Nonlinear Stochastic Control and Filtering with Engineering-Oriented Complexities

ENGINEERING SYSTEMS AND SUSTAINABILITY SERIES

Series Editor: Ganti Prasada Rao

Co-Editors: Andrew P. Sage, Heinz Unbehauen, Desineni S. Naidu, Hughes Garnier

Published Titles

Nonlinear Stochastic Control and Filtering with Engineering-Oriented Complexities
Guoliang Wei, Zidong Wang, and Wei Qian

Multi-Stage Flash Desalination: Modeling, Simulation, and Adaptive Control
Abraha Woldai

ENGINEERING SYSTEMS AND SUSTAINABILITY SERIES

Series Editor: Ganti Prasada Rao

Co-Editors: Heinz Unbehauen, D. Subbaram Naidu, Hugues Garnier, and Zidong Wang

Nonlinear Stochastic Control and Filtering with Engineering-Oriented Complexities

Guoliang Wei
Zidong Wang
Wei Qian

CRC Press is an imprint of the
Taylor & Francis Group, an **informa** business

CRC Press
Taylor & Francis Group
6000 Broken Sound Parkway NW, Suite 300
Boca Raton, FL 33487-2742

First issued in paperback 2020

ISBN 13: 978-0-367-57458-1 (pbk)
ISBN 13: 978-1-4987-6074-4 (hbk)

Visit the Taylor & Francis Web site at
http://www.taylorandfrancis.com

and the CRC Press Web site at
http://www.crcpress.com

Contents

List of Figures

List of Tables

Symbols

Symbol Description

Symbol	Description
$\mathbb{R}^n$	The n-dimensional Euclidean space.
$\mathbb{Z}$	The set of integers.
$\mathbb{Z}^+(\mathbb{Z}^-)$	The set of nonnegative (negative) integers.
$\lvert\cdot\rvert$	The Euclidean norm in $\mathbb{R}^n$.
$L_2[0,\infty)$	The space of square-integrable vector functions over $[0,\infty)$.
$\lVert\cdot\rVert_2$	The usual $L_2[0,\infty)$ norm.
$C([-\tau,0];\mathbb{R}^n)$	The family of continuous functions ϕ from $[-\tau,0]$ to $\mathbb{R}^n$ with the norm $\lVert\phi\rVert = \sup_{-\tau\leq\theta\leq 0}\lvert\phi(\theta)\rvert$.
$(\Omega,\mathcal{F},\{\mathcal{F}_t\}_{t\geq 0},\mathcal{P})$	The complete probability space with a filtration $\{\mathcal{F}_t\}_{t\geq 0}$ satisfying the usual conditions (i.e., the filtration contains all P-null sets and is right continuous).
$L^p_{\mathcal{F}_t}([-\tau,0];\mathbb{R}^n)$	The family of all $\mathcal{F}_t$-measurable $C([-\tau,0];\mathbb{R}^n)$-valued random variables $\xi = \{\xi(\theta) : -\tau \leq \theta \leq 0\}$, such that $\sup_{-\tau\leq\theta\leq 0}\mathbb{E}\lvert\xi(\theta)\rvert^p < \infty$, for $t \geq 0$.
$\omega(t)$	m-dimension Brownian motion defined on a complete probability space $(\Omega,\mathcal{F},\{\mathcal{F}_t\}_{t\geq 0},\mathcal{P})$.
$Pr\{\cdot\}$	The occurrence probability of the event "·".
$\mathbb{E}\{x\}$	The expectation of stochastic variable x.
$\mathbb{E}\{x\|y\}$	The expectation of x conditional on y, x, and y are all stochastic variables.
$\otimes$	The Kronecker product of matrices.
$\mathbb{R}^{n\times m}$	The set of all $n\times m$ real matrices.
I	The identity matrix of compatible dimension.
$X > Y$	The $X-Y$ is positive definite, where X and Y are symmetric matrices.
$X \geq Y$	The $X-Y$ is positive semi-definite, where X and Y are symmetric matrices.
M^T	The transpose matrix of M.
$\lambda_{\max}(M)$	The maximum eigenvalue of M, when M is symmetric.
$\lambda_{\min}(M)$	The minimum eigenvalue of M, when M is symmetric.
$\lVert M\rVert$	The spectral norm of M.
$\text{diag}_q\{\star\}$	$\text{diag}\{\star,\ldots,\star\}$ with q blocks of $\star$.
$*$	The ellipsis for terms induced by symmetry, in symmetric block matrices.

diag$\{M_1, ..., M_n\}$
The block diagonal matrix with diagonal blocks being the matrices $M_1, ..., M_n$.

$\mathfrak{M}_n[\,(M_1)_{i_1,j_1}, ..., (M_r)_{i_r,j_r}\,]$
The nth-order block square matrix whose all nonzero blocks are the i_1j_1th block M_1, the i_2j_2th block $M_2, ...,$ the i_rj_rth block M_r, and *all other* blocks are zero matrices.

Series Editor Foreword

Professor Andrew P. Sage of George Mason University, USA, my friend, colleague and collaborator for many decades, and also as a co-editor of this series, was a renowned scholar in the fields of systems, control, and estimation. His passing in October 2014 has been a great loss to all of us in the academic community. I remain grateful to him for his cooperation and advice. Professor Zidong Wang of Brunel University agreed to join as an honorary co-editor. I thank all the co-editors and Gagandeep Singh of CRC Press for their support.

This book

Human-engineered systems develop through technological innovations and designs towards meeting both growing human needs and challenges of associated complexities. In the context of man-made systems of control nonlinearities and uncertainties enter in many ways. Complexities such as time delays, jump parameters, nonlinearities, sensor faults and parameter uncertainties, etc. are ubiquitous. These have been tackled through various methods in the past several decades and the literature related to them is vast. The emerging complexities such as randomly occurring incomplete information implying missing measurements, randomly occurring nonlinearities, and coupling delays have been the focus of attention of control and signal processing communities in order to ensure the stability and reliability of the systems. The authors of this volume have devoted much time on these aspects and published significant results. They have tried to put together the state-of-the-art methods in this book dealing with the phenomena of engineering-oriented complexities both traditional and the newly emerging from a nonlinear stochastic control perspective.

Ganti Prasada Rao
Series Editor
Engineering Systems and Sustainability

Series Editor

GANTI PRASADA RAO earned a BE degree in electrical engineering from Andhra University, India in 1963, MTech (control systems engineering) and PhD degrees in electrical engineering in 1965 and 1970 respectively, both from the Indian Institute of Technology (IIT), Kharagpur, India. From July 1969 to October 1971, he was with the Department of Electrical Engineering, PSG College of Technology, Coimbatore, India as an assistant professor. In October 1971, he joined the Department of Electrical Engineering, IIT Kharagpur as an assistant professor and was a professor there from May 1978 to June 1997. From May 1978 to August 1980, he was the chairman of the Curriculum Development Cell (electrical engineering) established by the Government of India at IIT Kharagpur. From October 1975 to July 1976, he was with the Control Systems Centre, University of Manchester Institute of Science and Technology (UMIST), Manchester, England, as a Commonwealth Postdoctoral Research Fellow. During 1982–1983, 1985, 1991, 2003, 2004, 2007, and 2009, he visited the Lehrstuhl für Elektrische Steuerung und Regelung, Ruhr-Universität Bochum, Germany as a Research Fellow of the Alexander von Humboldt Foundation. He visited the Fraunhofer Institut für Rechnerarchitectkur und Softwaretchnik (FIRST) Berlin, in 2003, 2004, 2007 2009, and 2011. He was a visiting professor in 2003 at University Henri Poincare, Nancy, France, and Royal Society–sponsored visiting professor at Brunel University, UK in 2007. During 1992–1996 he was scientific advisor to the Directorate of Power and Desalination Plants, Water and Electricity Department, Government of Abu Dhabi and the International Foundation for Water Science and Technology where he worked in the field of desalination plant control. He is presently a member of the UNESCO-EOLSS Joint Committee.

Dr. Rao has authored/coauthored four books: *Piecewise Constant Orthogonal Functions and Their Applications to Systems and Control*, *Identification of Continuous Dynamical Systems–The Poisson Moment Functional (PMF) Approach* (with D.C. Saha), *Generalised Hybrid Orthogonal Functions and Their Applications in Systems and Control* (with A. Patra) all the three published by Springer in 1983, 1983 and 1996 respectively, and *Identification of Continuous Systems* (with H. Unbehauen) published by North Holland in 1987. He is co-editor (with N.K. Sinha) of *Identification of Continuous Systems - Methodology and Computer Implementation,* Kluwer, 1991. He has authored/coauthored over 150 research papers.

He has been on the editorial boards of *International Journal of Modeling and Simulation, Control Theory and Advanced Technology (C-TAT), Systems*

Science (Poland), *Systems Analysis Modeling and Simulation* (SAMS), *International Journal of Advances in Systems Science and Applications* (IJASSA) and *The Students' Journal of IETE (India)*. He was guest editor of three special issues: one of *C-TAT on Identification and Adaptive Control – Continuous Time Approaches,* Vol. 9, No. 1, March 1993, and *The Students' Journal of IETE on Control*, Vols. I&II, 1992–93, *System Science, Special Issue dedicated to Z. Bubnicki*, Volume 33, No. 3, 2007. He organized several invited sessions in IFAC Symposia on Identification and System Parameter Estimation, 1988, 1991, 1994, and the World Congress 1993. He was a member of the International Program Committees of these symposia during 1988–1997. He was a member of the IFAC Technical Committee on Modelling, Identification and Signal Processing in 1996, and was chairman of the Technical Committee of the 1989 National Systems Conference in India. He is co-editor (with Achim Sydow) of the book series "Numerical Insights Series" published by Taylor & Francis. He is a member of the International Advisory Boards of the International Institute of General Systems Science (IGSS), *Systems Science Journal* (Poland) and International Congresses of World Organisation of the Systems and Cybernetics (WOSC).

Since 1996, Prof. Rao has been closely associated with the development of the *Encyclopedia of Desalination and Water Resources* (DESWARE) (www.desware.net) and *Encyclopedia of Life Support Systems* (EOLSS), developed under the auspices of the UNESCO (www.eolss.net).

He earned several academic awards including the IIT Kharagpur Silver Jubilee Research Award 1985, The Systems Society of India Award 1989, International Desalination Association Best Paper Award 1995, and Honorary Professorship of the East China University of Science and Technology, Shanghai. He was elected to fellowship of the IEEE with the citation 'FOR DEVELOPMENT OF CONTINUOUS TIME IDENTIFICATION TECHNIQUES'. The International Foundation for Water Science and Technology has established the 'Systems and Information Laboratory' in the Electrical Engineering Department at the Indian Institute of Technology, Kharagpur, in his honor.

He is listed in several biographical volumes and is a Life Fellow of The Institution of Engineers (India), Life Member Systems Society of India, Indian Society for Technical Education, Fellow of The Institution of Electronics and Telecommunication Engineers (India), Life Fellow of IEEE (USA) and a Fellow of the Indian National Academy of Engineering.

Preface

Engineering-oriented complexities exist everywhere, including time delays, Markovian jump parameters, nonlinearities, sensor faults, and parameter uncertainties, etc. These complexities have been studied for several decades and could be referred to as the traditional engineer-oriented complexities. Recently, a class of newly emerged engineering-oriented complexities, i.e., randomly occurring incomplete information (such as missing measurements, randomly occurring nonlinearities, randomly occurring coupling delays, etc.), has gained some initial research interest in various communities, such as signal processing and control engineering. It should be mentioned that the occurrence of the engineering-oriented complexities inevitably affects system performance and may even lead to instability of controlled systems. So far, the analysis issues of linear systems with engineering-oriented complexities have attracted quite a lot research attention. However, the corresponding research on nonlinear stochastic systems with engineering-oriented complexities has not been fully investigated. Therefore, a seemingly important research problem is to develop new approaches addressing the analysis and synthesis problems for nonlinear stochastic systems with engineering-oriented complexities.

In this book, we focus on the phenomena of the engineering-oriented complexities, including the traditional engineer-oriented complexities and newly emerged engineering-oriented complexities from a nonlinear stochastic control perspective. The content of this book is conceptually divided mainly into three parts. In the early chapters of the book, a series of control and filtering approaches, especially robust control/filtering methods, is introduced for stochastic systems with traditional engineer-oriented complexities. Furthermore, some sufficient conditions for the existence of the desired controllers and filters are given by verifying the feasibility of the corresponding linear matrix inequalities or Hamilton–Jacobi inequalities. In the bulk of the book, the incomplete information with varying occurring probabilities is first introduced to represent a new complexity that belongs to the newly emerged engineering-oriented complexities. Accordingly, a probability-dependent gain-scheduling approach is proposed to design the filters and controllers. In the final chapters of the book, both the theory and methodology developed in previous parts are employed to deal with the state estimation/filtering problems for networked control systems, genetic regulatory networks, and complex networks.

The compendious framework and description of this book are given as follows. In Chapter 1, the research background and motivation are discussed, the research problems to be addressed in this book are proposed, and the outline of this book is given. Chapter 2 is concerned with the robust stability and stabilization problems for a class of stochastic time-delay interval systems with nonlinear disturbances. Robust stabilization and H_∞ control problems are investigated in Chapter 3 for a class of stochastic time-delay uncertain systems with Markovian switching and nonlinear disturbance. In Chapter 4, the design issues of the H_∞ state estimator and H_∞ output feedback controller are investigated for stochastic time-delay systems with nonlinear disturbances, sensor nonlinearities, and Markovian jumping parameters. Chapter 5 provides some analysis on H_∞ performance for a general class of nonlinear stochastic systems with time-delays, where the addressed systems are described by general stochastic functional differential equations. In Chapter 6, the filtering problem is studied for a class of discrete-time stochastic nonlinear time-delay systems with missing measurement and stochastic disturbances. Chapters 7, 8, and 9 consider the probability-dependent control and filtering problems for time-varying nonlinear systems with incomplete information by using the gain-scheduling technique. In Chapter 10, the filtering problem is addressed for a class of discrete-time stochastic nonlinear networked control systems with multiple random communication delays and random packet losses. The filtering problem is studied in Chapter 11 for a class of nonlinear genetic regulatory networks with state-dependent stochastic disturbances and state delays. In Chapter 12, the H_∞ state estimation problem is investigated for a class of discrete-time complex networks with probabilistic missing measurements and randomly occurring coupling delays. Based on the gain-scheduling technique, the H_∞ synchronization control problem is addressed in Chapter 13 for a class of dynamical networks with randomly varying nonlinearities.

Guoliang Wei, Zidong Wang, and Wei Qian
Authors

Authors

Guoliang Wei earned the BSc degree in mathematics from Henan Normal University, Xinxiang, China, in 1997, and the MSc degree in applied mathematics and the PhD degree in control engineering, both from Donghua University, Shanghai, China, in 2005 and 2008, respectively. He is currently a professor with the Department of Control Science and Engineering, University of Shanghai for Science and Technology, Shanghai, China. From March 2010 to May 2011, he was an Alexander von Humboldt research fellow in the Institute for Automatic Control and Complex Systems, University of Duisburg-Essen, Germany. From March 2009 to February 2010, he was a postdoctoral research fellow in the Department of Information Systems and Computing, Brunel University, Uxbridge, UK, sponsored by the Leverhulme Trust of the UK. From June to August 2007, he was a research assistant at the University of Hong Kong. From March to May 2008, he was a research assistant at the City University of Hong Kong. His research interests include nonlinear systems, stochastic systems, and bioinformatics. He has published more than 50 papers in refereed international journals. He is a very active reviewer for many international journals.

Zidong Wang is currently professor of dynamical systems and computing at Brunel University London in the United Kingdom. From January 1997 to December 1998, he was an Alexander von Humboldt research fellow with the Control Engineering Laboratory, Ruhr-University Bochum, Germany. From January 1999 to February 2001, he was a lecturer with the Department of Mathematics, University of Kaiserslautern, Germany. From March 2001 to July 2002, he was a university senior research fellow with the School of Mathematical and Information Sciences, Coventry University, UK. In August 2002, he joined the Department of Information Systems and Computing, Brunel University, UK, as a lecturer, and was then promoted to a reader in September 2003 and to a chair professor in July 2007. Professor Wang's research interests include dynamical systems, signal processing, bioinformatics, control theory and applications. He has published more than 200 papers in refereed international journals. According to the Web of Science, his publications have received more than 8000 citations (excluding self-citations) with an h-index 48. He was awarded the Humboldt research fellowship in 1996 from the Alexander von Humboldt Foundation, the JSPS research fellowship in 1998 from the Japan Society for the Promotion of Science, and the William Mong Visiting research fellowship in 2002, from the University of Hong Kong.

Professor Wang is an IEEE fellow for his contributions to networked control and complex networks. He has served or is serving as an associate editor for *IEEE Transactions on Automatic Control*, *IEEE Transactions on Neural Networks*, *IEEE Transactions on Signal Processing*, *IEEE Transactions on Systems, Man, and Cybernetics–Part C*, *IEEE Transactions on Control Systems Technology*, *Circuits, Systems & Signal Processing*, *Asian Journal of Control*, an action editor for *Neural Networks*, an editorial board member for *IET Control Theory and Applications*, *International Journal of Systems Science*, *Neurocomputing*, *International Journal of Computer Mathematics*, *International Journal of General Systems*, and an associate editor on the Conference Editorial Board for the IEEE Control Systems Society. He is a senior member of the IEEE, a fellow of the Royal Statistical Society, a member of the program committee for many international conferences, and a very active reviewer for many international journals. He was nominated as an appreciated reviewer for *IEEE Transactions on Signal Processing* in 2006–2008 and 2011, an appreciated reviewer for *IEEE Transactions on Intelligent Transportation Systems* in 2008; an outstanding reviewer for *IEEE Transactions on Automatic Control* in 2004 and for the journal *Automatica* in 2000.

Wei Qian earned his PhD degree in the State Key Lab of Industrial Control Technology from Zhejiang University in 2009, and he is now an associate professor at the School of Electrical Engineering and Automation in Henan Polytechnic University of China. His current research interests include time-delay systems, stochastic systems, networked control systems, etc. He has published more than 20 papers in refereed journals.

1

Introduction

CONTENTS

System complexities exist everywhere, with examples including time-delays, Markovian jump parameters, nonlinear disturbances, and parameter uncertainties. Such complexities are often referred to as traditional engineering phenomena that can be found in a variety of engineering systems, such as stochastic systems, linear parameter varying systems, genetic regulatory networks, networked control systems, and complex networks. Recently, with the rapid developments of networked systems, a new class of engineering-oriented complexities, namely, randomly occurring incomplete information, has been studied due to the fact that signal transmissions via networked systems are often hindered by the limited capacity of the devices. The phenomena falling in this group mainly include missing measurements [2, 172, 183], randomly occurring delays [168], randomly occurring sensor saturations [90, 182], randomly occurring nonlinearities [167, 192], and randomly occurring actuator faults [159, 190]. For more details about those randomly occurring incomplete information, we refer the readers to the survey paper [31] and [142].

When modeling the randomly occurring incomplete information, the Bernoulli distribution model has proven to be a flexible yet effective one that has been frequently employed, see, e.g., [186, 187]. Accordingly, great effort has been devoted to better describe the randomly occurring incomplete information in the real world. It is worth mentioning that the authors of this book have successively put forward the general probabilistic model and the Bernoulli distribution model with time-varying probabilities to enrich the research in this area, which have received a significant amount of attention. In addition, it should be noticed that the randomly occurring incomplete information, together with tractional engineering phenomena, may potentially cause a lot of undesired problems that would seriously degrade the system performance. Hence, there is a great need to develop new strategies to deal with the engineering-oriented complexities.

Fortunately, during the past several decades, a great number of theories and techniques have been developed to handle the engineering-oriented complexities. Robust control/filtering, especially, H_∞ control/filtering is the most used one, since it can achieve the guaranteed performances with respect to the external disturbances. Besides, the Razumikhin-type methodology is an effective tool to deal with the stability issues of time-delay systems. Nevertheless, when it comes to the systems with randomly varying incomplete information, neither the robust control/filtering method nor the Razumikhin-type approach can provide the satisfactory control/filtering scheme, which motivates us to come up with a novel probability-dependent gain-scheduled technique to provide an elegant solution.

In this chapter, we focus on the control and filtering problems for nonlinear stochastic systems with the engineering-oriented complexities and aim to give a brief survey on some recent advances in this area. The engineering-oriented complexities addressed in this book include the traditional engineering phenomena (such as time-delays, Markovian jump parameters, nonlinearities, and parameter uncertainties, etc.) and randomly occurring incomplete information (such as missing measurements, randomly occurring nonlinearities, etc.). Secondly, several typical systems that are often accompanied with engineering-oriented complexities are introduced, and a series of analysis and synthesis techniques handling the engineering-oriented complexities is also discussed. Finally, the organization structure of this book is shown in Figure 1.1.

1.1 Background, Motivation, and Research Problem

1.1.1 Engineering-Oriented Complexities

This section discusses several engineering-oriented complexities that are studied thoroughly in this book. In what follows, the engineering-oriented complexities addressed in this book are categorized into two groups. The first one includes the traditional engineering phenomena, such as time-delays, Markovian jump parameters, nonlinearities, and parameter uncertainties etc., while another group involves the randomly occurring incomplete information. We only focus on two types of randomly occurring incomplete information (i.e., missing measurements and randomly occurring nonlinearities) in this section. For more details about the randomly occurring incomplete information, we refer the readers to the survey papers [31] and [142] where comprehensive discussions have been given.

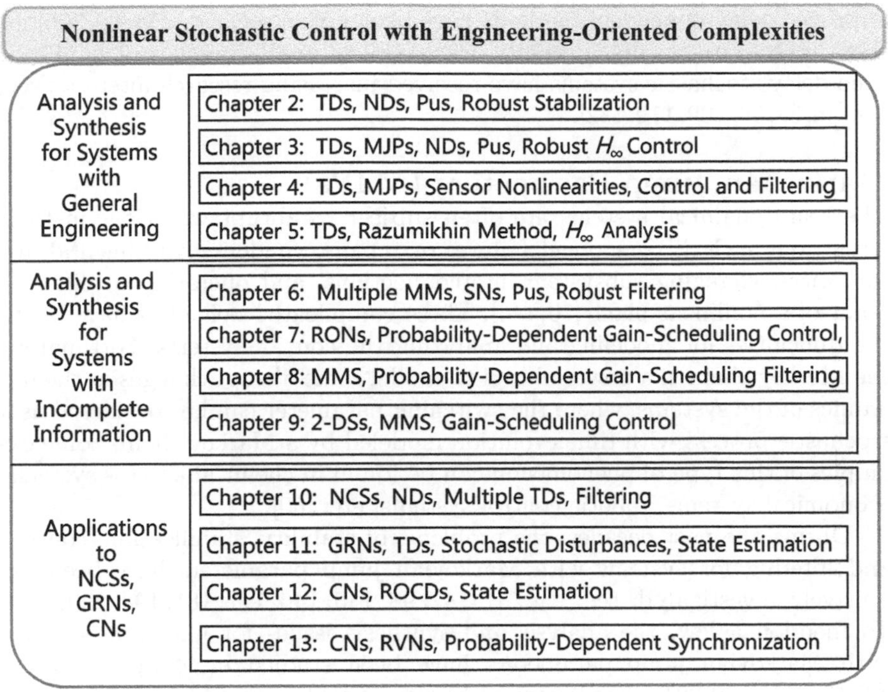

Nonlinear Stochastic Control with Engineering-Oriented Complexities

Analysis and Synthesis for Systems with General Engineering	Chapter 2: TDs, NDs, Pus, Robust Stabilization
	Chapter 3: TDs, MJPs, NDs, Pus, Robust H_∞ Control
	Chapter 4: TDs, MJPs, Sensor Nonlinearities, Control and Filtering
	Chapter 5: TDs, Razumikhin Method, H_∞ Analysis
Analysis and Synthesis for Systems with Incomplete Information	Chapter 6: Multiple MMs, SNs, Pus, Robust Filtering
	Chapter 7: RONs, Probability-Dependent Gain-Scheduling Control,
	Chapter 8: MMS, Probability-Dependent Gain-Scheduling Filtering
	Chapter 9: 2-DSs, MMS, Gain-Scheduling Control
Applications to NCSs, GRNs, CNs	Chapter 10: NCSs, NDs, Multiple TDs, Filtering
	Chapter 11: GRNs, TDs, Stochastic Disturbances, State Estimation
	Chapter 12: CNs, ROCDs, State Estimation
	Chapter 13: CNs, RVNs, Probability-Dependent Synchronization

FIGURE 1.1
The architecture of the book.

1.1.1.1 Traditional Engineering Phenomena

Time-Delays (TDs)

It is well known that time-delays are frequently encountered in practical systems, such as biology systems, engineering systems, traffic systems, and networked control systems [199]. The existence of time-delays may cause undesirable dynamic behaviors and add additional difficulties for the analysis and synthesis problems of nonlinear stochastic systems.

The commonly studied time-delays include constant delay, time-varying delay, distributed delay [99, 201], mixed time-delay [179, 185], sensor delay [175], randomly occurring time-delay [215], and randomly coupling time-delay in complex network [168]. In the past two decades, with the improvement of the computation capacity, the linear matrix inequalities (LMIs)-based methods have played an important role in the research on the general topic of time-delay systems; see, e.g., [12, 45, 123, 145, 146, 179, 186]. It is worth mentioning that, since the delay-dependent LMI techniques take the length of delays into account, the delay-dependent stability criteria are shown to be less conservative than the traditional delay-independent ones, especially when the time-delays are known and small; see, e.g., [21, 44, 54, 207, 221]. Accordingly, some improved delay-dependent methods have been given; see, e.g., [58–60, 206] for some up-to-date results. Recently, the stability analysis issues of nonlinear time-delay stochastic systems have received increasing research interests; see, e.g., [6, 38, 56, 109, 110, 132].

Markovian Jump Parameters (MJPs)

The real dynamical systems are often subject to abrupt variations in their structures, such as component failures or repairs, sudden environmental disturbances, changing subsystem interconnections, and operating in different point of a nonlinear plant [70, 144]. These complicated behaviors have led to the difficulties for modeling and analyzing the control systems. Fortunately, the Markov chain is a powerful tool to model some abrupt changes in the dynamics of the systems, where the switching parameter can be considered as a stochastic process with time evolution modeled by a Markov chain. Some examples of this type of phenomenon can be found in chemical process systems, economical systems, robotic control systems, etc. [50].

During the past decades, the problems of stability, stabilization, control, and filtering for systems with Markovian jump parameters have been extensively investigated; see, e.g., [14, 47, 98, 145, 146, 177, 181, 194, 220, 221]. To mention a few, the analysis and synthesis issues for uncertain systems with Markovian jump parameters have been studied in [47, 145, 146]. In [98,177,194,220,221], the filtering and control problems have been investigated for Markovian jump systems with or without nonlinear disturbances. The H_∞ control problem has been studied in [14], where the stochastic stability conditions have been proposed by using LMI technique. In [181], the stochastic stabilization problem has been investigated for systems with Markovian jump parameters and nonlinear exogenous disturbance inputs by solving a set of either LMIs or coupled quadratic matrix inequalities. So far, the analysis and synthesis problems have not yet been fully investigated for stochastic systems with Markovian jump parameters and time-delays.

Nonlinearities

It is well recognized that nonlinearities are inevitable and could not be eliminated thoroughly in many real-world systems, such as electrical circuits, mechanical systems, control systems, and other engineering systems [75]. In many papers, the nonlinearity is taken as the exogenous nonlinear disturbance (ND) input, which comes from the linearization process of an originally highly nonlinear plant [52, 170, 177]. In real-world applications, the nonlinearity is an inevitable feature for some sensors, which generally results from the harsh environments, such as uncontrollable elements (e.g., variations in flow rates, temperature, etc.) and aggressive conditions (e.g., corrosion, erosion, and fouling, etc.) [127], and this class of nonlinearities is often referred to as sensor nonlinearities. Because the sensor nonlinearities cannot be simply ignored and often lead to poor performances of the controlled system, the analysis and synthesis problems have received much research attention for various systems with sensor nonlinearities [15,54,78,97]. On the other hand, there are many different conditions for describing the nonlinearities. For example, sector-nonlinearity (also called sector-like nonlinearity), which is known to be quite general, includes the widely used Lipschitz condition as a special case. Moreover, the stochastic nonlinearity (SN) described by statistical means has drawn particular research focus, since it covers several well-studied nonlinearities in stochastic systems; see, e.g., [211, 219].

Parameter Uncertainties (PUs)

The parameter uncertainties serve as one important kind of complexity for system modeling and may arise from variations of the operating point, aging of the devices, identification errors, etc. Therefore, in the past decades, considerable attention has been devoted to address the analysis and synthesis problem for linear/nonlinear uncertain systems, and a large number of papers have been published; see [37, 52, 53, 100, 106, 125, 146, 160, 224] for some recent results. Besides, the robust H_∞ control problems have also been investigated for time-delay systems with parameter uncertainties; see, e.g., [47, 48, 147].

The parameter uncertainties can be generally classified into three types. The first one is the interval uncertainty, where the uncertain parameters take values in some certain intervals. The systems with interval uncertainties are known as interval systems, which have received considerable research attention to discuss the stability analysis and stabilization problems; see, e.g., [67, 114] and the references therein. The second one is the norm-bounded uncertainty, where the norm of the uncertainty matrix is often bounded by a real value. In fact, there are many physical systems in which the parameter uncertainties can be modeled in this manner, e.g., systems satisfying "matching conditions" [76]. During the past decades, the stabilization, control, and filtering problems have been extensively investigated for systems with norm-bounded uncertainties; for example, robust filtering [195], robust stabilization [76], robust H_∞ control [200], robust Kalman filtering [202], and nonfragile filtering [86]. Recently, the third type of uncertainties, namely, polytopic uncertainties, has gained increasing attention; see, e.g., [65, 208, 209], where the uncertainty has been described by a convex combination of N vertices. This representation is quite general and can encompass the well-known case of interval uncertainties; see [152, 189] for some latest results.

1.1.1.2 Randomly Occurring Incomplete Information

Missing Measurements (MMs)

In practical systems, due to various reasons, such as sensor temporal failure or network transmission delay/loss, the measurement signals may be subject to information missing. In case of information missing (called packet loss), the measurement output may contain noise only at certain time points. Such a measurement missing phenomenon usually occurs in a probabilistic way and has attracted considerable attention during the past few years; see, e.g., [149,

186] and the references therein. The Bernoulli distributed model has become a popular approach to model the measurement missing phenomenon, which is specified by a conditional probability distribution. As early as in [18,122], such a model has been employed to synthesize the recursive filtering problems for systems with missing measurements. Recently, with the rapid developments of networked control systems, renewed effort has been devoted to this effective and flexible model; see, e.g., [186,187,228].

It should be pointed out that, in all aforementioned literature, the probability 0 is used to stand for an entire signal missing, and the probability 1 denotes the intactness (i.e., there is no signal missing at all), and all the sensors have the same missing probability. Such a description, however, does have its limitations, since it cannot cover some practical cases, for example, the case when only partial information is missing and the case when the individual sensor has different missing probabilities. Note that the latter case has been dealt with in [63,64], where the minimum variance linear state estimators have been designed for linear systems with multiple sensors with different failure/delay rates. In [66,195], a more general model describing the multiple missing measurements has been proposed and the filtering problems have been studied for stochastic systems. Recently, a probability-dependent gain-scheduling approach has been developed in [87,193] to address the missing measurement with time-varying probabilities.

Randomly Occurring Nonlinearities (RONs)

Nowadays, the so-called randomly occurring nonlinearities (RONs) are prevalent in the networked control systems, such as the internet-based three-tank system for leakage fault diagnosis [31]. The RONs are a new kind of nonlinear disturbances that may appear intermittently in a random way due probably to random failures and repairs of the components, intermittently switching in the interconnections of subsystems, etc. The RONs have been initially introduced in [169,185], where the global synchronization problems for complex networks have been discussed. Since then, a series of papers has been published concerning the control and filtering problems for systems with RONs; see, e.g., [28,94,108,143,168]. For example, in [108], a game theory approach has been applied to obtain the H_∞ controller when RONs occur. The dissipative control problem has investigated in [28] for systems with simultaneous presence of RONs, state saturations as well as multiple missing measurements. It should be noted that the RONs in [28,94,108,143,169,185] have been assumed to satisfy a time-invariant Bernoulli distribution. Such an assumption will limit the application scope, since RONs usually appear with time-varying probabilities. Very recently, [168] has extended the RONs model in [169,185] to the time-varying ones and designed the time-varying gain-scheduled controller based on the time-varying probabilities. Furthermore, [167] has considered the case that the RONs are varying between different kinds of nonlinearities according to the Bernoulli distribution, where the addressed phenomenon is named as randomly varying nonlinearities (RVNs).

1.1.2 Systems with Engineering-Oriented Complexities

It is well known that the engineering-oriented complexities ubiquitously exist in most of real systems. This book covers several typical systems that often encounter the engineering-oriented complexities, such as stochastic systems, linear parameter-varying systems, networked control systems, genetic regulatory networks and complex networks. In the following, we will introduce each of them in order to stir the reader's interest.

Stochastic Systems

Owing to pervasive existence of stochastic perturbations in reality, stochastic models have been successfully utilized to describe many practical systems, such as mechanical systems, economic systems, and biological systems, etc. It's notable that the stochasticity is one of the main sources of degrading the system performance, and it also poses significant challenges for the analysis and synthesis of practical systems. It is not surprising that, over the past few decades, the study of stabilization, control, and filtering problems for stochastic systems has been paid much attention from many researchers and a large number of results have been reported in the literature; see, e.g., [7, 14,41,73,74,88,147,184,188,220]. Among them, there are two most stunning achievements developed specifically for the stochastic systems. One of them is the linear quadratic Gaussian control [7], and another one is the famed Kalman filter, which was developed by Kalman in 1960s [73]. Both of them have found numerous applications in a variety of areas.

Even though many fundamental results for deterministic systems have been extended to stochastic systems, the analysis and synthesis problems for stochastic systems still require new tools to pursue better solutions. In this respect, the Itô-type stochastic systems, which are characterized by stochastic differential equations, have stirred a lot of research interests; see, e.g., [40,47,204], particularly, the book [113].

Recently, due to the increasing complexities of the systems, the analysis and synthesis issues for stochastic systems with one or multiple other engineering-oriented complexities have become popular, and some latest results have been reported; see, e.g. [14,29,48,100,143,156,168,173,176,180,186, 192,194,205,221]. To be specific, many researchers have studied the stochastic systems with uncertain parameters [48,176], with nonlinearities [168,192], with Markovian jump parameters [14,173], with time-delays [156,186,194,205], and with randomly occurring incomplete information [29,143].

Linear Parameter-Varying Systems (LPVSs)

Linear parameter-varying systems (LPVSs) are a very special class of systems, whose system matrices are functions of measurable time-varying parameters, and the measurements of these parameters provide real-time information according to the variations of the plant's characteristics. In the past few years, the research on LPVSs has become a promising topic from both theoretical and engineering viewpoints. For example, the sensor-fault-tolerant control problem has been considered in [1] for LPVSs, while the analysis and synthesis issues have been addressed in [197] for LPVSs with parameter-varying time-delays. Moreover, the designs of robust controller/filter for LPVSs have also received a great deal of attention, and some representative results have appeared in the literature; see, e.g. [4,16,24,155,163].

On the other hand, for the purpose of designing a controller/filter with less conservatism for LPVSs, it's natural to construct the novel Lyapunov functions with scheduling parameters, which are usually called the parameter-dependent Lyapunov functions. Very recently, the parameter-dependent Lyapunov function approach has been applied to deal with the gain-scheduled control/filtering problems, and some results have been reported in the literature [5,153,171,229]. It should be noticed that, in order to design an appropriate controller/filter for LPVSs, the gain-scheduling approach has been proved to be an effective one in this process. The idea of the gain-scheduling approach is to design controller/filter gains as functions of the scheduling parameters, which are supposed to be available in real time and can be utilized to adjust the controller/filter with hope to get the better performance. Therefore, the gain-scheduling control and filtering problems for LPVSs have stirred a great deal of interest in these years; see, e.g. [153, 171, 229]. In addition, several successful applications of the gain-scheduling approach for LPVSs have been discussed in the survey paper [136].

Two-Dimension Systems (2-DSs)

In the past few decades, two-dimension systems (2-DSs) have been receiving considerable attention due to their wide applications in many engineering fields, such as the analysis of satellite weather photos, multidimensional digital filtering, multivariable network realizability, electron heating systems, etc. [72, 198]. From a physical perspective, 2-D system is a class of system that depends on two variables. For example, an image is a generalization of a temporal signal where it is defined over two spatial dimensions instead of a single temporal dimension [135]. It can essentially be governed by the Roesser model, with space coordinates i and j taking the place of time t. There are a great number of results focusing on the modeling of 2-DSs, for instance, the state-space models of 2-DSs have been investigated in [72, 130]. As is well known, the Roesser model is one of the most important models of 2-D linear systems, which has been widely used due to its practicality. Meanwhile, the design of controllers and filters, stability analysis, and stabilization for 2-DSs have been extensively studied by researchers, and a large number of results have been reported; see, e.g. [27, 84, 93, 105, 118].

Networked Control Systems (NCSs)

With the successful applications of networks in the complex dynamical processes, such as advanced aircraft, spacecraft, automotive, and manufacturing processes, the networked control systems (NCSs) have received considerable attention in the literature; see, e.g., [51, 80, 139, 162] and references therein. Note that the devices are connected by the shared medium, and the transmission capacity of the communication network is limited, which in turn affect the number of bits or packets per second transmitted via the network. Consequently, a series of networked phenomena has appeared. For example, the communication delays and missing measurements are commonly encountered. Therefore, in the past decades, the filtering and control problems with communication delays and/or missing measurements have been extensively studied for NCSs; see, e.g., [68, 117, 124, 138, 140].

In the literature, the communication delays occurring in networked systems have received considerable research attention, and many methods have been proposed to deal with the communication delays; see, e.g., [80,117,124,133,164, 175, 213, 218, 226]. On the other hand, the data-missing problem emerging in the NCSs is another important research issue to be addressed. During the last few years, there has been a growing research interest on how to model the data-missing phenomenon in network-based control and filtering problems [138, 140, 186, 210, 223]. It is worth mentioning that, in [57], a unified measurement model has been proposed to account for both the communication delays and the data missing by extending the results in [175, 210, 213]. Recently, the nonlinear Itô-type stochastic NCSs have also attracted considerable attention, see e.g., [22, 137, 165].

Genetic Regulatory Networks (GRNs)

It is well known that the encoded information contained in genes is necessary for the organism to develop within a changing external environment. Genetic regulatory networks are the mechanisms that have evolved to regulate the expression of genes, where the expression level of a gene is regulated negatively or positively by its own production (protein). DNA microarray technology [128] has made it possible to measure gene expression levels on a genomic scale, and has therefore been extensively applied to gene transcription analysis. Theoretical analysis and experimental investigation on genetic regulatory networks (GRNs) have quickly become an attractive area of research in the biological and biomedical sciences, and received great attention over the past decade [19, 36, 49, 85, 85, 134, 203].

Recently, there has been much interest to reconstruct models for GRNs, for example, boolean network models [69], linear differential equation models [20, 25, 26], and single negative feedback loop networks [36]. In biological systems or artificial genetic networks, time-delays exist due primarily to the slow processes of transcription, translation, translocation, and the finite switching speed of amplifiers. It has been pointed out in [19, 150, 151] that the time-delays may play an important role in dynamics of genetic networks, and mathematical models without addressing the delay effects may even provide wrong predictions of the mRNA and protein concentrations [150, 151]. It has also been shown in [119], by mathematically modeling recent data, that the observed oscillatory expression and activity of three proteins are most likely driven by transcriptional delays. Very recently, the asymptotic stability has been discussed in [134] for the delayed GRNs with SUM regulatory logic, where time-delays are assumed to be time-varying and belong to given intervals.

On the other hand, the stochastic noise (fluctuations) in real-world gene expression data is of great importance when we construct the models for GRNs [23, 161]. In general, the stochastic noise arises in gene expression in one of two ways, namely, internal noise and external noise. The internal random fluctuations in genetic networks are inevitable, as chemical reactions are probabilistic [126] and the external noise originates in the random variation of the externally set control parameters [55]. Recently, a stochastic nonlinear dynamic model has been developed in [17] for GRNs under intrinsic fluctuation and extrinsic noise, and a method has been proposed to determine the robust stability under intrinsic fluctuations and identify the genes that are significantly affected by extrinsic noises.

In practice, for the ultimate goal of identifying genes of interest and designing drugs, biologists would be interested in knowing the steady-state values of the actual network states, that is, the concentrations of the mRNA and protein. Unfortunately, due to the inherent state delay and state-dependent noises, the actual network measurements are far from the true network states, and any subsequent analysis based only on the network measurements would probably have little value in application. This gives rise to the following filtering research issue: given a gene regulatory network that contains both transmission delays and intrinsic fluctuations, how does one estimate the network states, such that the estimation error could exponentially converge to zero in the mean square sense? Such a filtering issue has been addressed in [196] for a linear GRN with stochastic disturbances in terms of the variance-constrained index, where the time-delays and regulation nonlinearities have been ignored, which has later solved by [178].

Complex Networks (CNs)

Complex networks consist of a group of highly inter-related nodes and can be used to model various practical systems in the real world. In many practical situations, it is very common that a large number of complicated systems can be approximatively described by complex networks for theoretical analysis, such as neural systems, social networks, food webs, power grids, and so on. Over the past decades, complex networks have received particularly research attention and an increasing interest since two fundamental academic papers have been published with the discoveries of the "small-world" and "scale-free" properties [9, 71]. Very recently, due to the arrival of the epoch of big data, complex networks have become a hot research area in the simultaneous presence of opportunities and challenges. How to handle the large scale of network data trends is a crucial technical problem to be solved.

On the other hand, as a complex and interesting phenomenon, the synchronization of dynamical networks has attracted continual research attention in the past few years; see, e.g., [39,99]. Several effective control approaches have been proposed to synchronize the complex networks, such as impulsive control and pinning control [89,102,157]. In recent years, the synchronization of complex networks has attracted considerable research interests, since it is a universal behavior in the natural world and commonly exists in many systems, such as the large-scale and complex networks of chaotic oscillators [95,101,103,104], the coupled systems exhibiting spatiotemporal chaos and autowaves [129,227], and the array of coupled neural networks [216]. In addition, the synchronization has promising prospects in a wide range of practical application fields, including parallel image processing [62], pattern storage and retrieval [79], and secure communications.

Parallel to the synchronization problem, the state estimation problem for complex networks is also important. During the past few decades, significant progress has been made to design various state estimators for complex networks; see, e.g., [29,141,168]. For example, H_∞ estimators have been designed for complex networks with randomly occurring coupling delays (ROCDs) in [168], with randomly occurring sensor saturations and randomly varying sensor delays simultaneously in [29], and with uncertain inner coupling and incomplete measurements in [141].

1.1.3 Analysis and Synthesis for Engineering-Oriented Complexities

Since the engineering-oriented complexities would degrade the system performance, hence, there is a great need to apply existing techniques and develop novel methods to tackle these complexities. In this section, we introduce three methods, including robust (H_∞) control and filtering, gain-scheduled control and filtering, and Razumikhin-type methodology. It is worth mentioning that the probability-dependent gain-scheduled approach developed in this book is very useful for the case when the probabilities of incomplete information are time-varying, which has been recognized by many researchers.

Robust (H_∞) Control and Filtering

As is well known, the design of robust controller and filter has been one of the research mainstreams in control community [8]. The aim of the robust control and filtering problems is to design controllers and filters that can tolerate with the uncertainties or disturbances. The last several decades have witnessed a significant development of robust control and filtering theory, and one striking branch is the H_∞ control and filtering theory, which uses the H_∞ performance index to evaluate the disturbance attenuation level. Generally speaking, the H_∞ control/filtering problem can be described as follows: Given a dynamic system with exogenous input and measured output, design a controller/filter to control/estimate a controlled output, such that the L_2 gain of the mapping from the exogenous input to the controlled output (output error) is minimized or no larger than some prescribed level in terms of the H_∞ norm. It is well known that the existence of a solution to the H_∞ problem is in fact associated with the solvability of an appropriate algebraic Riccati equation (for the linear cases) or a so-called Hamilton–Jacobi equation (for the nonlinear ones).

In the past two decades, the H_∞ analysis and synthesis problems have been extensively studied for linear/nonlinear systems since the original work has been published in [225]. The standard H_∞ control problem has been completely solved in [34] for linear systems by deriving simple state-space formulas. For nonlinear systems, the H_∞ performance evaluation can be conducted through analyzing the L_2 gain of the relationship from the external disturbance to the system output. Moreover, a bounded real lemma has been given in [10,35,82] for stochastic nonlinear systems. When both the Markovian jump parameters and time-delays appear in the stochastic systems, the H_∞ filtering problem has been studied in [194], and some useful stochastic stability conditions have been proposed by employing the LMI technique. In [177], the robust H_∞ filter design problem has been investigated for stochastic time-delay systems with missing measurements.

Gain-Scheduled Control and Filtering

The gain-scheduling approach is one of the most popular ways to design the controller or filter, whose gains can be updated by a set of tuning parameters in order to optimize the closed-loop system's performance. On the other hand, the randomly occurring incomplete information often occurs with time-varying probabilities, which can also be considered as a tuning parameter for the controller or filter. Under such consideration, a novel gain-scheduling approach, namely, probability-dependent gain-scheduling approach, has been proposed to deal with the analysis and synthesis problems for systems with randomly occurring incomplete information.

It's worth mentioning that, by utilizing the probability-dependent gain-scheduling approach, the gains of the designed gain-scheduled controller/filter are time-varying consisting of two parts, one of which is constant matrix that can be obtained by solving a set of LMIs or Riccati inequalities, and the other part is the time-varying parameter that is measurable in real time. The controller/filter gains are scheduled along with the time-varying parameters and, therefore, have much less conservatism than the conventional ones with constant (fixed) gains only. With the development of the related research in the past several years, the probability-dependent gain-scheduled controller/filter scheme has been shown to be a very useful method to address the analysis and synthesis problems for systems with randomly occurring incomplete information.

Since firstly introduced in [193], the probability-dependent gain-scheduling approach has received increasing research attention, and a great number of results have been reported; see, e.g. [87,107,192,193]. In [192], the probability-dependent gain-scheduled state feedback control problem has been studied for a class of discrete-time stochastic delayed systems with randomly occurring nonlinearities. In [87], the output feedback control problem has been investigated for systems with missing measurement and discrete distributed delays. Parallel to the control issues, the probability-dependent gain-scheduling filtering problems have also been considered. In [180], an elegant result has been presented for systems with missing measurements, while a robust H_∞ deconvolution filter has been designed in [107] for systems with the randomly occurring sensor delays.

Razumikhin-Type Methodology

When it comes to the stability issues of time-delay systems, the corresponding Lyapunov–Krasovskii functional requires the state variable in the interval with regard to the length of the delay, and the Lyapunov–Krasovskii theorem cannot be applied directly. Fortunately, in [199], this difficulty can be solved by using the Razumikhin-type theorem.

The past several decades have witnessed a rapid progress on the stability issues of time-delay systems; see, e.g., [111,112,115,116,158]. For example, [111] has explored the exponential stability issues of stochastic functional differential equations with the help of Razumikhin-type theorem. Subsequently, a more general result on neutral stochastic differential delay equations has been presented in [112], which covered the result in [111] as a special case. Furthermore, the connections between the Razumikhin-type theorem and the input-to-state stability small gain theorem have been revealed in [158]. Recently, by utilizing the Razumikhin-type theory, the H_∞ analysis problem has been studied in [147] for stochastic nonlinear differential systems, and the stability analysis problem has been investigated in [115] for a class of stochastic delay interval systems.

1.2 Outline

- In Chapter 1, the research background, motivation and research problems are first introduced, which mainly involve the stochastic nonlinear systems, complex networks, networked control systems, parameter-varying systems, control and filtering, and gain-scheduling technique. In addition, the outline of the book is listed.

- In Chapter 2, the robust stability and stabilization problems are investigated for a class of stochastic time-delay interval systems with nonlinear disturbances by using the delay-dependent analysis technique. Based on the Itô's differential formula and the Lyapunov stability theory, sufficient conditions for the solvability of the addressed problems are given in terms of the LMI technique.

- In Chapter 3, the robust stabilization and robust H_∞ control problems are studied for a class of time-delay uncertain stochastic systems with

Markovian switching and nonlinear disturbance. Here, the nonlinear disturbances include the time-delay term and are also mode dependent, hence the description of the nonlinearities addressed in this chapter is more general than those in the literature. A state feedback controller is designed such that, for all nonlinear disturbances, Markovian switching and admissible uncertainties, the closed-loop system is stochastically stable with a prescribed disturbance attenuation level.

- In Chapter 4, the design problems of H_∞ output feedback controller and filter are addressed for time-delay stochastic systems with nonlinear disturbances, sensor nonlinearities, and Markovian jump parameters, respectively. A delay-dependent approach is developed to design the H_∞ controller and filter for stochastic delay jump systems, such that, for both the nonlinear disturbances and sensor nonlinearities, the augmented systems are stochastically stable with a prescribed disturbance attenuation level.

- In Chapter 5, the H_∞ analysis problem is discussed for a general class of nonlinear stochastic systems with time-delays, where the addressed systems are described by general stochastic differential equations. By using the Razumikhin-type method, we first establish sufficient conditions to guarantee the internal stability of the time-delay stochastic systems, and then deal with the H_∞ analysis problem in order to quantify the disturbance attenuation level of the addressed nonlinear stochastic time-delays system. General conditions are derived under which the L_2 gain of the system is less than or equal to a given constant. Subsequently, some easy-to-test criteria are given that can be used to check the stability of the addressed systems.

- In Chapter 6, the filtering problem is addressed for a class of discrete-time stochastic nonlinear time-delay systems with missing measurements and stochastic disturbances. The sensor measurement missing is assumed to be random and different for individual sensor, which is modeled by a set of random variables satisfying certain probabilistic distributions on the interval [0 1]. Such probabilistic distributions could be any commonly used discrete distributions. By using the LMI method, sufficient conditions are derived to ensure the existence of the desired filters, and the filter gain is characterized in terms of the solution to a set of LMIs.

- In Chapter 7, by constructing the probability-dependent Lyapunov functions, the gain-scheduled control problem is investigated for a class of discrete-time stochastic delayed systems with randomly occurring sector-nonlinearities. The sector-nonlinearities are assumed to occur according to a time-varying Bernoulli distribution with known conditional probability in real time. The aim of the addressed gain-scheduled control problem is to design a controller with scheduled gains such that, for all RONs, time-delays, and external noise disturbances, the closed-loop system is exponentially mean-square stable. Note that the designed gain-scheduled controller is based on the measured time-varying probability and is therefore less conservative than the conventional controllers with constant gains. It is shown that the time-varying controller gains can be derived in terms of the measurable probability by solving a convex optimization problem via the semidefinite program method.

- In Chapter 8, the gain-scheduled filtering problem is studied for a class of discrete-time systems with missing measurements, nonlinear disturbances, and external stochastic noises. The measurement missing phenomenon is assumed to occur in a random way, and the missing probability is time-varying with securable upper and low bounds that can be measured in real time. The aim of the addressed gain-scheduled filtering problem is to design a filter, such that, for all missing measurements, nonlinear disturbances, and external noise disturbances, the error dynamics is exponentially mean-square stable. The desired filter is equipped with time-varying gains based primarily on the time-varying missing probability and is therefore less conservative than the traditional filters with fixed gains. It is shown that the filter parameters can be derived in terms of the measurable probability via the semidefinite program method.

- In Chapter 9, the static output feedback control problem is studied for a class of 2-D uncertain stochastic nonlinear systems with time-delays, missing measurements, and multiplicative noises. A time-varying Bernoulli distribution model is proposed to describe the changing characteristic of the occurring probability of missing measurements, and the time-varying gains of the desired gain-scheduled controller include constant gains and time-varying probability that can adapt to the missing measurements with time-varying probability. By employing the probability-dependent gain-scheduled method, the designed static output feedback controller possesses less conservatism than the traditional one that is with constant gains only.

- In Chapter 10, the filtering problem is addressed for a class of discrete-time stochastic nonlinear networked systems with multiple random communication delays and random packet losses. The communication delay and packet losses, which are frequently encountered in communication networks with limited signal transmission capacity, are modeled by a stochastic mechanism that combines a certain set of indicator functions dependent on the same stochastic variable. A linear filter is designed, such that, for all random incomplete measurement phenomenon, stochastic disturbances as well as sector nonlinearities, the filtering error dynamics is exponentially mean-square stable.

- Chapter 11 is concerned with the filtering problem for a class of nonlinear genetic regulatory networks with state-dependent stochastic disturbances and state delays. The feedback regulation is described by a sector-like nonlinear function. The true concentrations of the mRNA and protein are estimated by designing a linear filter with guaranteed exponential stability of the filtering augmented systems. By using the LMI technique, sufficient conditions are first derived for ensuring the exponential mean square stability with a prescribed decay rate for the gene regulatory model, and then the filter gain is characterized in terms of the solution to an LMI, which can be easily solved by using available software packages.

- In Chapter 12, the H_∞ state estimation problem is investigated for a class of discrete-time complex networks with randomly occurring phenomena. The proposed randomly occurring phenomena include both probabilistic missing measurements and randomly occurring coupling delays, which are described by two random variables satisfying individual probability distributions. The purpose of the addressed H_∞ state estimation problem is to design a state estimator, such that, for all nonlinear disturbances, missing measurements as well as coupling delays, the dynamics of the augmented systems is guaranteed to be exponentially mean-square stable with a given H_∞ performance level. By constructing a novel Lyapunov–Krasovskii functional and utilizing convex optimization method as well as Kronecker product, we derive the sufficient conditions for the existence of the desired state estimator.

- In Chapter 13, the H_∞ synchronization control problem is investigated for a class of dynamical networks with randomly varying nonlinearities. The time-varying nonlinearities of each node are modeled to be randomly switched between two different nonlinear functions by utilizing a Bernoulli distributed variable satisfying a randomly varying conditional probability distribution. A probability-dependent gain scheduling method is adopted to handle the time-varying characteristic of the switching probability. Attention is focused on the design of a sequence of gain-scheduled controllers, such that the controlled networks are exponentially mean-square stable, and the H_∞ synchronization performance is achieved in the simultaneous presence of randomly varying nonlinearities and external energy bounded disturbances. In view of semidefinite programming method, controller parameters are derived in terms of the solutions to a series of LMIs that can be easily solved by using the MATLAB toolbox.

2

Robust Stabilization for Stochastic Time-Delay Interval Systems

CONTENTS

In this chapter, we deal with the robust stability and stabilization problems for a class of stochastic time-delay interval systems with nonlinear disturbances by developing delay-dependent analysis techniques. The robust stability analysis problem is first dealt with, where the aim is to derive sufficient conditions, such that the system is asymptotically stable in the mean-square sense, dependent on the length of the time-delays, for all admissible nonlinear disturbances as well as interval time-varying uncertain parameters. Then, we tackle the robust stabilization problem where a memoryless state feedback controller is designed to stabilize the closed-loop system. By using Itô's differential formula and the Lyapunov stability theory, sufficient conditions for the solvability of these problems are derived in term of linear matrix inequalities, which can be easily checked by resorting to available software packages. A numerical example is exploited to demonstrate the effectiveness of the obtained results.

The remainder of this chapter is organized as follows. In Section 2.1, the robust stability and stabilization problems for a class of stochastic time-delay interval systems with nonlinear disturbances is formulated. Section 2.2 deals with the robust stability analysis problem in a unified linear matrix inequality (LMI) framework. In Section 2.3, an LMI design procedure is developed for the state feedback controller, and a delay-dependent LMI technique is developed in order to obtain a less conservative condition. An illustrative example is presented to show the effectiveness of the proposed algorithm in Section 2.4. Section 2.5 provides summary remarks.

2.1 Problem Formulation

For a matrix $D_{n_1 \times n_2}$, define the following matrix interval:

$$D_I = [\underline{D}, \bar{D}] = \{D = [d_{ij}]_{n_1 \times n_2} : \underline{d}_{ij} \leq d_{ij} \leq \bar{d}_{ij}, 1 \leq i \leq n_1, 1 \leq j \leq n_2\},$$

where $\underline{D} = [\underline{d}_{ij}]_{n_1 \times n_2}$ and $\bar{D} = [\bar{d}_{ij}]_{n_1 \times n_2}$ satisfy $\underline{d}_{ij} \leq \bar{d}_{ij}$ for all $1 \leq i \leq n_1$, $1 \leq j \leq n_2$.

Consider the following stochastic time-delay interval system with nonlinear disturbance:

$$dx(t) = [Ax(t) + A_d x(t-\tau) + Bu(t) + f(x(t), x(t-\tau))]\,dt + Ex(t)d\omega(t) \quad (2.1)$$

$$x(t) = \phi(t), \quad \forall t \in [-\tau, 0], \quad (2.2)$$

where $x(t) \in \mathbb{R}^n$ is the state, $u(t) \in \mathbb{R}^p$ is the control input, $f(\cdot, \cdot)$ is an unknown nonlinear exogenous disturbance input, and $\omega(t)$ is a one-dimensional Brownian motion satisfying

$$\mathbb{E}\{d\omega(t)\} = 0, \quad \mathbb{E}\{d\omega^2(t)\} = dt.$$

Furthermore, τ is a real constant time-delay satisfying $0 \leq \tau < \infty$, and $\phi(t) \in C([-\tau, 0]; \mathbb{R}^n)$ is the initial function. The system matrices $A \in A_I$, $A_d \in A_{dI}$, $B \in B_I$, and $E \in E_I$, where $A_I = [\underline{A}, \bar{A}] = \{A = [a_{ij}]_{n \times n}\}$, $A_{dI} = [\underline{A}_d, \bar{A}_d] = \{A_d = [a_{dij}]_{n \times n}\}$, $B_I = [\underline{B}, \bar{B}] = \{B = [b_{ij}]_{n \times p}\}$, and $E_I = [\underline{E}, \bar{E}] = \{E = [e_{ij}]_{n \times n}\}$.

By setting

$$\begin{aligned}
A_0 &= \frac{1}{2}(\underline{A} + \bar{A}), \quad \tilde{A} = (\tilde{a}_{ij}) = \frac{1}{2}(\bar{A} - \underline{A}), \\
A_{d0} &= \frac{1}{2}(\underline{A}_d + \bar{A}_d), \; \tilde{A}_d = (\tilde{a}_{dij}) = \frac{1}{2}(\bar{A}_d - \underline{A}_d), \\
B_0 &= \frac{1}{2}(\underline{B} + \bar{B}), \quad \tilde{B} = (\tilde{b}_{ij}) = \frac{1}{2}(\bar{B} - \underline{B}), \\
E_0 &= \frac{1}{2}(\underline{E} + \bar{E}), \quad \tilde{E} = (\tilde{e}_{ij}) = \frac{1}{2}(\bar{E} - \underline{E}),
\end{aligned}$$

we can rewrite A, A_d, B, and E as follows:

$$\begin{cases}
A = A_0 + A^\delta = A_0 + \Sigma_{i,\,j=1}^n e_i a_{ij}^\delta e_j^T, \; |a_{ij}^\delta| \leq \tilde{a}_{ij}, \\
A_d = A_{d0} + A_d^\delta = A_{d0} + \Sigma_{i,\,j=1}^n e_i a_{dij}^\delta e_j^T, \; |a_{dij}^\delta| \leq \tilde{a}_{dij}, \\
B = B_0 + B^\delta = B_0 + \Sigma_{i=1}^n \Sigma_{j=1}^p e_i b_{ij}^\delta h_j^T, \; |b_{ij}^\delta| \leq \tilde{b}_{ij}, \\
E = E_0 + E^\delta = E_0 + \Sigma_{i,\,j=1}^n e_i e_{ij}^\delta e_j^T, \; |e_{ij}^\delta| \leq \tilde{e}_{ij},
\end{cases} \quad (2.3)$$

where $e_k \in \mathbb{R}^n$ or $h_k \in \mathbb{R}^p$ denotes the column vector, with the kth element being 1 and others being 0.

Remark 2.1 *In practice, the interval uncertainties described in (2.3) are frequently encountered in many engineering systems, which may result from the variation of operating points, aging of the devices, identification errors, etc. For example, when modeling a real world plant, we often use an interval to estimate a certain parameter so as to allow for some margin for error in the parameter identification. In the past few years, the control problems for systems with interval uncertainty have attracted considerable research attentions; see [67, 114] and references therein.*

In this chapter, the nonlinear disturbances are assumed to satisfy the following boundedness condition.

Assumption 2.1 *There exist real constant matrices $G_1 \in \mathbb{R}^{n\times n}$ and $G_2 \in \mathbb{R}^{n\times n}$, such that the unknown nonlinear vector function $f(\cdot,\cdot)$ satisfies:*

$$|f(x(t), x(t-\tau))| \le |G_1 x(t)| + |G_2 x(t-\tau)|. \tag{2.4}$$

Remark 2.2 *The exogenous nonlinear time-varying disturbance has been dealt with in many papers, such as [177]. In Assumption 2.1, the nonlinear disturbance $f(x(t), x(t-\tau))$ in the system (2.1)–(2.2) involves the delayed term, which is more general than that studied in [177]. To the best of the authors' knowledge, there has been few research efforts reported in the literature on using delay-dependent technique to deal with the robust stabilization problem for stochastic time-delay interval systems, including such kind of nonlinear exogenous disturbances.*

For the sake of simplicity, we denote:

$$\begin{cases}
F_0 = [E_0^T \ E_0^T],\ H_1 = [G_1^T \ G_1^T],\ H_2 = [G_2^T \ G_2^T] \\
J_1 = \bar{\tau}^{-1} Z - \epsilon_2 I - \Sigma_4 - \Sigma_5,\ J_2 = \mathrm{diag}\{X - \Sigma_6,\ \bar{\tau}^{-1} I - \Sigma_7\}, \\
J_3 = \mathrm{diag}\{2\epsilon_1 I \ 2\epsilon_2 I\} \\
\mathcal{X} = [\underbrace{X, ..., X}_{n}],\ \mathcal{Y} = [\underbrace{Y^T, ..., Y^T}_{n}]^T,\ \mathcal{I} = [\underbrace{I, ..., I}_{n}],\ \mathcal{Z} = [\underbrace{Z, ..., Z}_{n}] \\
\Upsilon = [\mathcal{X}\ \mathcal{X}\ \mathcal{I}\ \mathcal{X}\ \mathcal{X}\ \mathcal{X}],\ U = \mathrm{diag}\{U_1, U_2, U_3, U_4, U_6, U_7\} \\
U_l = \mathrm{diag}\{\eta_{l11}, ..., \eta_{l1n}, ..., \eta_{ln1}, ..., \eta_{lnn}\},\ (l = 1, 2, 4, 5, 6, 7), \\
U_l = \mathrm{diag}\{\eta_{l11}, ..., \eta_{ln1}, ..., \eta_{l1n}, ..., \eta_{lnn}\},\ (l = 3, 8), \\
U_l = \mathrm{diag}\{\eta_{l11}, ..., \eta_{l1p}, ..., \eta_{ln1}, ..., \eta_{lnp}\},\ (l = 9, 10), \\
\Sigma_1 = \Sigma_{i,j=1}^n \eta_{1ij} \tilde{a}_{ij}^2 e_i e_i^T,\ \Sigma_2 = \Sigma_{i,j=1}^n \eta_{2ij} \tilde{a}_{dij}^2 e_i e_i^T, \\
\Sigma_3 = \Sigma_{i,j=1}^n \eta_{3ij} \tilde{a}_{dij}^2 e_j e_j^T,\ \Sigma_4 = \Sigma_{i,j=1}^n \eta_{4ij} \tilde{a}_{ij}^2 e_i e_i^T \\
\Sigma_5 = \Sigma_{i,j=1}^n \eta_{5ij} \tilde{a}_{dij}^2 e_i e_i^T,\ \Sigma_6 = \Sigma_{i,j=1}^n \eta_{6ij} \tilde{e}_{ij}^2 e_i e_i^T, \\
\Sigma_7 = \Sigma_{i,j=1}^n \eta_{7ij} e_{ij}^2 e_i e_i^T,\ \Sigma_8 = \Sigma_{i,j=1}^n \eta_{8ij} \tilde{a}_{dij}^2 e_j e_j^T \\
\Sigma_9 = \Sigma_{i=1}^n \Sigma_{j=1}^p \eta_{9ij} \tilde{b}_{ij}^2 h_i h_i^T,\ \Sigma_{10} = \Sigma_{i=1}^n \Sigma_{j=1}^p \eta_{10ij} \tilde{b}_{ij}^2 h_i h_i^T \\
\Phi = (A_0 + A_{d0})X + X(A_0 + A_{d0})^T + S + \epsilon_1 I + \bar{\tau} T + \Sigma_1 + \Sigma_2, \\
\Psi = (A_0 + A_{d0})X + X(A_0 + A_{d0})^T + B_0 Y + Y^T B_0^T + S + \epsilon_1 I \\
\qquad + \bar{\tau} T + \Sigma_1 + \Sigma_2 + \Sigma_9.
\end{cases} \tag{2.5}$$

Observe the system (2.1)–(2.2), and let $x(t;\xi)$ denote the state trajectory from the initial data $x(\theta) = \xi(\theta)$ on $-\tau \leq \theta \leq 0$ in $L^2_{\mathcal{F}_0}([-\tau, 0]; \mathbb{R}^n)$. Obviously, $x(t, 0) \equiv 0$ is the trivial solution of system (2.1)–(2.2) corresponding to the initial data $\xi = 0$.

Before formulating the problem to be coped with, we first introduce the following stability concepts for (2.1)–(2.2).

Definition 1 For the stochastic time-delay interval system (2.1)–(2.2) with $u(t) = 0$ and every $\xi \in L^2_{\mathcal{F}_0}([-\tau, 0]; \mathbb{R}^n)$, the trivial solution is said to be *mean-square asymptotically stable* if

$$\lim_{t \to \infty} \mathbb{E}|x(t)|^2 = 0.$$

Definition 2 The stochastic time-delay interval system (2.1)–(2.2) with the state feedback controller $u(t) = Kx(t)$ is said to be *robustly stochastically stabilizable* if there exists a gain matrix $K \in R^{p \times n}$, such that the closed-loop system is mean-square asymptotically stable.

The purpose of this chapter is to design a state feedback controller, such that the stochastic time-delay interval system (2.1)–(2.2) with nonlinear disturbance is stochastically stabilized by developing delay-dependent techniques.

2.2 Robust Stability Analysis

First, let us give the following lemmas that will be used in the proof of our main results.

Lemma 2.1 *(Schur Complement) [13] Given the constant matrices Σ_1, Σ_2, and Σ_3, where $\Sigma_1 = \Sigma_1^T$ and $0 < \Sigma_2 = \Sigma_2^T$. Then $\Sigma_1 + \Sigma_3^T \Sigma_2^{-1} \Sigma_3 < 0$ if and only if*

$$\begin{bmatrix} \Sigma_1 & \Sigma_3^T \\ \Sigma_3 & -\Sigma_2 \end{bmatrix} < 0,$$

or equivalently,

$$\begin{bmatrix} -\Sigma_2 & \Sigma_3 \\ \Sigma_3^T & \Sigma_1 \end{bmatrix} < 0.$$

Lemma 2.2 *[170] Let $\mathcal{A}$, $\mathcal{D}$, $\mathcal{H}$, $\mathcal{W}$, and F be real matrices of appropriate dimension, such that $\mathcal{W} > 0$ and $F^T F \leq I$. We have the following:*

1) For a scalar $\varepsilon > 0$ and two vectors x, $y \in \mathbb{R}^n$,

$$2x^T \mathcal{D} F \mathcal{H} y \leq \varepsilon^{-1} x^T \mathcal{D}\mathcal{D}^T x + \varepsilon y^T \mathcal{H}^T \mathcal{H} y; \tag{2.6}$$

2) For any scalar $\varepsilon > 0$ such that $\mathcal{W} - \varepsilon \mathcal{D}\mathcal{D}^T > 0$,

$$(\mathcal{A} + \mathcal{D}F\mathcal{H})^T \mathcal{W}^{-1} (\mathcal{A} + \mathcal{D}F\mathcal{H}) \leq \mathcal{A}^T (\mathcal{W} - \varepsilon \mathcal{D}\mathcal{D}^T)^{-1} \mathcal{A} + \varepsilon^{-1} \mathcal{H}^T \mathcal{H}. \tag{2.7}$$

For presentation convenience, we define the following new state variable

$$y(t) = Ax(t) + A_d x(t-\tau) + Bu(t) + f(x(t), x(t-\tau)), \tag{2.8}$$

and then the system (2.1) can be represented as

$$dx(t) = y(t)dt + Ex(t)d\omega(t). \tag{2.9}$$

In the following theorem, a delay-dependent LMI approach is developed to solve the robust stability analysis problem for the stochastic time-delay interval system (2.1)–(2.2), with $u(t) = 0$, and a sufficient condition is derived to ensure the solvability of the problem.

Theorem 1 Consider the system (2.1)–(2.2), with $u(t) \equiv 0$. If there exist positive definite matrices $X > 0$, $S > 0$, $Z > 0$, $T > 0$, and positive scalars $\epsilon_1 > 0$, $\epsilon_2 > 0$, $\eta_{lij} > 0$, $(i, j = 1, ..., n,\ l = 1, ..., 8)$, such that the following linear matrix inequalities

$$\begin{bmatrix} \Phi & 0 & A_{d0} & XA_0^T & XF_0 & XH_1 & 0 & \Upsilon & 0 \\ * & -S & 0 & XA_{d0}^T & 0 & 0 & XH_2 & 0 & \mathcal{X} \\ * & * & -I+\Sigma_3 & 0 & 0 & 0 & 0 & 0 & 0 \\ * & * & * & -J_1 & 0 & 0 & 0 & 0 & 0 \\ * & * & * & * & -J_2 & 0 & 0 & 0 & 0 \\ * & * & * & * & * & -J_3 & 0 & 0 & 0 \\ * & * & * & * & * & * & -J_3 & 0 & 0 \\ * & * & * & * & * & * & * & -U & 0 \\ * & * & * & * & * & * & * & * & -U_5 \end{bmatrix} < 0 \tag{2.10}$$

$$\begin{bmatrix} -T & A_{d0}Z & 0 \\ * & -Z+\Sigma_8 & \mathcal{Z} \\ * & * & -U_8 \end{bmatrix} < 0 \tag{2.11}$$

hold, where Φ, F_0, H_1, H_2, J_1, J_2, J_3, Υ, U, U_5, U_8, Σ_3, Σ_8, $\mathcal{X}$, and $\mathcal{Z}$ are all defined in (2.5), then the system (2.1)–(2.2), with $u(t) \equiv 0$ is mean-square asymptotically stable.

Proof 1 *Recalling the Newton–Leibniz formula and (2.9), we can write that, for $t \geq \tau$,*

$$\begin{aligned} x(t-\tau) &= x(t) - \int_{t-\tau}^{t} dx(s) \\ &= x(t) - \left[\int_{t-\tau}^{t} y(s)ds + \int_{t-\tau}^{t} Ex(s)d\omega(s)\right]. \end{aligned} \tag{2.12}$$

Then, it is easy to know from (2.12) that the following system is equivalent to

(2.1)–(2.2), with $u(t) = 0$:

$$\begin{aligned} dx(t) &= [(A + A_d)x(t) - A_d \int_{t-\tau}^{t} y(s)ds - A_d \int_{t-\tau}^{t} Ex(s)d\omega(s) \\ &\quad + f(x(t), x(t-\tau))]dt + Ex(t)d\omega(t). \quad (2.13) \\ x(t) &= \psi(t), \quad t \in [-2\tau, 0], \quad r(0) = r_0, \quad (2.14) \end{aligned}$$

where $\psi(t)$ is the initial function. Hence, it suffices to prove the mean-square asymptotic stability of the above system.

Now, let $P = X^{-1} > 0$, $Q = PSP > 0$, $R = Z^{-1} > 0$, and define the following Lyapunov–Krasovskii functional candidate for the system (2.13):

$$\begin{aligned} V(x(t), t) &= x^T(t)Px(t) + \int_{t-\tau}^{t} x^T(s)Qx(s)ds + \int_{t-\tau}^{t} \int_{s}^{t} y^T(\beta)Ry(\beta)d\beta ds \\ &\quad + \int_{t-\tau}^{t} \int_{s}^{t} |Ex(\beta)|^2 d\beta ds. \quad (2.15) \end{aligned}$$

Noticing the fact of

$$\tau x^T(t)Wx(t) = \int_{t-\tau}^{t} x^T(t)Wx(t)ds,$$

it can be derived by Itô's differential formula [81] that

$$dV(x(t), t) = \mathcal{L}V(x(t), t)dt + 2x^T(t)PEx(t)d\omega(t), \quad (2.16)$$

where

$$\begin{aligned} \mathcal{L}V(x(t), t) &= x^T(t)[(A + A_d)^T P + P(A + A_d) + Q + E^T PE + \tau W]x(t) \\ &\quad -2x^T(t)PA_d \left(\int_{t-\tau}^{t} y(s)ds + \int_{t-\tau}^{t} Ex(s)d\omega(s) \right) \\ &\quad +2x^T(t)Pf(x(t), x(t-\tau)) - x^T(t-\tau)Qx(t-\tau) \\ &\quad +\tau y^T(t)Ry(t) - \int_{t-\tau}^{t} y^T(s)Ry(s)ds + \tau x^T(t)E^T Ex(t) \\ &\quad - \int_{t-\tau}^{t} |Ex(s)|^2 ds - \int_{t-\tau}^{t} x^T(t)Wx(t)ds, \quad (2.17) \end{aligned}$$

with $W = PTP > 0$.

Noting (2.4) and Lemma 2.2, we can calculate that

$$\begin{aligned} & 2x^T(t)Pf(x(t), x(t-\tau)) \\ \leq\ & \epsilon_1 x^T(t)P^2 x(t) + \epsilon_1^{-1} f^T(x(t), x(t-\tau))f(x(t), x(t-\tau)) \\ \leq\ & \epsilon_1 x^T(t)P^2 x(t) + \epsilon_1^{-1}(|G_{1i}x(t)| + |G_{2i}x(t-\tau)|)^2 \\ \leq\ & \epsilon_1 x^T(t)P^2 x(t) + 2\epsilon_1^{-1}[x^T(t)G_1^T G_1 x(t) \\ & +x^T(t-\tau)G_2^T G_2 x(t-\tau)]. \quad (2.18) \end{aligned}$$

Again, we can obtain from Lemma 2.2 that

$$-2x^T(t)PA_d\int_{t-\tau}^{t}Ex(s)d\omega(s) \le x^T(t)PA_dA_d^TPx(t) + |\int_{t-\tau}^{t}Ex(s)d\omega(s)|^2. \tag{2.19}$$

Moreover,

$$\mathbb{E}|\int_{t-\tau}^{t}Ex(s)d\omega(s)|^2 \le \int_{t-\tau}^{t}\mathbb{E}|Ex(s)|^2ds. \tag{2.20}$$

Using Lemma 2.2 and (2.8), we have

$$\begin{aligned}
& \tau y^T(t)Ry(t) \\
= & [Ax(t)+A_dx(t-\tau)+f(x(t),x(t-\tau))]^T(\tau R) \\
& [Ax(t)+A_dx(t-\tau)+f(x(t),x(t-\tau))] \\
\le & [Ax(t)+A_dx(t-\tau)]^T[(\tau R)^{-1}-\epsilon_2 I]^{-1}[Ax(t)+A_dx(t-\tau)] \\
& +\epsilon_2^{-1}f^T(x(t),x(t-\tau))f(x(t),x(t-\tau)) \\
\le & [Ax(t)+A_dx(t-\tau)]^T[(\tau R)^{-1}-\epsilon_2 I]^{-1}[Ax(t)+A_dx(t-\tau)] \\
& +2\epsilon_2^{-1}[x^T(t)G_1^TG_1x(t)+x^T(t-\tau)G_2^TG_2x(t-\tau)].
\end{aligned} \tag{2.21}$$

Substituting (2.18)–(2.21) into (2.17) and taking expectation lead to

$$\mathbb{E}\mathcal{L}V(x(t),t) \le \mathbb{E}\{\bar{x}^T(t)\Omega\bar{x}(t)\} + \int_{t-\tau}^{t}\mathbb{E}\{\bar{x}^T(t,s)\Pi\bar{x}(t,s)\}, \tag{2.22}$$

where

$$\Omega := \begin{bmatrix} \Omega_1+\Delta & 0 \\ 0 & \Omega_2 \end{bmatrix} + \begin{bmatrix} A^T \\ A_d^T \end{bmatrix}[(\tau R)^{-1}-\epsilon_2 I)]^{-1}[A \ \ A_d], \tag{2.23}$$

$$\Pi := \begin{bmatrix} -W & -PA_d \\ -A_d^TP & -R \end{bmatrix}, \tag{2.24}$$

with

$$\begin{aligned}
\bar{x}(t) &= [x^T(t) \ \ x^T(t-\tau)]^T, \ \bar{x}(t,s)=[x^T(t) \ \ y^T(s)]^T, \\
\Omega_1 &= (A+A_d)^TP+P(A+A_d)+Q+\epsilon_1P^2+\tau W, \\
\Omega_2 &= 2(\epsilon_1^{-1}+\epsilon_2^{-1})G_2^TG_2-Q, \\
\Delta &= PA_dA_d^TP+E^TPE+\tau E^TE+2(\epsilon_1^{-1}+\epsilon_2^{-1})G_1^TG_1.
\end{aligned}$$

It remains to show that $\Omega<0$ *and* $\Pi<0$*. By Schur complement lemma, it is easily seen that* $\Omega<0$ *if and only if*

$$\begin{bmatrix}
\Omega_1 & 0 & PA_d & A^T & E^T & E^T & H_1 & 0 \\
* & -Q & 0 & A_d^T & 0 & 0 & 0 & H_2 \\
* & * & -I & 0 & 0 & 0 & 0 & 0 \\
* & * & * & \epsilon_2 I-\tau^{-1}Z & 0 & 0 & 0 & 0 \\
* & * & * & * & -P^{-1} & 0 & 0 & 0 \\
* & * & * & * & * & -\tau^{-1}I & 0 & 0 \\
* & * & * & * & * & * & -J_3 & 0 \\
* & * & * & * & * & * & * & -J_3
\end{bmatrix} < 0, \tag{2.25}$$

where H_1, H_2 are defined in (2.5).

On the other hand, we note that premultiplying and postmultiplying (2.25) by

$$\mathrm{diag}(X, X, I, I, I, I, I, I)$$

yields

$$\bar{\Omega} = \begin{bmatrix} \bar{\Omega}_1 & 0 & A_d & XA^T & XE^T & XE^T & XH_1 & 0 \\ * & -S & 0 & XA_d^T & 0 & 0 & 0 & XH_2 \\ * & * & -I & 0 & 0 & 0 & 0 & 0 \\ * & * & * & \epsilon_2 I - \tau^{-1}Z & 0 & 0 & 0 & 0 \\ * & * & * & * & -X & 0 & 0 & 0 \\ * & * & * & * & * & -\tau^{-1}I & 0 & 0 \\ * & * & * & * & * & * & -J_3 & 0 \\ * & * & * & * & * & * & * & -J_3 \end{bmatrix} < 0, \tag{2.26}$$

with

$$\bar{\Omega}_1 = X(A + A_d)^T + (A + A_d)X + S + \epsilon_1 I + \tau T.$$

Similarly, premultiplying and postmultiplying $\Pi < 0$ *by* $\mathrm{diag}(X, Z)$ *result in*

$$\bar{\Pi} = \begin{bmatrix} -T & -A_d Z \\ -ZA_d^T & -Z \end{bmatrix} < 0. \tag{2.27}$$

Note that we use the shorthand $\mathfrak{M}_n[(M_1)_{i_1,j_1}, (M_2)_{i_2,j_2}, ..., (M_r)_{i_r,j_r}]$ *to represent an nth-order block square matrix whose all nonzero blocks are the* i_1j_1*th block* M_1*, the* i_2j_2*th block* M_2*, ... , the* i_rj_r*th block* M_r*, and all other blocks are zero matrices. Then, the matrix* $\bar{\Omega}$ *can be further rearranged as*

$$\begin{aligned} \bar{\Omega} &= \begin{bmatrix} \bar{\Omega}_{10} & 0 & A_{d0} & XA_0^T & XE_0^T & XE_0^T & XH_1 & 0 \\ * & -S & 0 & XA_{d0}^T & 0 & 0 & 0 & XH_2 \\ * & * & -I & 0 & 0 & 0 & 0 & 0 \\ * & * & * & \epsilon_2 I - \tau^{-1}Z & 0 & 0 & 0 & 0 \\ * & * & * & * & -X & 0 & 0 & 0 \\ * & * & * & * & * & -\tau^{-1}I & 0 & 0 \\ * & * & * & * & * & * & -J_3 & 0 \\ * & * & * & * & * & * & * & -J_3 \end{bmatrix} \\ &\quad + \mathfrak{M}_8\left[(\bar{\Omega}_1^\delta)_{1,1}\right] + \mathfrak{M}_8\left[(A_d^\delta)_{1,3},\ ((A_d^\delta)^T)_{3,1}\right] \\ &\quad + \mathfrak{M}_8\left[(X(A^\delta)^T)_{1,4},\ (A^\delta X)_{4,1}\right] + \mathfrak{M}_8\left[(X(A_d^\delta)^T)_{2,4},\ (A_d^\delta X)_{4,2}\right] \\ &\quad + \mathfrak{M}_8\left[(X(E^\delta)^T)_{1,5},\ (E^\delta X)_{5,1}\right] + \mathfrak{M}_8\left[(X(E^\delta)^T)_{1,6},\ (E^\delta X)_{6,1}\right] \\ &:= \Phi_0 + \Phi_1 + \Phi_2 + \Phi_3 + \Phi_4 + \Phi_5 + \Phi_6, \end{aligned} \tag{2.28}$$

where

$$\begin{aligned} \bar{\Omega}_{10} &= X(A_0 + A_{d0})^T + (A_0 + A_{d0})X + S + \epsilon_1 I + \tau T, \\ \bar{\Omega}_1^\delta &= X(A^\delta + A_d^\delta)^T + (A^\delta + A_d^\delta)X. \end{aligned}$$

It follows from Lemma 2.2 and (2.3) that, for any real scalars $\eta_{lij} > 0$ $(i,\ j = 1, ..., n; l = 1, 2)$, *the following holds:*

$$
\begin{aligned}
\Phi_1 &= [X,0,0,0,0,0,0,0]^T[(A^\delta + A_d^\delta)^T,0,0,0,0,0,0,0] \\
&\quad +[(A^\delta + A_d^\delta)^T,0,0,0,0,0,0,0]^T[X,0,0,0,0,0,0,0] \\
&= \Sigma_{i,\,j=1}^n \left([X,0,0,0,0,0,0,0]^T[(e_i a_{ij}^\delta e_j^T)^T,0,0,0,0,0,0,0]\right. \\
&\quad +[(e_i a_{ij}^\delta e_j^T)^T,0,0,0,0,0,0,0]^T[X,0,0,0,0,0,0,0] \\
&\quad +[X,0,0,0,0,0,0,0]^T[(e_i a_{dij}^\delta e_j^T)^T,0,0,0,0,0,0,0] \\
&\quad \left.+[(e_i a_{dij}^\delta e_j^T)^T,0,0,0,0,0,0,0]^T[X,0,0,0,0,0,0,0]\right) \\
&= \Sigma_{i,\,j=1}^n \left([e_j^T X,0,0,0,0,0,0,0]^T[(e_i a_{ij}^\delta)^T,0,0,0,0,0,0,0]\right. \\
&\quad +[(e_i a_{ij}^\delta)^T,0,0,0,0,0,0,0]^T[e_j^T X,0,0,0,0,0,0,0] \\
&\quad +[e_j^T X,0,0,0,0,0,0,0]^T[(e_i a_{dij}^\delta)^T,0,0,0,0,0,0,0] \\
&\quad \left.+[(e_i a_{dij}^\delta)^T,0,0,0,0,0,0,0]^T[e_j^T X,0,0,0,0,0,0,0]\right) \\
&\leq \Sigma_{i,\,j=1}^n \left(\eta_{1ij}^{-1}[e_j^T X,0,0,0,0,0,0,0]^T[e_j^T X,0,0,0,0,0,0,0]\right. \\
&\quad +\eta_{1ij}\tilde{a}_{ij}[e_i^T,0,0,0,0,0,0,0]^T[e_i^T,0,0,0,0,0,0,0] \\
&\quad +\eta_{2ij}^{-1}[e_j^T X,0,0,0,0,0,0,0]^T[e_j^T X,0,0,0,0,0,0,0] \\
&\quad \left.+\eta_{2ij}\tilde{a}_{dij}[e_i^T,0,0,0,0,0,0,0]^T[e_i^T,0,0,0,0,0,0,0]\right) \\
&= \mathfrak{M}_8\left[(\Sigma_1 + \Sigma_2)_{1,1}\right] + [\mathcal{X},0,0,0,0,0,0,0]^T U_1^{-1} [\mathcal{X},0,0,0,0,0,0,0] \\
&\quad + [\mathcal{X},0,0,0,0,0,0,0]^T U_2^{-1} [\mathcal{X},0,0,0,0,0,0,0] \\
&= \mathfrak{M}_8\left[(\Sigma_1 + \Sigma_2 + \mathcal{X}U_1^{-1}\mathcal{X} + \mathcal{X}U_2^{-1}\mathcal{X})_{1,1}\right], \qquad (2.29)
\end{aligned}
$$

where $\mathcal{X}$, Σ_1, Σ_2, U_1, *and* U_2 *are defined in (2.5).*

Similarly, for any scalars $\eta_{lij} > 0$, $(i,\ j = 1, ..., n,\ l = 3, 4, 5, 6, 7, 8)$, *we have*

$$
\begin{aligned}
\Phi_2 &\leq \mathfrak{M}_8\left[(\Sigma_3)_{3,3}\right] + [\mathcal{I},0,0,0,0,0,0,0]^T U_3^{-1} [\mathcal{I},0,0,0,0,0,0,0], \\
&= \mathfrak{M}_8\left[(U_3^{-1})_{1,1},\ \ (\Sigma_3)_{3,3}\right] \qquad (2.30) \\
\Phi_3 &\leq \mathfrak{M}_8\left[(\Sigma_4)_{4,4}\right] + [\mathcal{X},0,0,0,0,0,0,0]^T U_4^{-1} [\mathcal{X},0,0,0,0,0,0,0], \\
&= \mathfrak{M}_8\left[(\mathcal{X}U_4^{-1}\mathcal{X})_{1,1},\ \ (\Sigma_4)_{4,4}\right] \qquad (2.31) \\
\Phi_4 &\leq \mathfrak{M}_8\left[(\Sigma_5)_{4\times 4}\right] + [0,\mathcal{X},0,0,0,0,0,0]^T U_5^{-1} [0,\mathcal{X},0,0,0,0,0,0], \\
&= \mathfrak{M}_8\left[(\mathcal{X}U_5^{-1}\mathcal{X})_{2,2},\ \ (\Sigma_5)_{4,4}\right] \qquad (2.32) \\
\Phi_5 &\leq \mathfrak{M}_8\left[(\Sigma_6)_{5\times 5}\right] + [\mathcal{X},0,0,0,0,0,0,0]^T U_6^{-1} [\mathcal{X},0,0,0,0,0,0,0], \\
&= \mathfrak{M}_8\left[(\mathcal{X}U_6^{-1}\mathcal{X})_{1,1},\ \ (\Sigma_6)_{5,5}\right] \qquad (2.33) \\
\Phi_6 &\leq \mathfrak{M}_8\left[(\Sigma_7)_{6\times 6}\right] + [\mathcal{X},0,0,0,0,0,0,0]^T U_7^{-1} [\mathcal{X},0,0,0,0,0,0,0], \\
&= \mathfrak{M}_8\left[(\mathcal{X}U_7^{-1}\mathcal{X})_{1,1},\ \ (\Sigma_7)_{6,6}\right] \qquad (2.34) \\
\bar{\Pi} &\leq \begin{bmatrix} -T & A_{d0}Z \\ ZA_{d0}^T & -Z + \Sigma_8 \end{bmatrix} + \begin{bmatrix} 0 \\ \mathcal{Z} \end{bmatrix} U_8^{-1} \begin{bmatrix} 0 \\ \mathcal{Z} \end{bmatrix}^T, \qquad (2.35)
\end{aligned}
$$

where Σ_3, Σ_4, Σ_5, Σ_6, Σ_7, Σ_8, U_3, U_4, U_5, U_6, U_7, U_8, $\mathcal{I}$, *and* $\mathcal{Z}$ *are defined in (2.5).*

According to Schur complement lemma, after tedious but straightforward calculations, it is followed from the conditions (2.10), (2.11), and (2.26)–(2.35) that

$$\bar{\Omega} < 0, \bar{\Pi} < 0.$$

Obviously, from the relationship between Ω *and* $\bar{\Omega}$*, the relationship between* Π *and* $\bar{\Pi}$*, and the inequality (2.22), we can obtain*

$$\Omega < 0, \quad \Pi < 0.$$

Therefore, we can conclude that

$$\mathbb{E}\mathcal{L}V(x(t), t) < 0,$$

which indicates that the trivial solution of (2.13) is asymptotically stable in the mean-square sense. This completes the proof.

Remark 2.3 *In Theorem 1, it is shown that the unforced stochastic time-delay interval system with nonlinear disturbances is mean-square asymptotically stable if two LMIs (2.10) and (2.11) are satisfied, and the stability criteria are dependent on the length of time-delays. Note that, by MATLAB toolbox, the feasibility of the LMIs (2.10) and (2.11) can be checked easily and the maximum allowable bound of the time-delays* τ *with which the stochastic delayed interval system (2.1)–(2.2) is mean-square asymptotically stable can be determined.*

2.3 Delay-Dependent Robust Stabilization

In this section, we aim to propose a design procedure for the state feedback controller that can robustly stochastically stabilize the addressed stochastic delayed interval systems with nonlinear disturbances. Again, a delay-dependent LMI technique will be developed in order to obtain a less conservative condition. The main result of this chapter is given in the following theorem.

Theorem 2 Consider the system (2.1)–(2.2). If there exist positive definite matrices $X > 0$, $S > 0$, $Z > 0$, $T > 0$, a matrix Y, and positive scalars $\epsilon_1 > 0$, $\epsilon_2 > 0$, $\eta_{lij} > 0$, $(l = 1, 2, ..., 8)$, $\eta_{9im} > 0$, $\eta_{10im} > 0$ $(i, j = 1, ..., n;\ m =$

$1, ..., p$), such that (2.11) and the following linear matrix inequality

$$\begin{bmatrix} \Psi & 0 & A_{d0} & \mathbb{A}_0 & XF_0 & XH_1 & 0 & \Upsilon & 0 & \mathcal{Y}^T & \mathcal{Y}^T \\ * & -S & 0 & \mathbb{A}_{d0} & 0 & 0 & XH_2 & 0 & \mathcal{X} & 0 & 0 \\ * & * & -\mathbb{I} & 0 & 0 & 0 & 0 & 0 & 0 & 0 & 0 \\ * & * & * & -\mathbb{J}_1 & 0 & 0 & 0 & 0 & 0 & 0 & 0 \\ * & * & * & * & -J_2 & 0 & 0 & 0 & 0 & 0 & 0 \\ * & * & * & * & * & -J_3 & 0 & 0 & 0 & 0 & 0 \\ * & * & * & * & * & * & -J_3 & 0 & 0 & 0 & 0 \\ * & * & * & * & * & * & * & -U & 0 & 0 & 0 \\ * & * & * & * & * & * & * & * & -U_5 & 0 & 0 \\ * & * & * & * & * & * & * & * & * & -U_9 & 0 \\ * & * & * & * & * & * & * & * & * & * & -U_{10} \end{bmatrix} < 0, \tag{2.36}$$

holds, where $\mathbb{A}_0 := XA_0^T + Y^T B_0^T$, $\mathbb{A}_{d0} := XA_{d0}^T$, $\mathbb{I} := I - \Sigma_3$, $\mathbb{J}_1 := J_1 - \Sigma_{10}$, and F_0, H_1, H_2, J_1, J_2, J_3, Υ, U, U_5, U_8, U_9, U_{10}, Σ_3, Σ_8, Σ_{10}, $\mathcal{X}$, $\mathcal{Y}$, and Ψ are all defined in (2.5), then with the state feedback controller given by

$$u(t) = Kx(t), \quad K = YX^{-1}, \tag{2.37}$$

the closed-loop system is robustly stochastically stable.

Proof 2 *Applying the controller (2.37) and Newton–Leibniz formula (2.12) to the stochastic interval system (2.1), which is equivalent to the result by replacing A with $A_c = A + BK$ in (2.13), we have*

$$dx(t) = [(A_c + A_d)x(t) - A_d \int_{t-\tau}^{t} y(s)ds - A_d \int_{t-\tau}^{t} Ex(s)d\omega(s) + f(x(t), x(t-\tau))]dt + Ex(t)d\omega(t), \tag{2.38}$$

$$x(t) = \psi(t), \quad t \in [-2\tau, 0], \quad r(0) = r_0. \tag{2.39}$$

The Lyapunov–Krasovskii function is chosen as:

$$V(x(t),t) = x^T(t)Px(t) + \int_{t-\tau}^{t} x^T(s)Qx(s)ds + \int_{t-\tau}^{t}\int_{s}^{t} y^T(\beta)Ry(\beta)d\beta ds + \int_{t-\tau}^{t}\int_{s}^{t} |Ex(\beta)|^2 d\beta ds. \tag{2.40}$$

Similar to the proof of Theorem 1, we obtain

$$\mathbb{E}\mathcal{L}V(x(t),t) \le \mathbb{E}\{\bar{x}(t)^T \Gamma \bar{x}(t)\} + \int_{t-\tau}^{t} \mathbb{E}\{\bar{x}(t,s)^T \Pi \bar{x}(t,s)\}, \tag{2.41}$$

where Π is defined in (2.24) and

$$\Gamma := \begin{bmatrix} \Gamma_1 + \Delta & 0 \\ 0 & \Omega_2 \end{bmatrix} + \begin{bmatrix} A_c^T \\ A_d^T \end{bmatrix} [(\tau R)^{-1} - \epsilon_2 I]^{-1} [A_c \ A_d], \tag{2.42}$$

with

$$\Gamma_1 = P(A_c + A_d) + (A_c + A_d)^T P + Q + \epsilon_1 P^2 + \tau W.$$

Along the similar line as that in the proof of Theorem 1, we can know from (2.5), (2.36), and the expression of K in (2.37) that

$$\Gamma < 0, \quad \Pi < 0,$$

and, therefore,

$$\mathbb{E}\mathcal{L}V(x(t), t) < 0,$$

which implies that the trivial solution of the closed-loop system (2.1)–(2.2) is robustly stochastically stable. The proof is completed.

2.4 An Illustrative Example

In this section, to illustrate the usefulness and flexibility of the theory developed in the previous section, we present a simple numerical example. Attention is focused on the design of a stabilizing controller for a class of stochastic time-delay interval system with nonlinear disturbance.

The system data of (2.1)–(2.2) are as follows:

$$\underline{A} = \begin{bmatrix} -3.5 & 0.9 \\ -0.1 & -4.3 \end{bmatrix}, \quad \bar{A} = \begin{bmatrix} -2.5 & 1.1 \\ 0.1 & -3.7 \end{bmatrix},$$

$$\underline{A}_d = \begin{bmatrix} 1 & 1 \\ 0 & 1 \end{bmatrix}, \quad \bar{A}_d = \begin{bmatrix} 1.4 & 1 \\ 0 & 1.6 \end{bmatrix},$$

$$\underline{B} = \begin{bmatrix} -1.4 & 0 \\ 0 & -1.3 \end{bmatrix}, \quad \bar{B} = \begin{bmatrix} 1.6 & 0 \\ 0 & 1.7 \end{bmatrix},$$

$$\underline{E} = \begin{bmatrix} 0 & -0.1 \\ -0.1 & 0.8 \end{bmatrix}, \quad \bar{E} = \begin{bmatrix} 2 & 0.1 \\ 0.1 & 2.2 \end{bmatrix},$$

$$G_1 = \begin{bmatrix} 0.5 & 0 \\ 0 & 0.1 \end{bmatrix}, \quad G_2 = \begin{bmatrix} 0.2 & 0 \\ 0 & 0.5 \end{bmatrix}.$$

Using MATLAB LMI control toolbox to solve the LMIs (2.5) and (2.36), we obtain the maximum allowable bound of the time-delays as $\bar{\tau} = 2.2793$. Hence, we have the conclusion that the stochastic interval delay system is robustly stabilizable when $\tau \leq 2.2793$.

The solutions of the LMIs (2.5) and (2.36) in the case of $\tau = 1.0$ are given as follows

$$X = \begin{bmatrix} 1.6840 & 0.0697 \\ 0.0687 & 0.9025 \end{bmatrix}, \quad S = \begin{bmatrix} 16.2992 & -4.8426 \\ -4.8426 & 7.8962 \end{bmatrix},$$

$$T = \begin{bmatrix} 39.6446 & -3.2759 \\ -3.2759 & 65.3736 \end{bmatrix}, \quad Z = \begin{bmatrix} 184.2647 & 3.9623 \\ 3.9623 & 172.2201 \end{bmatrix},$$

$$Y = \begin{bmatrix} -79.1243 & -3.0749 \\ -2.9078 & -85.4815 \end{bmatrix}, \quad K = \begin{bmatrix} -46.9938 & 0.2207 \\ 2.1987 & -94.8815 \end{bmatrix},$$

$$\epsilon_1 = 15.3141, \quad \epsilon_2 = 26.0197,$$

$$U_1 = \mathrm{diag}(44.9998, 201.1568, 201.2179, 54.4045),$$

$$U_2 = \mathrm{diag}(155.5229, 222.5990, 222.6691, 54.4045),$$

$$U_3 = \mathrm{diag}(11.1924, 222.6691, 222.5990, 4.2363),$$

$$U_4 = \mathrm{diag}(121.2648, 216.7939, 216.8615, 118.4990),$$

$$U_5 = \mathrm{diag}(196.4067, 222.5903, 222.6697, 118.4997),$$

$$U_6 = \mathrm{diag}(0.6582, 45.2894, 45.3510, 0.6294),$$

$$U_7 = \mathrm{diag}(0.3798, 37.7572, 37.7977, 0.6979),$$

$$U_8 = \mathrm{diag}(0.6070, 0.7385, 1.3274, 0.5459),$$

$$U_9 = \mathrm{diag}(414.7423, 634.7105, 458.4166, 380.3232),$$

$$U_{10} = \mathrm{diag}(448.2585, 634.7105, 458.4166, 521.1614).$$

According to Theorem 2, with the designed controller gain K, the closed-loop system is asymptotically stable in the mean-square sense for all admissible interval uncertainties and nonlinear disturbances.

Figure 2.1 and Figure 2.3 give the evolution of states $x_1(t)$ and $x_2(t)$ for the uncontrolled systems. Figure 2.2 and Figure 2.4 depict the results of the evolution of states $x_1(t)$ and $x_2(t)$ for the controlled systems. The simulation results have illustrated our theoretical analysis.'

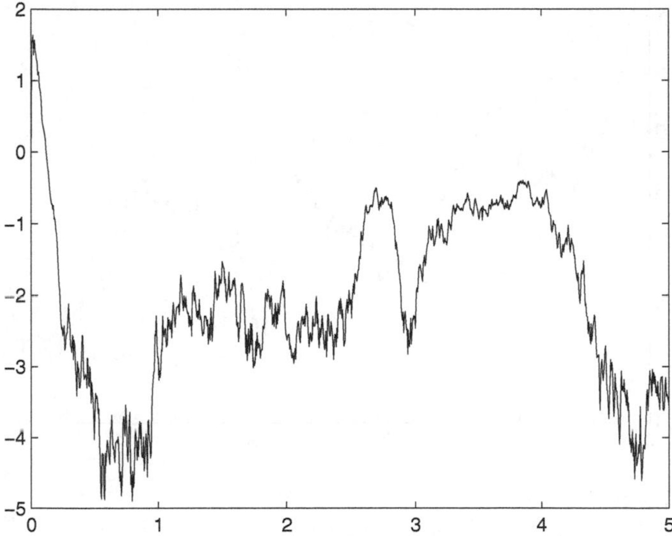

FIGURE 2.1
State evolution $x_1(t)$ of uncontrolled systems.

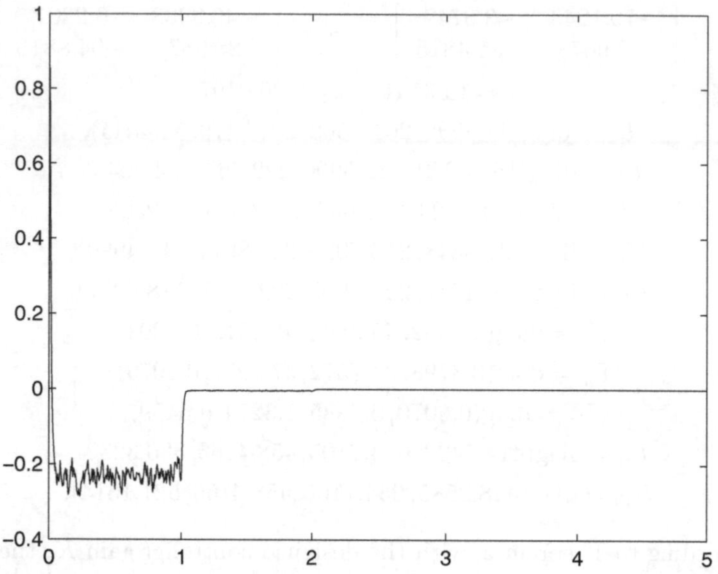

FIGURE 2.2
State evolution $x_1(t)$ of controlled systems.

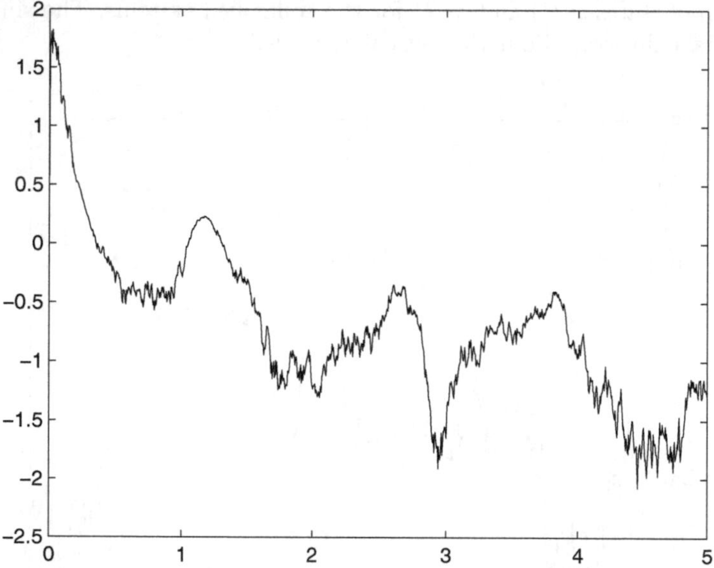

FIGURE 2.3
State evolution $x_2(t)$ of uncontrolled systems.

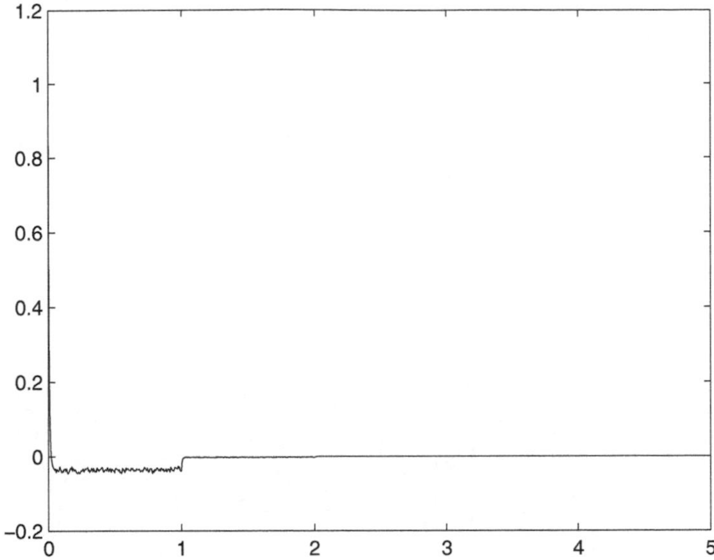

FIGURE 2.4
State evolution $x_2(t)$ of controlled systems.

2.5 Summary

In this chapter, we have investigated the robust stability analysis problem as well as the robust stabilization problem for a class of stochastic time-delay interval systems with nonlinear disturbances. A delay-dependent LMI approach has been developed to derive sufficient conditions under which the controlled system is mean-square asymptotically stable, where the conditions are dependent on the length of the time-delays. A numerical example has been employed to illustrate the effectiveness of the obtained results.

3

Robust H_∞ Control for Markov Systems with Nonlinear Disturbance

CONTENTS

In this chapter, the robust stabilization and robust H_∞ control problems for a class of stochastic time-delay uncertain systems with Markovian switching and nonlinear disturbance are investigated. Here, the nonlinear disturbance includes the time-delay term and is also mode dependent, hence, the description of the nonlinearities addressed in this chapter is more general than those in the literature. Our goal is to design a state feedback controller, which might also involve time-delay, such that for the considered nonlinear disturbances, Markovian switching and all admissible uncertainties, the closed-loop system is stochastically stable with a given disturbance attenuation level γ, independent of the time-delays. Sufficient criteria will be derived based on an LMI method, and a numerical example is given to illustrate the effectiveness of our main results.

The rest of this chapter is arranged as follows. Section 3.1 formulates the robust stabilization and robust H_∞ control problems for a class of stochastic time-delay uncertain systems with Markovian switching and nonlinear disturbance. In Section 3.2, robust stabilization and robust H_∞ control problems are analyzed. H_∞ control problem for the same system is studied, then an LMI approach is developed to ensure the solvability of the problem and a direct design method is provided of the desired controllers in Section 3.3. In Section 3.4, an illustrative numerical example is provided to show the effectiveness and usefulness of the proposed approach. Section 3.5 gives our summary.

3.1 Problem Formulation

In this chapter, $\{r(t), t \geq 0\}$ is a right-continuous Markov chain on the probability space taking values in a finite state space $S = \{1, 2, \cdots, N\}$ with the following transition probabilities:

$$P\{r(t+\Delta t) = j | r(t) = i\} = \begin{cases} \gamma_{ij}\Delta t + o(\Delta t) & \text{if } i \neq j, \\ 1 + \gamma_{ii}\Delta t + o(\Delta t) & \text{if } i = j, \end{cases} \tag{3.1}$$

where $\Delta t > 0$ and $\lim_{\Delta t \to 0} o(\Delta t)/\Delta t = 0$. Here $\gamma_{ij} \geq 0$ is the transition rate from i to j if $i \neq j$, while $\gamma_{ii} = -\sum_{j \neq i} \gamma_{ij}$.

Consider the following uncertain stochastic time-delay systems with nonlinear disturbance and Markovian switching:

$$\begin{aligned} dx(t) &= \big[(A(r(t)) + \Delta A(t, r(t)))x(t) + (A_d(r(t)) + \Delta A_d(t, r(t)))x(t-\tau) \\ &\quad + (B_u(r(t)) + \Delta B_u(t, r(t)))u(t) + B_v(r(t))v(t) \\ &\quad + f(x(t), x(t-\tau), r(t))\big]dt \\ &\quad + \big[(E(r(t)) + \Delta E(t, r(t)))x(t) \\ &\quad + (E_d(r(t)) + \Delta E_d(t, r(t)))x(t-\tau)\big]d\omega(t), \end{aligned} \tag{3.2}$$

$$z(t) = C(r(t))x(t) + D_u(r(t))u(t), \tag{3.3}$$

$$x(t) = \phi(t), \; r(t) = r(0), \quad \forall t \in [-\tau, 0], \tag{3.4}$$

where $x(t) \in \mathbb{R}^n$ is the state, $u(t) \in \mathbb{R}^r$ is the control input, $v(t) \in \mathbb{R}^p$ is the disturbance input that belongs to $L_2[0, \infty)$, $f(\cdot, \cdot, \cdot)$ is an unknown nonlinear exogenous disturbance input, $Z(t) \in \mathbb{R}^q$ is the controlled output, $w(t)$ is a one-dimensional Brownian motion satisfying $\mathbb{E}\{dw(t)\} = 0$, and $\mathbb{E}\{dw(t)^2\} = dt$, τ represents a real constant time-delay that satisfies $0 \leq \tau < \infty$, and $\phi(t) \in C([-\tau, 0]; \mathbb{R}^n)$ is an initial function.

For a fixed mode, $A(r(t))$, $A_d(r(t))$, $B_u(r(t))$, $B_v(r(t))$, $E(r(t))$, $E_d(r(t))$, $C(r(t))$, and $D_u(r(t))$ are known constant matrices with appropriate dimensions, and $\Delta A(t, r(t))$, $\Delta A_d(t, r(t))$, $\Delta B_u(t, r(t))$, $\Delta E(t, r(t))$, and $\Delta E_d(t, r(t))$ are real-valued matrix functions representing norm-bounded parameter uncertainties and satisfy

$$\begin{aligned} &[\Delta A(t, r(t)) \;\; \Delta A_d(t, r(t)) \;\; \Delta B_u(t, r(t)) \;\; \Delta E(t, r(t)) \;\; \Delta E_d(t, r(t))] \\ &\quad = M(r(t))F(t, r(t))[N_a(r(t)) \;\; N_{ad}(r(t)) \;\; N_u(r(t)) \;\; N_e(r(t)) \;\; N_{ed}(r(t))], \end{aligned}$$

where, for a fixed mode, $M(r(t))$, $N_a(r(t))$, $N_{ad}(r(t))$, $N_u(r(t))$, $N_e(r(t))$, and $N_{ed}(r(t))$ are constant matrices of appropriate dimensions, and $F(t, r(t))$ is an unknown matrix function satisfying

$$F^T(t, r(t))F(t, r(t)) \leq I, \; \forall \, t \leq 0; \; r(t) = i \in S. \tag{3.5}$$

Here, S is the set of the operation modes of systems (3.2)–(3.4).

The state feedback controller considered in this chapter is of the form

$$u(t) = K(r(t))x(t) + K_d(r(t))x(t-\tau), \tag{3.6}$$

where $K(r(t))$ and $K_d(r(t))$ are controller parameters to be designed.

Remark 3.1 Note that delayed feedback term is included in (3.6), which is to reflect the input delay phenomena. It has been shown in [121, 224] that the controller with delay compensation can have more robustness than the memoryless controllers.

In the sequel, we denote the matrix associated with the ith mode by

$$\Gamma_i \triangleq \Gamma(r(t) = i),$$

where the matrix Γ could be A, A_d, B_u, B_v, C, D_u, E, E_d, N_a, N_{ad}, N_u, N_e, N_{ed}, K, or K_d.

Throughout the chapter, we will make an assumption on the exogenous nonlinear time-varying disturbance term $f(x(t), x(t-\tau), r(t))$. Note that the description on the nonlinearities given below is more general than that studied in [52, 170, 177].

Assumption 3.1 *There exist known real constant and mode-dependent matrices $G_1(r(t)), G_2(r(t)) \in \mathbb{R}^{n\times n}$ for a fixed mode, such that the unknown nonlinear vector function $f(\cdot,\cdot,\cdot)$ satisfies the following bounded condition:*

$$|f(x(t), x(t-\tau), r(t))| \leq |G_1(r(t))x(t)| + |G_2(r(t))x(t-\tau)|. \tag{3.7}$$

We now introduce the following stability concepts for (3.2)–(3.4), which will be needed for formulating the problem to be dealt with in this chapter.

Observe the system (3.2)–(3.4), and let $x(t;\xi)$ denote the state trajectory from the initial data $x(\theta) = \xi(\theta)$ on $-\tau \leq \theta \leq 0$ in $L^2_{\mathcal{F}_0}([-\tau, 0]; \mathbb{R}^n)$. Obviously, $x(t, 0) \equiv 0$ is the trivial solution of system (3.2)–(3.4) corresponding to the initial datum $\xi = 0$.

Definition 3 For the stochastic uncertain time-delay jumping system (3.2)–(3.4), with $u(t) = 0$, $v(t) = 0$, and every $\xi \in L^2_{\mathcal{F}_0}([-\tau, 0]; \mathbb{R}^n)$, the trivial solution is said to be *mean-square asymptotically stable* if

$$\lim_{t\to\infty} \mathbb{E}|x(t)|^2 = 0; \tag{3.8}$$

and is said to be *mean-square exponentially stable* if there exist scalars $\alpha > 0$ and $\beta > 0$, such that

$$\mathbb{E}|x(t,\xi)|^2 \leq \alpha e^{\beta t} \sup_{-\tau\leq\theta\leq 0} \mathbb{E}|\xi(\theta)|^2. \tag{3.9}$$

Definition 4 Given a scalar $\gamma > 0$, the stochastic uncertain time-delay jumping system (3.2)–(3.4) is said to be *robustly stochastically stable with disturbance attenuation* γ if it is mean-square exponentially stable and, under zero initial condition, the condition $\|z(t)\|_{\mathbb{E}_2} < \gamma\|v(t)\|_2$ holds for all nonzero $v(t) \in L_2[0, \infty)$, where

$$\|z(t)\|_{\mathbb{E}_2} = \left(\mathbb{E}\left\{\int_0^{\infty} |z(t)|^2 dt\right\}\right)^{1/2}.$$

The purpose of this chapter is to design a state feedback controller of the type (3.6), such that the uncertain stochastic time-delay jumping system with nonlinear disturbance (3.2)–(3.4) is stochastically stable with disturbance attenuation $\gamma > 0$. An LMI approach will be established to explore the analysis and design issues of the problem addressed.

3.2 Robust Stabilization

In the following theorem, the LMI method is used to solve the robust stabilization problem for the stochastic system (3.2)–(3.4), with $v(t) = 0$, where a sufficient condition is developed to guarantee the solvability.

Theorem 3 Consider the system (3.2)–(3.4), with $v(t) \equiv 0$. If there exist sequences of scalars $\varepsilon_{1i} > 0$, $\varepsilon_{2i} > 0$, positive definite matrices $X_i > 0$, and matrices Y_i, K_{di}, such that the following LMIs

$$\begin{bmatrix} \Omega_i & \mathbb{A}_{di} & X_i & X_i E_i^T & \Lambda_i & X_i G_{1i}^T & \mathbb{N}_{ai} & X_i N_{ei} \\ * & -I & 0 & E_{di}^T & 0 & 0 & \mathbb{N}_{adi} & N_{edi}^T \\ * & * & -\mu I & 0 & 0 & 0 & 0 & 0 \\ * & * & * & \mathbb{X}_i & 0 & 0 & 0 & 0 \\ * & * & * & * & -J_i & 0 & 0 & 0 \\ * & * & * & * & * & -1/2I & 0 & 0 \\ * & * & * & * & * & * & -\varepsilon_{1i} I & 0 \\ * & * & * & * & * & * & * & -\varepsilon_{2i} I \end{bmatrix} < 0, \ \forall i \in S \tag{3.10}$$

hold, where

$$\begin{aligned} \Omega_i &:= A_i X_i + B_{ui} Y_i + X_i A_i^T + Y_i^T B_{ui}^T + \varepsilon_{1i} M_i M_i^T + I, \\ \mathbb{A}_{di} &:= A_{di} + B_{ui} K_{di}, \mathbb{X}_i := \varepsilon_{2i} M_i M_i^T - X_i, \\ \mathbb{N}_{ai} &:= X_i N_{ai}^T + Y_i^T N_{ui}, \mathbb{N}_{adi} := N_{adi}^T + K_{di} N_{ui}^T, \\ \Lambda_i &:= [X_i, \cdots, X_i]_{n \times (N-1)n}, \ \mu := \left(2 \max_{i \in S} \|G_{2i}\|^2 + 1\right)^{-1}, \\ J_i &:= diag\{\gamma_{i1}^{-1} X_1, \cdots, \gamma_{ii-1}^{-1} X_{i-1}, \gamma_{ii+1}^{-1} X_{i+1}, \cdots, \gamma_{iN}^{-1} X_N\}, \end{aligned}$$

then, with the state feedback controller of the form

$$u(t) = K_i x(t) + K_{di} x(t-\tau), \tag{3.11}$$

where $K_i = Y_i X_i^{-1}$ and K_{di} is determined by (3.10), the closed-loop system is robustly exponentially stable in the mean-square sense.

Proof 3 *First, the closed-loop system can be easily obtained as follows by substituting the controller (3.11) into system (3.2) with $v(t) = 0$:*

$$\begin{aligned} dx(t) &= [(A_{ci} + \Delta A_{ci}(t))x(t) + (A_{dci} + \Delta A_{dci}(t))x(t-\tau) \\ &\quad + f(x(t), x(t-\tau), i)]dt + [(E_i + \Delta E_i(t))x(t) \\ &\quad + (E_{di} + \Delta E_{di}(t))x(t-\tau)]d\omega(t), \end{aligned} \tag{3.12}$$

where

$$\begin{aligned} A_{ci} &= A_i + B_{ui}K_i,\ \Delta A_{ci}(t) = M_i F_i(t) N_{ci}, \\ N_{ci} &= N_{ai} + N_{ui}K_i,\ \Delta A_{dci}(t) = M_i F_i(t) N_{dci}, \\ A_{dci} &= A_{di} + B_{ui}K_{di},\ N_{dci} = N_{adi} + N_{ui}K_{di}. \end{aligned} \tag{3.13}$$

Now, let

$$P_i = X_i^{-1}, \quad Q = \mu^{-1} I. \tag{3.14}$$

It is clear from the definition of μ that $Q \geq I + 2G_{2i}^T G_{2i}$. Furthermore, from (3.10), it is easy to see that

$$P_i^{-1} - \varepsilon_{2i} M_i M_i^T > 0. \tag{3.15}$$

Define the following Lyapunov functional candidate for system (3.12):

$$V(x(t), t, i) = x^T(t) P_i x(t) + \int_{t-\tau}^{t} x^T(\theta) Q x(\theta) d\theta. \tag{3.16}$$

It can be derived by Itô's formula [81] that:

$$\begin{aligned} dV(x(t), t, i) &= \mathcal{L}V(x(t), t, i)dt + 2x^T(t) P_i[(E_i + \Delta E_i(t))x(t) \\ &\quad + (E_{di} + \Delta E_{di}(t))x(t-\tau)]d\omega(t), \end{aligned} \tag{3.17}$$

where

$$\begin{aligned} \mathcal{L}V(x(t), t, i) &= x^T(t)\left[P_i A_{ci} + A_{ci}^T P_i + \sum_{i=1}^{N} \gamma_{ij}(t) P_j + Q\right] x(t) \\ &\quad + 2x^T(t) P_i [A_{dci} x(t-\tau) + \Delta A_{ci}(t) x(t) + \Delta A_{dci}(t) x(t-\tau) \\ &\quad + f(x(t), x(t-\tau), i)] - x^T(t-\tau) Q x(t-\tau) \\ &\quad + [(E_i(t)) + \Delta E_i(t))x(t) + (E_{di} + \Delta E_{di}(t))x(t-\tau)]^T P_i \\ &\quad \cdot [(E_i(t)) + \Delta E_i(t))x(t) + (E_{di} + \Delta E_{di}(t))x(t-\tau)]. \end{aligned} \tag{3.18}$$

Based on Assumption 3.1, it is not difficult to obtain from (3.7) that

$$\begin{aligned} & 2x^T(t)P_i f(x(t), x(t-\tau), i) \\ \leq\ & x^T(t)P_i^2 x(t) + f(x(t), x(t-\tau), i)^T f(x(t), x(t-\tau), i) \\ \leq\ & x^T(t)P_i^2 x(t) + (|G_{1i}x(t)| + |G_{2i}x(t-\tau)|)^2 \\ \leq\ & x^T(t)P_i^2 x(t) + 2x^T(t)G_{1i}^T G_{1i}x(t) \\ & +2x^T(t-\tau)G_{2i}^T G_{2i}x(t-\tau). \end{aligned} \tag{3.19}$$

Furthermore, from (2.6) in Lemma 2.2, we have

$$\begin{aligned} & 2x^T(t)P_i[\Delta A_{ci}(t)x(t) + \Delta A_{dci}(t)x(t-\tau)] \\ =\ & 2x^T(t)P_i M_i F_i(t)[N_{ci}x(t) + N_{dci}x(t-\tau)] \\ \leq\ & \varepsilon_{1i}x^T(t)P_i M_i M_i^T P_i x(t) + \varepsilon_{1i}^{-1}[N_{ci}x(t) + N_{dci}x(t-\tau)]^T \\ & \cdot[N_{ci}x(t) + N_{dci}x(t-\tau)]. \end{aligned} \tag{3.20}$$

Again, noting (2.7) in Lemma 2.2 and (3.15), we can obtain

$$\begin{aligned} & [\tilde{E}_i + M_i F_i(t)\tilde{N}_i]^T P_i[\tilde{E}_i + M_i F_i(t)\tilde{N}_i] \\ \leq\ & \tilde{E}_i^T(P_i^{-1} - \varepsilon_{2i}M_i M_i^T)^{-1}\tilde{E}_i + \varepsilon_{2i}^{-1}\tilde{N}_i^T\tilde{N}_i, \end{aligned} \tag{3.21}$$

where

$$\tilde{E}_i := [E_i \ \ E_{di}], \quad \tilde{N}_i := [N_{ei} \ \ N_{edi}].$$

Substituting (3.19)–(3.21) into (3.18) results in

$$\mathcal{L}V(x(t), t, i) \leq \begin{bmatrix} x^T(t) & x^T(t-\tau) \end{bmatrix} \Pi_i \begin{bmatrix} x(t) \\ x(t-\tau) \end{bmatrix}, \tag{3.22}$$

where

$$\begin{aligned} \Pi_i \ :=\ & \begin{bmatrix} \Psi_i & P_i A_{di} + P_i B_{ui} K_{di} \\ A_{di}^T P_i + K_{di}^T B_{ui}^T P_i & -Q + 2G_{2i}^T G_{2i} \end{bmatrix} \\ & +\varepsilon_{1i}^{-1}\begin{bmatrix} N_{ci}^T \\ N_{cdi}^T \end{bmatrix}[N_{ci} \ \ N_{cdi}] \\ & +\tilde{E}_i^T(P_i^{-1} - \varepsilon_{2i}M_i M_i^T)^{-1}\tilde{E}_i + \varepsilon_{2i}^{-1}\tilde{N}_i^T\tilde{N}_i, \end{aligned} \tag{3.23}$$

with

$$\begin{aligned} \Psi_i \ :=\ & [A_i + B_{ui}K_i]^T P_i + P_i[A_i + B_{ui}K_i] + P_i^2 + \sum_{i=1}^{N}\gamma_{ij}P_j \\ & +2G_{1i}^T G_{1i} + \varepsilon_{1i}P_i M_i M_i^T P_i + Q. \end{aligned} \tag{3.24}$$

It can be seen from the relation $P_i = X_i^{-1}$ that premultiplying and post-

multiplying (3.10) by diag$(P_i, I, I, I, I, I, I, I)$ yield

$$\begin{bmatrix} \bar{\Omega}_i & & I & E_i^T & \bar{\Lambda}_i & G_{1i}^T & \bar{\mathbb{N}}_{ai} & N_{ei} \\ * & -I & 0 & E_{di}^T & 0 & 0 & \mathbb{N}_{adi} & N_{edi}^T \\ * & * & -\mu I & 0 & 0 & 0 & 0 & 0 \\ * & * & * & \mathbb{P}_i & 0 & 0 & 0 & 0 \\ * & * & * & * & -J_i & 0 & 0 & 0 \\ * & * & * & * & * & -1/2I & 0 & 0 \\ * & * & * & * & * & * & -\varepsilon_{1i}I & 0 \\ * & * & * & * & * & * & * & -\varepsilon_{2i}I \end{bmatrix} < 0, \tag{3.25}$$

where

$$\begin{aligned} \bar{\Omega}_i &:= [A_i + B_{ui}K_i]^T P_i + P_i[A_i + B_{ui}K_i] + P_i^2 + \varepsilon_{1i}M_iM_i^T \\ \bar{\mathbb{A}}_{di} &:= P_iA_{di} + P_iB_{ui}K_{di}, \mathbb{P}_i := \varepsilon_{2i}M_iM_i^T - P_i^{-1}, \\ \bar{\mathbb{N}}_{ai} &:= N_{ai}^T + K_i^T N_{ui}, \\ \bar{\Lambda}_i &:= [I, \cdots, I]_{n\times(N-1)n}, \end{aligned}$$

which, by Schur complement lemma, implies that $\Pi_i < 0$. Therefore, it follows directly from (3.18) that for all

$$[x^T(t) \;\; x^T(t-\tau)]^T \neq 0, \tag{3.26}$$

we have

$$\mathcal{L}V(x(t), t, i) \leq 0. \tag{3.27}$$

Defining

$$\lambda_P = \max_{i\in S} \lambda_{\max}(P_i), \;\; \lambda_p = \min_{i\in S} \lambda_{\min}(P_i), \;\; \lambda_\Pi = \min_{i\in S}(-\lambda_{\max}(\Pi_i))$$

and letting δ be the unique positive root to the equation:

$$\delta(\lambda_P + \mu^{-1}he^{\delta h}) = \lambda_\Pi + \min(1, \lambda_\Pi e^{\delta h}), \tag{3.28}$$

we are now ready to prove the mean-square exponential stability of the closed-loop system. To do this, we modify the Lyapunov functional candidate (3.16) as

$$V_1(x(t), t, i) = e^{\delta t}\left(x^T(t)P_ix(t) + \int_{t-\tau}^{t} x^T(s)Qx(s)ds\right), \tag{3.29}$$

and then obtain the following equation

$$\begin{aligned} &\mathbb{E}V_1(x(t), t, i) \\ =\;& \mathbb{E}V_1(x(0), 0, r(0)) + \mathbb{E}\int_0^t e^{\delta s}\left[\delta x^T(s)P_ix(s)\right. \\ &\left. + \delta\int_{s-\tau}^{s} x^T(\beta)Qx(\beta)d\beta + \mathcal{L}V(x(s), s, r(s))\right] ds. \end{aligned} \tag{3.30}$$

It follows from (3.18) that

$$\begin{aligned}
&\mathbb{E}V_1(x(t),t,i)\\
\leq\ & (\lambda_P+\mu^{-1}h)\mathbb{E}\|\xi\|^2+(\delta\lambda_P-\lambda_\Pi)\mathbb{E}\int_0^t e^{\delta s}|x(s)|^2ds\\
&+\delta\mathbb{E}\int_0^t e^{\delta s}\int_{s-\tau}^s x^T(\beta)Qx(\beta)d\beta ds-\lambda_\Pi\mathbb{E}\int_0^t e^{\delta s}|x(s-\tau)|^2ds.
\end{aligned}$$

Noticing the definition of δ and the two facts of

$$\begin{aligned}
&\int_{s-\tau}^s x^T(\beta)Qx(\beta)d\beta ds\leq\mu^{-1}\int_{-\tau}^t|x(\beta)|^2\left(\int_{\max(\beta,0)}^{\min(\beta+\tau,t)}e^{\delta s}ds\right)d\beta\\
\leq\ & \mu^{-1}\int_{-\tau}^t|x(\beta)|^2\tau e^{\delta(\beta+\tau)}d\beta\leq\mu^{-1}\tau e^{\delta\tau}\left(\frac{\|\xi\|^2}{\delta}+\int_0^t e^{\delta s}|x(s)|^2ds\right)
\end{aligned}$$

and

$$\begin{aligned}
&-\lambda_\Pi\mathbb{E}\int_0^t e^{\delta s}|x(s-\tau)|^2ds=-\lambda_\Pi e^{\delta\tau}\mathbb{E}\int_{-\tau}^{t-\tau}e^{\delta s}|x(s)|^2ds\\
\leq\ & -\min(1,\lambda_\Pi e^{\delta\tau})\mathbb{E}\int_0^t e^{\delta s}|x(s)|^2ds+\mathbb{E}\int_{t-\tau}^t e^{\delta s}|x(s)|^2ds,
\end{aligned}$$

we have

$$\mathbb{E}V_1(x(t),t,i)\leq(\lambda_P+\mu^{-1}\tau(1+e^{\delta\tau}))\mathbb{E}\|\xi\|^2+\mathbb{E}\int_{t-\tau}^t e^{\delta s}|x(s)|^2ds. \tag{3.31}$$

Moreover, since $Q=\mu^{-1}I\geq I$, we obtain

$$\mathbb{E}V_1(x(t),t,i)\geq e^{\delta t}\lambda_P\mathbb{E}|x(t)|^2+\mathbb{E}\int_{t-\tau}^t e^{\delta s}|x(s)|^2ds. \tag{3.32}$$

It follows from (3.31), (3.32) that

$$e^{\delta t}\lambda_P\mathbb{E}|x(t)|^2\leq(\lambda_P+\mu^{-1}\tau(1+e^{\delta\tau}))\mathbb{E}\|\xi\|^2,$$

or

$$\lim_{t\to\infty}\sup(1/t)\log(\mathbb{E}|x(t,\xi)|^2)\leq-\delta,$$

which indicates that the trivial solution of (3.10) is exponentially stable in the mean-square sense. This completes the proof of this theorem.

Remark 3.2 It is shown in Theorem 3 that the addressed robust stabilization problem is solvable if a set of LMIs is feasible. As the Markovian jumping parameters, modeling uncertainties, time-delays, and nonlinear disturbances are simultaneously considered, the stability analysis results presented in Theorem 3 are quite general, which have recovered many existing ones as special cases, such as those in [48,177,181]. Besides, the proposed controller structure with delay compensation offers more robustness than the traditional memoryless controller.

3.3 H_∞ Control

In this section, based on the stability analysis results obtained in the previous section, we continue to study the H_∞ control problem for the same class of stochastic delay systems with Markovian switching and nonlinear disturbances. Again, we aim to develop an LMI approach to establish sufficient conditions that ensure the solvability of the problem and provide a direct design method of the desired controllers.

The main results in this section are given in the following theorem.

Theorem 4 Consider the system (3.2)–(3.4) for a given scalar $\gamma > 0$. If there exist sequences of scalars $\varepsilon_{1i} > 0$, $\varepsilon_{2i} > 0$, positive matrices $X_i > 0$, and matrices Y_i, K_{di} satisfying the following LMIs

$$\begin{bmatrix} \Omega_i & A_{di}+B_{ui}K_{di} & X_i & X_iE_i^T & \Lambda_i & X_iG_{1i}^T & X_iN_{ai}^T+Y_i^TN_{ui} & X_iN_{ei} & B_{vi} & X_iC_i^T+Y_i^TD_{ui}^T \\ * & -I & 0 & E_{di}^T & 0 & 0 & N_{adi}^T+K_{di}N_{ui}^T & N_{edi}^T & 0 & C_{di}^T+K_{di}^TD_{ui}^T \\ * & * & -\mu I & 0 & 0 & 0 & 0 & 0 & 0 & 0 \\ * & * & * & \varepsilon_{2i}M_iM_i^T-X_i & 0 & 0 & 0 & 0 & 0 & 0 \\ * & * & * & * & -J_i & 0 & 0 & 0 & 0 & 0 \\ * & * & * & * & * & -1/2I & 0 & 0 & 0 & 0 \\ * & * & * & * & * & * & -\varepsilon_{1i}I & 0 & 0 & 0 \\ * & * & * & * & * & * & * & -\varepsilon_{2i}I & 0 & 0 \\ * & * & * & * & * & * & * & * & -\gamma^2 I & 0 \\ * & * & * & * & * & * & * & * & * & -I \end{bmatrix} < 0, \quad (3.33)$$

$\forall i \in S$,

where

$$\begin{aligned} \Omega_i &:= A_iX_i + B_{ui}Y_i + X_iA_i^T + Y_i^TB_{ui}^T + \varepsilon_{1i}M_iM_i^T + I \\ \Lambda_i &:= [X_i, \cdots, X_i]_{n\times(N-1)n} \\ J_i &:= diag\{\gamma_{i1}^{-1}X_1, \cdots, \gamma_{ii-1}^{-1}X_{i-1}, \gamma_{ii+1}^{-1}X_{i+1}, \cdots, \gamma_{iN}^{-1}X_N\} \\ \mu &:= \left(2\max_{i\in S}\|G_{2i}\|^2+1\right)^{-1}, \end{aligned}$$

then with the following state feedback controller

$$u(t) = K_i x(t) + K_{di} x(t-\tau), \tag{3.34}$$

where $K_i = Y_i X_i^{-1}$ and K_{di} is determined by (3.33), the closed-loop system of (3.2)–(3.4) and (3.34) is robustly exponentially stable in the mean-square sense with a disturbance attenuation $\gamma > 0$.

Proof 4 *Applying the controller (3.34) to (3.2)–(3.4), we obtain the closed-loop system as follows:*

$$\begin{aligned} dx(t) &= [(A_{ci} + \Delta A_{ci}(t))x(t) + (A_{dci} + \Delta A_{dci}(t))x(t-\tau) \\ &\quad + f(x(t), x(t-\tau), i) + B_{vi}v(t)]dt \\ &\quad + [(E_i + \Delta E_i(t))x(t) + (E_{di} + \Delta E_{di}(t))x(t-\tau)]d\omega(t), \quad (3.35) \\ z(t) &= (C_i + D_{ui}K_i)x(t) + (C_{di} + D_{ui}K_{di})x(t-\tau), \quad (3.36) \end{aligned}$$

where A_{ci}, $\Delta A_{ci}(t)$, A_{dci}, and $\Delta A_{dci}(t)$ are the same as in (3.13).

Obviously, if the inequalities (3.33) hold, then the inequalities (3.10) hold as well, and therefore it follows from Theorem 3 that the system in (3.35) and (3.36) is mean-square exponentially stable.

Continue to choose the Lyapunov functional candidate $V(x(t), t, i)$ given in (3.16) and we have

$$\begin{aligned} &\mathcal{L}V(x(t), t, i) \\ = \; & x^T(t)\left[A_{ci}^T P_i + P_i A_{ci} + \sum_{i=1}^{N} \gamma_{ij}(t) P_j + Q\right] x(t) \\ & + 2x^T(t)P_i\left[A_{dci}x(t-\tau) + \Delta A_{ci}^T(t)x(t) + \Delta A_{dci}(t)x(t-\tau)\right. \\ & \left. + f(x(t), x(t-\tau), i)\right] - x^T(t-\tau)Qx(t-\tau) + 2x^T(t)P_i B_{vi} v(t) \\ & + [(E_i + \Delta E_i(t))x(t) + (E_{di} + \Delta E_{di}(t))x(t-\tau)]^T P_i \\ & \cdot [(E_i + \Delta E_i(t))x(t) + (E_{di} + \Delta E_{di}(t))x(t-\tau)] \\ \leq \; & \begin{bmatrix} x^T(t) & x^T(t-\tau) \end{bmatrix} \Pi_i \begin{bmatrix} x(t) \\ x(t-\tau) \end{bmatrix} + 2x^T(t)P_i B_{vi} v(t), \end{aligned} \tag{3.37}$$

where Π_i is defined in (3.23).

Next, let us assume the zero initial condition, i.e., $x(t) = 0$ for $t \in [-\tau, 0]$, and define

$$J(t) = \mathbb{E}\left\{\int_0^t [z^T(s)z(s) - \gamma^2 v^T(s)v(s)]ds\right\}. \tag{3.38}$$

From Dynkin's formula [81] and the fact that $x(0) = 0$, we have

$$\mathbb{E}\{V(x(t), t, r(t))\} = \mathbb{E}\left\{\int_0^t \mathcal{L}V(x(s), s, r(s))ds\right\}. \tag{3.39}$$

From (3.38) and (3.39), it is easy to find that

$$\begin{aligned} & J(t) \\ = \;& \mathbb{E}\left\{\int_0^t [z^T(s)z(s) - \gamma^2 v^T(s)v(s) + \mathcal{L}V(x(s), s, r(s))]ds\right\} \\ & -\mathbb{E}\{V(x(t), t, r(t))\} \\ \leq \;& \mathbb{E}\left\{\int_0^t [z^T(s)z(s) - \gamma^2 v^T(s)v(s) + \mathcal{L}V(x(s), s, r(s))]ds\right\}. \end{aligned} \tag{3.40}$$

By denoting

$$\eta(t) := [x^T(t) \;\; x^T(t-\tau) \;\; v^T(t)], \tag{3.41}$$

we can obtain from (3.37) that

$$z^T(s)z(s) - \gamma^2 v^T(s)v(s) + \mathcal{L}V(x(s), s, i) \leq \eta(s)^T \Gamma_i \eta(s), \tag{3.42}$$

where

$$\begin{aligned} \Gamma_i \;:=\; & \begin{bmatrix} \Psi_i & P_iA_{di} + P_iB_{ui}K_{di} & P_iB_{vi} \\ A_{di}^TP_i + K_{di}^TB_{ui}^TP_i & -Q + 2G_{2i}^TG_{2i} & 0 \\ B_{vi}^TP_i & 0 & -\gamma^2 I \end{bmatrix} \\ & +\varepsilon_{1i}^{-1}\begin{bmatrix} N_{ci}^T \\ N_{cdi}^T \\ 0 \end{bmatrix}[N_{ci} \;\; N_{cdi} \;\; 0] \\ & + \begin{bmatrix} C_i^T + K_i^TD_{ui}^T \\ C_{di}^T + K_{di}^TD_{ui}^T \\ 0 \end{bmatrix}[C_i + D_{ui}K_i \;\; C_{di} + D_{ui}K_{di} \;\; 0] \\ & +\hat{E}_i^T(P_i^{-1} - \varepsilon_{2i}M_iM_i^T)^{-1}\hat{E}_i + \varepsilon_{2i}^{-1}\hat{N}_i^T\hat{N}_i, \end{aligned} \tag{3.43}$$

with

$$\begin{aligned} \Psi_i \;:=\; & [A_i + B_{ui}K_i]^TP_i + P_i[A_i + B_{ui}K_i] + P_i^2 + Q \\ & + \sum_{i=1}^N \gamma_{ij}(t)P_j + \varepsilon_{1i}P_iM_iM_i^TP_i + 2G_{1i}^TG_{1i} \end{aligned} \tag{3.44}$$

and

$$\hat{E}_i := [E_i,\, E_{di},\, 0],\; \hat{N}_i = [N_{ei}\, N_{edi},\, 0].$$

Premultiplying and postmultiplying (3.33) by $diag(P_i, I, I, I, I, I, I, I, I, I)$

result in

$$
\begin{bmatrix}
\bar{\Omega}_i & \bar{A}_{di} & I & E_i^T & \bar{\Lambda}_i & G_{1i}^T & N_{ai}^T + K_i^T N_{ui} & N_{ei}^T & P_i B_{vi} & C_i^T + K_i^T D_{ui}^T \\
* & -I & 0 & E_{di}^T & 0 & 0 & N_{adi}^T + K_{di}^T N_{ui}^T & N_{edi}^T & 0 & C_{di}^T + K_{di}^T D_{ui}^T \\
* & * & -\mu I & 0 & 0 & 0 & 0 & 0 & 0 & 0 \\
* & * & * & \varepsilon_{2i} M_i M_i^T - P_i^{-1} & 0 & 0 & 0 & 0 & 0 & 0 \\
* & * & * & * & -J_i & 0 & 0 & 0 & 0 & 0 \\
* & * & * & * & * & -1/2I & 0 & 0 & 0 & 0 \\
* & * & * & * & * & * & -\varepsilon_{1i} I & 0 & 0 & 0 \\
* & * & * & * & * & * & * & -\varepsilon_{2i} I & 0 & 0 \\
* & * & * & * & * & * & * & * & -\gamma^2 I & 0 \\
* & * & * & * & * & * & * & * & * & -I
\end{bmatrix} < 0, \quad (3.45)
$$

where

$$
\begin{aligned}
\bar{\Lambda}_i &:= [I, \cdots, I]_{n \times (N-1)n} \\
\bar{\Omega}_i &:= [A_i + B_{ui} K_i]^T P_i + P_i [A_i + B_{ui} K_i] + P_i^2 + \varepsilon_{1i} P_i M_i M_i^T P_i, \\
\bar{A}_{di} &:= P_i A_{di} + P_i B_{ui} K_{di}.
\end{aligned}
$$

By the Schur complement, the inequality (3.45) implies that

$$\Gamma_i < 0. \quad (3.46)$$

To this end, we know from (3.40), (3.42), and (3.46) that, for all $t > 0$, $J(t) < 0$, *and, therefore,*

$$\mathbb{E}\left\{\int_0^t [z^T(s) z(s)\right\} ds \leq \gamma^2 \left\{\int_0^t v^T(s) v(s) ds\right\},$$

which shows

$$\|z(t)\|_{E_2} < \gamma \|v(t)\|_2. \quad (3.47)$$

From Definition 4, we have the conclusion that the closed-loop system (3.35)–(3.36) is stochastically stable with a disturbance attenuation rate γ, *which ends the proof.*

It is worth mentioning that the results given in Theorem 4 are general enough to cover many cases studied in the literature. For example, if we drop out the uncertain parameters, input delay, and nonlinear disturbance terms in the system (3.2)–(3.4), then the system reduces to the usual stochastic delay system with Markovian switching. In this case, the following corollary follows directly from Theorem (4).

Corollary 3.1 Consider the system (3.2)–(3.4) without nonlinear disturbance, input delay, and uncertain parameters. Given a scalar $\gamma > 0$, if there exist sequences of positive matrices $X_i > 0$ and matrices Y_i satisfying the following linear matrix inequalities

$$\begin{bmatrix} \Omega_i & X_i & A_{di} & B_{vi} & X_iE_i^T & \Lambda_i & X_iC^T + Y_iD_{ui}^T \\ * & -I & 0 & 0 & E_{di}^T & 0 & 0 \\ * & * & -I & 0 & 0 & 0 & 0 \\ * & * & * & -\gamma^2 & 0 & 0 & 0 \\ * & * & * & * & -X_i & 0 & 0 \\ * & * & * & * & * & -J_i & 0 \\ * & * & * & * & * & * & -I \end{bmatrix} < 0, \; \forall\, i \in S, \quad (3.48)$$

where

$$\begin{aligned} \Omega_i &= A_iX_i + B_{ui}Y_i + X_iA_i^T + Y_i^TB_{ui}^T, \\ \Lambda_i &= [X_i, \cdots, X_i]_{n\times(N-1)n}, \\ J_i &= diag\{\gamma_{i1}^{-1}X_1, \cdots, \gamma_{ii-1}^{-1}X_{i-1}, \gamma_{ii+1}^{-1}X_{i+1}, \cdots, \gamma_{iN}^{-1}X_N\}, \end{aligned}$$

then the closed-loop system is stochastically stable with a disturbance attenuation rate γ by the state feedback control law

$$u(t) = K_ix(t), \;\; K_i = Y_iX_i^{-1}, \, i \in S. \quad (3.49)$$

Remark 3.3 In the main results given in this chapter, a single time-invariant delay is considered, which is to avoid unnecessarily complicated notations. We like to point out that it would be straightforward to extend the present results to the multiple time-varying delay cases (see, e.g., [222,224]) within the same LMI framework.

3.4 An Illustrative Example

In this section, we will illustrate the usefulness and flexibility of the theory developed in this chapter by designing a robust stabilizing controller for an uncertain stochastic time-delay jumping system with nonlinear disturbances, which is assumed to have two modes.

The system data of (3.2)–(3.4) are given as follows:

$$\begin{bmatrix} \gamma_{11} & \gamma_{12} \\ \gamma_{21} & \gamma_{22} \end{bmatrix} = \begin{bmatrix} -1.5 & 1.5 \\ 0.8 & -0.8 \end{bmatrix}, \;\; \tau = 0.1, \; \gamma = 1.3$$

Mode (I):

$$A_1 = \begin{bmatrix} 1 & 0 \\ 0 & -2 \end{bmatrix}, \quad A_{d1} = \begin{bmatrix} 0.1 & 0 \\ 0 & 0.3 \end{bmatrix}, \quad E_1 = \begin{bmatrix} 0.1 & 1 \\ -1 & 1 \end{bmatrix}$$

$$E_{d1} = \begin{bmatrix} 0.2 & 0 \\ -1 & 1 \end{bmatrix}, \quad G_{11} = \begin{bmatrix} 0.5 & 0 \\ 0 & 0.1 \end{bmatrix}, \quad G_{21} = \begin{bmatrix} 0.1 & 0 \\ 0 & 0.5 \end{bmatrix}$$

$$B_{u1} = I_2, \quad B_{v1} = 0.5I_2, \quad D_{u1} = [0.2 \ 0.1],$$

$$C_1 = [0.5 \ 0.2], \quad C_{d1} = [0.3 \ 0.1]$$

$$M_1 = [0.1 \ 0.1]^T, \quad N_{a1} = [0.2 \ 0.1], \quad N_{ad1} = [0.15 \ 0.2]$$

$$N_{u1} = [0.3 \ 0.1], \quad N_{e1} = [0.4 \ 0.3], \quad N_{ed1} = [0.1 \ 0.1].$$

Mode (II):

$$A_2 = \begin{bmatrix} -3 & 0 \\ 0 & -0.5 \end{bmatrix}, \quad A_{d2} = \begin{bmatrix} 0.5 & 0 \\ 0 & 0.5 \end{bmatrix}, \quad E_2 = \begin{bmatrix} 1 & 0 \\ 0 & -0.5 \end{bmatrix}$$

$$E_{d2} = \begin{bmatrix} 1 & 0 \\ -1 & 0.2 \end{bmatrix}, \quad G_{12} = \begin{bmatrix} 0.3 & 0 \\ 0 & 0.32 \end{bmatrix}, \quad G_{22} = \begin{bmatrix} 0.3 & 0 \\ 0 & 0.4 \end{bmatrix}$$

$$B_{u2} = 0.5I_2, \quad B_{v2} = I_2, \quad D_{u2} = [0.5 \ 0.5],$$

$$C_2 = [1.5 \ 0], \quad C_{d2} = [0.3 \ 0.1]$$

$$M_2 = [0.2 \ 0.2]^T, \quad N_{a2} = [0.13 \ 0.21], \quad N_{ad2} = [0.15 \ 0.23]$$

$$N_{u2} = [0.33 \ 0.16], \quad N_{e2} = [0.24 \ 0.31], \quad N_{ed2} = [0.12 \ 0.31].$$

Using MATLAB LMI Toolbox to solve the LMIs (3.33), we obtain a solution as follows:

$$X_1 = \begin{bmatrix} 5.1212 & -3.3772 \\ -3.3772 & 11.0210 \end{bmatrix}, \quad Y_1 = \begin{bmatrix} -136.8106 & 212.2489 \\ 202.2063 & -451.7080 \end{bmatrix}$$

$$K_1 = \begin{bmatrix} -17.5636 & 13.8765 \\ 15.6105 & -36.2025 \end{bmatrix}, \quad K_{d1} = \begin{bmatrix} 0.6065 & 0.6586 \\ 0.1293 & 0.1782 \end{bmatrix}$$

$$X_2 = \begin{bmatrix} 6.3615 & -4.8212 \\ -4.8212 & 5.0922 \end{bmatrix}, \quad Y_2 = \begin{bmatrix} -254.3627 & 250.5542 \\ 241.6184 & -244.2887 \end{bmatrix}$$

$$K_2 = \begin{bmatrix} -9.5404 & 40.1706 \\ 5.7493 & -42.5295 \end{bmatrix}, \quad K_{d2} = \begin{bmatrix} 3.2552 & -5.1699 \\ -2.5316 & 5.3148 \end{bmatrix}$$

$$\varepsilon_{11} = 221.8908, \varepsilon_{12} = 11.9560, \varepsilon_{21} = 48.6809, \varepsilon_{22} = 1.5683.$$

Therefore, according to Theorem 4, the closed-loop system of (3.2)–(3.4), with the controller (3.34) is robustly exponentially stable in the mean-square sense with a disturbance attenuation $\gamma = 1.3$.

Figure 3.1, Figure 3.3, Figure 3.5, and Figure 3.7 give the evolution of states $x_{11}(t)$, $x_{12}(t)$, $x_{21}(t)$ and $x_{22}(t)$ for the uncontrolled systems, and Figure 3.2, Figure 3.4, Figure 3.6, and Figure 3.8 depict the results of the evolution of states $x_{11}(t)$, $x_{12}(t)$, $x_{21}(t)$ and $x_{22}(t)$ for the controlled systems, where $x_{ij}(t)$ $(i, j = 1, 2)$ denote the state evolution when the mode changes from i to j. The simulation results have illustrated our theoretical analysis.

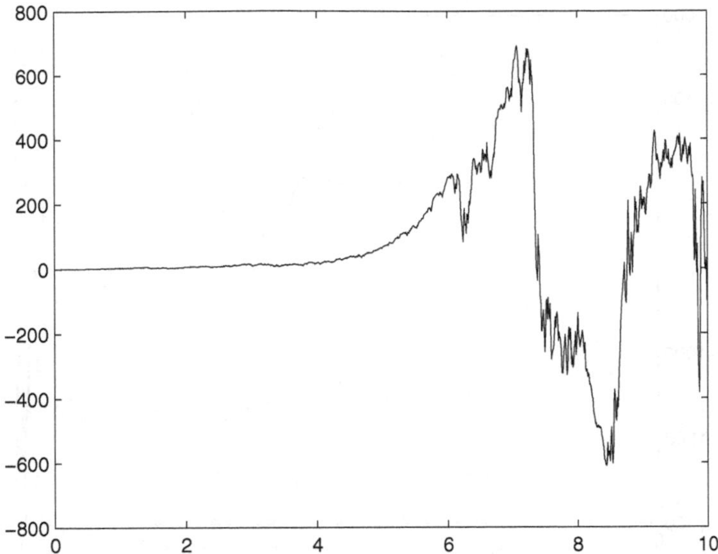

FIGURE 3.1
State evolution $x_{11}(t)$ of uncontrolled systems.

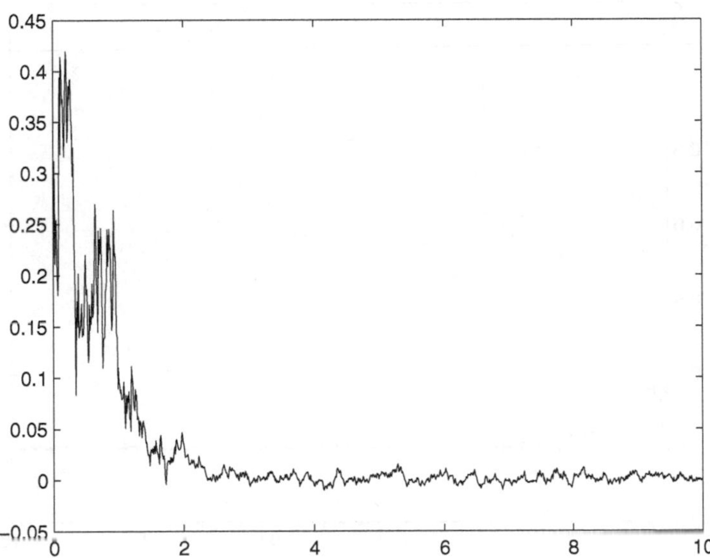

FIGURE 3.2
State evolution $x_{11}(t)$ of controlled systems.

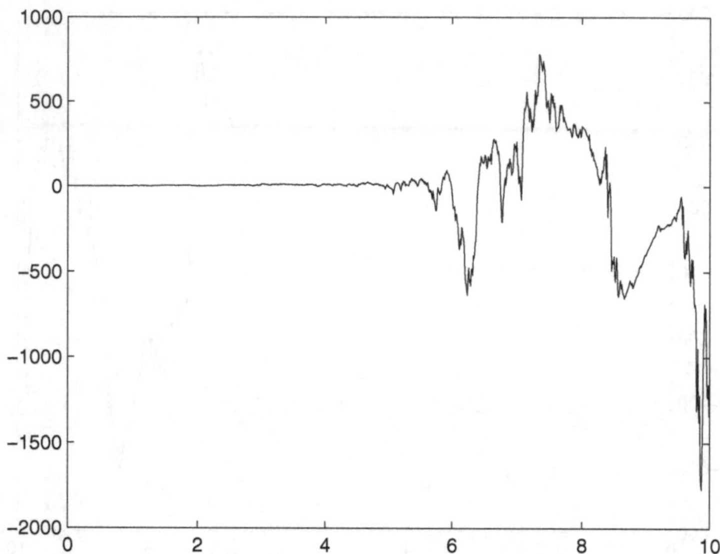

FIGURE 3.3
State evolution $x_{12}(t)$ of uncontrolled systems.

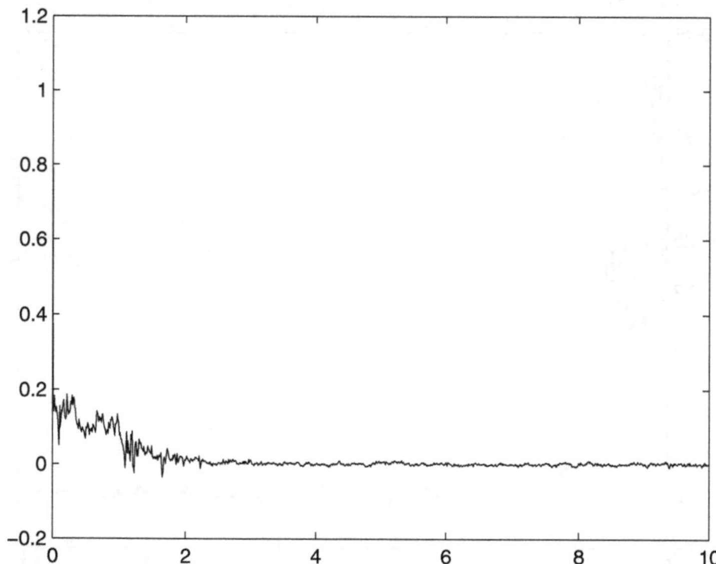

FIGURE 3.4
State evolution $x_{12}(t)$ of controlled systems.

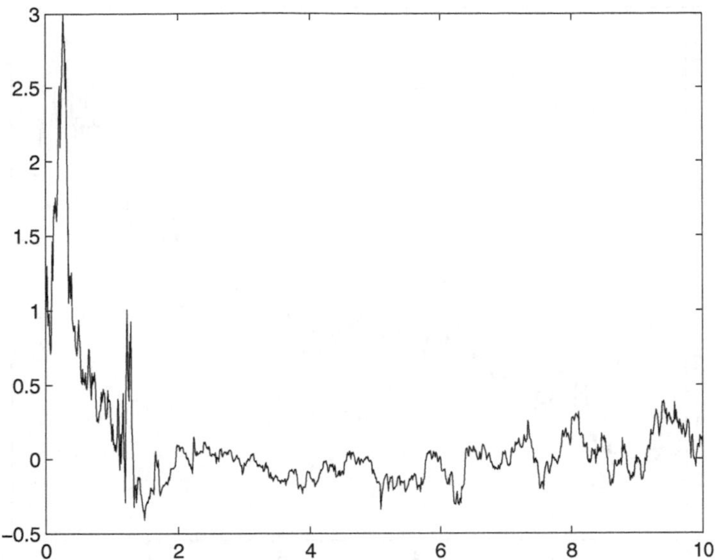

FIGURE 3.5
State evolution $x_{21}(t)$ of uncontrolled systems.

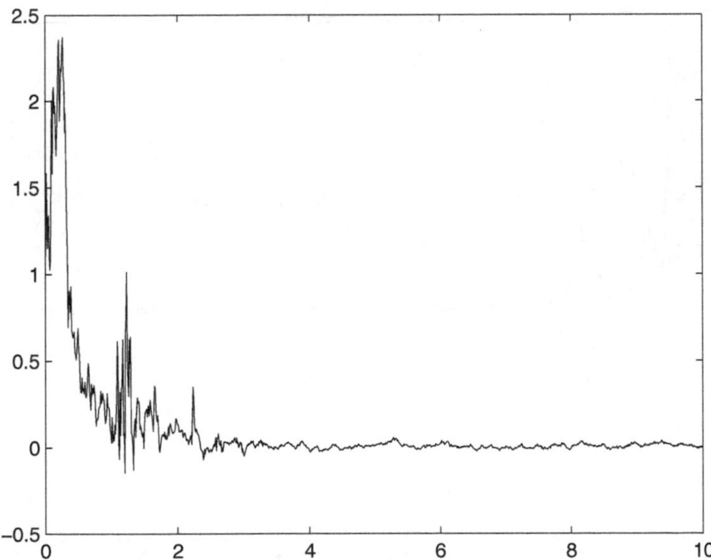

FIGURE 3.6
State evolution $x_{21}(t)$ of controlled systems.

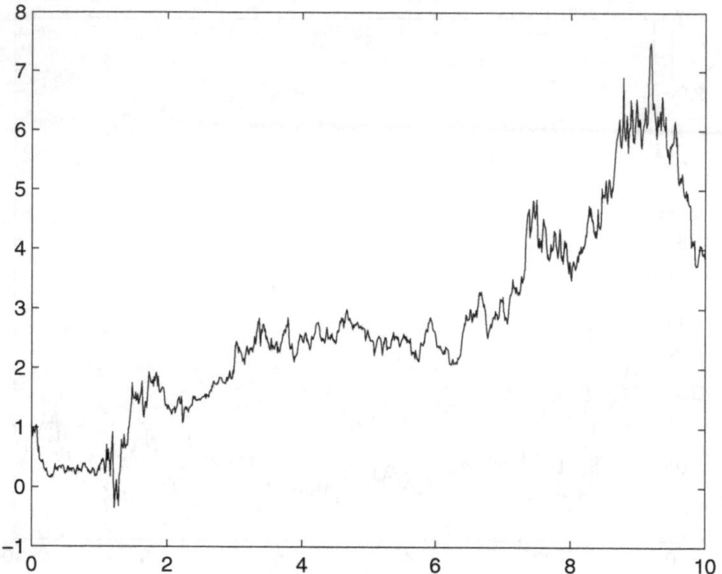

FIGURE 3.7
State evolution $x_{22}(t)$ of uncontrolled systems.

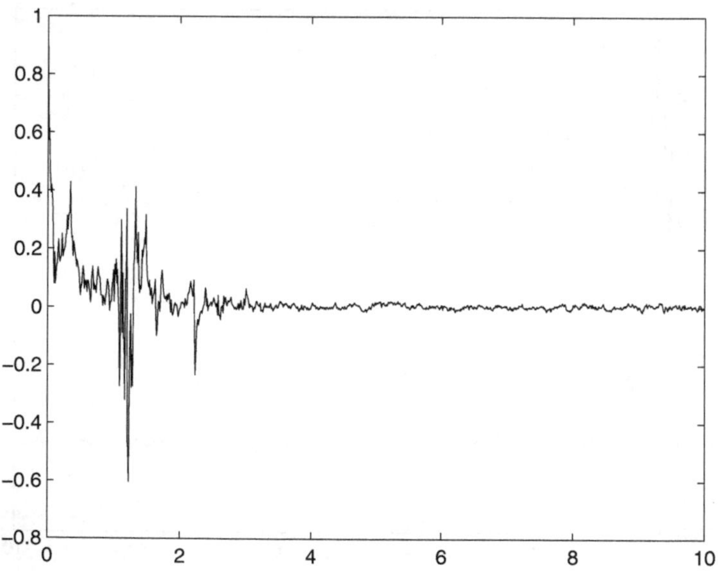

FIGURE 3.8
State evolution $x_{22}(t)$ of controlled systems.

3.5 Summary

In this chapter, we have investigated the robust stabilization problem and H_∞ control problem for a class of uncertain stochastic time-delay systems with Markovian switching and nonlinear disturbances, which involve the delay compensation term in the feedback structure. By using the linear matrix inequality technique, we have established the H_∞ controller design procedure that guarantees the stochastic stability of the closed-loop system with a given disturbance rejection attenuation rate $\gamma > 0$. We have also shown that many existing results could be covered as special cases of our main results. Finally, a numerical example has been exploited to illustrate the effectiveness of the results.

4

H_∞ *Filtering and Control for Markov Systems with Sensor Nonlinearities*

CONTENTS

In this chapter, we aim to solve the H_∞ filter and output feedback H_∞ controller design problems for a class of stochastic time-delay systems with nonlinear disturbances, sensor nonlinearity, and Markovian jumping parameters. Both the filter and controller analysis and synthesis problems are tackled. A delay-dependent approach is developed to design the H_∞ filter and output feedback H_∞ controller for the stochastic delay jumping systems such that, for the addressed nonlinear disturbances and sensor nonlinearities, the resulting system is stochastically stable with a prescribed disturbance rejection attenuation level γ. By using Itô's differential formula and the Lyapunov stability theory, sufficient conditions for the solvability of the filter and controller design problem are derived in term of linear matrix inequalities (LMIs). These conditions are dependent on the information of the time-delays, which can be easily checked by resorting to available software packages. A numerical example and the corresponding simulation results are exploited to demonstrate the effectiveness of the proposed filter and controller design method.

The remainder of this chapter is organized as follows. In Section 4.1, the H_∞ filter design problem for a class of stochastic time-delay systems with nonlinear disturbances, sensor nonlinearity, and Markovian jumping parameters is formulated. In Section 4.2, state estimator analysis and synthesis problems are investigated separately. In Section 4.3, a simulation example is presented to illustrate the usefulness and flexibility of the filter design method developed.

In Section 4.4, the output feedback H_∞ controller design problem for a class of stochastic time-delay systems with nonlinear disturbances, sensor nonlinearity, and Markovian jumping parameters is formulated. In Section 4.5, H_∞ performance and output feedback controller synthesis problems are investigated separately. In Section 4.6, a numerical example is presented to illustrate the usefulness and flexibility of the controller design method developed. In Section 4.7, summary remarks are provided.

4.1 Problem Formulation

Let $\{r(t), t \geq 0\}$ be a right-continuous Markov chain that has been introduced in (3.1) of Chapter 3.

Consider the following stochastic time-delay system with both the sensor nonlinearity and Markovian switching:

$$\begin{aligned}(\Sigma) : dx(t) &= [A(r(t))x(t) + A_d(r(t))x(t-\tau) + B_1(r(t))v(t) + D(r(t))u(t) \\ &\quad + f(x(t), x(t-\tau), r(t))]\, dt + E_d(r(t))x(t-\tau)d\omega(t), \qquad (4.1)\\ y(t) &= \psi(\mu) + B_2(r(t))v(t), \qquad (4.2)\\ z(t) &= L(r(t))x(t), \qquad (4.3)\\ x(t) &= \phi(t),\ r(t) = r(0), \quad \forall t \in [-\tau, 0], \qquad (4.4)\end{aligned}$$

where $x(t) \in \mathbb{R}^n$ is the state, $y(t) \in \mathbb{R}^r$ is the measured output, $u(t) \in \mathbb{R}^s$ is the control input, $z(t) \in \mathbb{R}^q$ is the controlled output, $v(t) \in \mathbb{R}^p$ is the disturbance input that belongs to $L_2[0, \infty)$. $f(\cdot,\cdot,\cdot)$ is an unknown nonlinear exogenous disturbance input, $\psi(\cdot)$ represents the sensor nonlinearity, and $\mu = C(r(t))x(t) + C_d(r(t))x(t-\tau)$. $\omega(t)$ is a one-dimensional Brownian motion satisfying $\mathbb{E}\{d\omega(t)\} = 0$ and $\mathbb{E}\{d\omega^2(t)\} = dt$. The constant τ is a real time-delay satisfying $0 \leq \tau < \infty$, and $\phi(t) \in L^p_{\mathcal{F}_0}([-\tau, 0]; \mathbb{R}^n)$ is an initial function. For a fixed mode $r(t) \in S$, $A(r(t))$, $A_d(r(t))$, $B_1(r(t))$, $B_2(r(t))$, $E_d(r(t))$, $C(r(t))$, $C_d(r(t))$, $D(r(t))$, and $L(r(t))$ are constant matrices with appropriate dimensions.

Assumption 4.1 *For a fixed system mode, there exist known real constant mode-dependent matrices $M_1(r(t)) \in \mathbb{R}^{n\times n}$ and $M_2(r(t)) \in \mathbb{R}^{n\times n}$, such that the unknown nonlinear vector function $f(\cdot,\cdot,\cdot)$ satisfies the following boundedness condition:*

$$|f(x(t), x(t-\tau), r(t))| \leq |M_1(r(t))x(t)| + |M_2(r(t))x(t-\tau)|. \qquad (4.5)$$

Remark 4.1 *Exogenous nonlinear time-varying disturbances, which may exist in many real-world systems, have been dealt with in many papers, such as [177, 194]. In Assumption 4.1, the nonlinear disturbance term $f(x(t), x(t-$*

$\tau), r(t))$ in (4.5) contains the delayed term, which is more general than that studied in [177, 194]. Note that the H_∞ filtering problem for stochastic delayed jumping systems with such kind of nonlinear exogenous disturbances has not been thoroughly investigated in the literature.

Assumption 4.2 *The nonlinear function $\psi(\cdot)$ in stochastic systems (4.1)–(4.4) represents the sector nonlinearities satisfying the following sector condition:*

$$(\psi(\mu) - K_1(r(t))\mu)^T (\psi(\mu) - K_2(r(t))\mu) \leq 0, \ \forall \, \mu \in \mathbb{R}^n, \tag{4.6}$$

where the matrices $K_1(r(t)) \geq 0$ and $K_2(r(t)) \geq 0$ $(K_2(r(t)) > K_1(r(t)))$ are given mode-dependent constant diagonal matrices.

Remark 4.2 *As in [78], it is customary to say that the nonlinear function belongs to a sector $[K_1(r(t)), K_2(r(t))]$. The nonlinear description in (4.6) is quite general and includes the usual Lipschitz condition as a special case. Note that both the control analysis and model reduction problems for systems with sector nonlinearities have been intensively studied; see, e.g., [54, 77, 83].*

For technical convenience, the nonlinear function $\psi(u)$ can be decomposed into a combination of a linear part and a nonlinear part as

$$\psi(\mu) = \psi_s(\mu) + K_1(r(t))\mu, \tag{4.7}$$

where the nonlinear part $\psi_s(\mu)$ belongs to the set Ψ_s given by

$$\Psi_s = \{\psi_s(\mu) : \psi_s^T(\mu)(\psi_s(\mu) - K(r(t))\mu) \leq 0\}, \tag{4.8}$$

with $K(r(t)) = K_2(r(t)) - K_1(r(t)) > 0$.

In this chapter, in order to estimate $z(t)$, we are interested in designing a filter of the following structure:

$$\begin{aligned} (\Sigma_f) : d\hat{x}(t) &= F(r(t))\hat{x}(t)dt + G(r(t))y(t)dt && (4.9)\\ \hat{z}(t) &= H(r(t))\hat{x}(t), && (4.10) \end{aligned}$$

where $\hat{x}(t) \in \mathbb{R}^n$, $\hat{z}(t) \in \mathbb{R}^q$, and $F(r(t))$, $G(r(t))$, and $H(r(t))$ are filter parameters to be determined.

Note that the set S consists of different operation modes of the system (4.1)–(4.4) for each possible values of $r(t) = i, \ i \in S$. In the sequel, we denote the matrix associated with the ith mode by

$$W_i \triangleq W(r(t) = i),$$

where the matrix W could be A, A_d, B_1, B_2, E_d, C, C_d, D, L, M_1, M_2, K_1, K_2, K, F, G, H, A_c, B_c, or C_c.

Let the filter estimation error be $e(t) = z(t) - \hat{z}(t)$. By augmenting the state variables

$$\xi(t) = \begin{bmatrix} x(t) \\ \hat{x}(t) \end{bmatrix}, \quad \xi_\tau = \begin{bmatrix} x(t-\tau) \\ \hat{x}(t-\tau) \end{bmatrix},$$

and combining (Σ) and (Σ_f), we obtain the filtering error dynamics as follows:

$$\begin{aligned}(\Sigma_e) : d\xi(t) &= \left[\bar{A}_i\xi(t) + \bar{A}_{di}N\xi_\tau + \bar{B}_i v(t) + N^T f(x(t), x(t-\tau), i)\right. \\ &\quad \left. + \bar{G}_i\psi_s(u)\right] dt + \bar{E}_{di}N\xi_\tau d\omega(t) & (4.11) \\ e(t) &= \bar{L}_i\xi(t), & (4.12)\end{aligned}$$

where

$$\begin{aligned}\bar{A}_i &= \begin{bmatrix} A_i & 0 \\ G_iK_{1i}C_i & F_i \end{bmatrix}, \; \bar{A}_{di} = \begin{bmatrix} A_{di} \\ G_iK_{1i}C_{di} \end{bmatrix}, \; \bar{B}_i = \begin{bmatrix} B_{1i} \\ G_iB_{2i} \end{bmatrix}, \\ \bar{G}_i &= \begin{bmatrix} 0 \\ G_i \end{bmatrix}, \; \bar{E}_{di} = \begin{bmatrix} E_{di} \\ 0 \end{bmatrix}, \; \bar{L}_i = \begin{bmatrix} L_i & -H_i \end{bmatrix}, \; N = \begin{bmatrix} I & 0 \end{bmatrix}.\end{aligned}$$

And then, we are interested in designing an output feedback controller of the following structure:

$$\begin{aligned}(\Sigma_c) : dx_c(t) &= A_c(r(t))x_c(t)dt + B_c(r(t))y(t)dt & (4.13) \\ u(t) &= C_c(r(t))x_c(t), & (4.14)\end{aligned}$$

where $x_c(t) \in \mathbb{R}^n$ is the controller state, and $A_c(r(t))$, $B_c(r(t))$, and $C_c(r(t))$ are controller parameters to be determined.

Augmenting the state variables

$$\theta(t) = \begin{bmatrix} x(t) \\ x_c(t) \end{bmatrix}, \quad \theta_\tau := \theta(t-\tau) = \begin{bmatrix} x(t-\tau) \\ x_c(t-\tau) \end{bmatrix},$$

and combining (Σ) and (Σ_c), we obtain the following dynamics:

$$\begin{aligned}(\Sigma_k) : d\theta(t) &= \left[\tilde{A}_i\theta(t) + \tilde{A}_{di}N\theta_\tau + \tilde{B}_i v(t) + N^T f(t) + \tilde{G}_i\psi_s(\mu)\right] dt \\ &\quad + \tilde{E}_{di}N\theta_\tau d\omega(t) & (4.15) \\ z(t) &= \tilde{L}_i\theta(t), & (4.16)\end{aligned}$$

where

$$\begin{aligned}\tilde{A}_i &= \begin{bmatrix} A_i & D_iC_{ci} \\ B_{ci}K_{1i}C_i & A_{ci} \end{bmatrix}, \; \tilde{A}_{di} = \begin{bmatrix} A_{di} \\ B_{ci}K_{1i}C_{di} \end{bmatrix}, \\ \tilde{B}_i &= \begin{bmatrix} B_{1i} \\ B_{ci}B_{2i} \end{bmatrix}, \; \tilde{E}_{di} = \begin{bmatrix} E_{di} \\ 0 \end{bmatrix}, \; \tilde{G}_i = \begin{bmatrix} 0 \\ B_{ci} \end{bmatrix}, \\ \tilde{L}_i &= \begin{bmatrix} L_i & 0 \end{bmatrix}, \; N = \begin{bmatrix} I & 0 \end{bmatrix}.\end{aligned}$$

For the purpose of presentation simplification, we define new state variables

$$\begin{aligned}\eta_f(t) &= \bar{A}_i\xi(t) + \bar{A}_{di}N\xi_\tau + \bar{B}_i v(t) + N^T f(t) + \bar{G}_i\psi_s(\mu), \\ \eta_c(t) &= \tilde{A}_i\theta(t) + \tilde{A}_{di}N\theta_\tau + \tilde{B}_i v(t) + N^T f(t) + \tilde{G}_i\psi_s(\mu), & (4.17)\end{aligned}$$

and then the systems (4.11) and (4.13) can, respectively, be rewritten as

$$d\xi(t) = \eta_f(t)dt + \bar{E}_{di}N\xi_\tau d\omega(t), \quad (4.18)$$
$$d\theta(t) = \eta_c(t)dt + \bar{E}_{di}N\theta_\tau d\omega(t). \quad (4.19)$$

Observe the system (4.11)–(4.12), (4.15)–(4.16), and let $\vartheta(t;\zeta)$ denote the state trajectory from the initial datum $\vartheta(\rho) = \zeta(\rho)$ on $-\tau \leq \rho \leq 0$ in $L^2_{\mathcal{F}_0}([-\tau,0];\mathbb{R}^n)$. Obviously, $\vartheta(t,0) \equiv 0$ is the trivial solution of systems (4.11)–(4.12), or (4.15)–(4.16) corresponding to the initial datum $\zeta = 0$.

Before formulating the problem to be investigated, we first introduce the following stability concepts for the augmented system (4.11)–(4.12) and (4.15)–(4.16).

Definition 5 For the systems (4.11)–(4.12), (4.15)–(4.16), and every $\zeta \in L^2_{\mathcal{F}_0}$ $([-\tau,0];\mathbb{R}^n)$, the trivial solution is said to be *mean-square asymptotically stable* if

$$\lim_{t\to\infty} \mathbb{E}|\vartheta(t)|^2 = 0; \quad (4.20)$$

and is said to be *mean-square exponentially stable* if there exist scalars $\alpha > 0$ and $\beta > 0$, such that

$$\mathbb{E}|\vartheta(t,\zeta)|^2 \leq \alpha e^{\beta t} \sup_{-2\tau\leq\rho\leq 0} \mathbb{E}|\zeta(\rho)|^2. \quad (4.21)$$

Definition 6 Given a scalar $\gamma > 0$, the augmented systems (4.11)–(4.12) ((4.15)–(4.16)) with sensor nonlinearity are said to be *stochastically stable with disturbance attenuation level* γ if it is mean-square exponentially stable and, under zero initial conditions, $\|e(t)\|_{\mathbb{E}_2} < \gamma\|v(t)\|_2$ ($\|z(t)\|_{\mathbb{E}_2} < \gamma\|v(t)\|_2$) holds for all nonzero $v(t) \in L_2[0,\infty)$, where

$$\|\cdot\|_{\mathbb{E}_2} := \left(\mathbb{E}\left\{\int_0^\infty |\cdot|^2 dt\right\}\right)^{1/2}.$$

The purpose of this chapter is first to design an H_∞ filter of the form (4.9)–(4.10) for the system (4.1)–(4.4), with $u(t) = 0$, such that, for all admissible time-delays, exogenous nonlinear disturbances, sensor nonlinearities, and Markovian jumping parameters, the filtering error system (4.11)–(4.12) is stochastically stable, with disturbance attenuation level γ, where the criteria depend on the length of time-delays, and then to obtain an output feedback H_∞ controller of the form (4.13)–(4.14) for the system (4.1)–(4.4), such that, for all admissible time-delays, exogenous nonlinear disturbances, sensor nonlinearities, and Markovian jumping parameters, the augmented system (4.15)–(4.16) is stochastically stable, with disturbance attenuation level γ, where the criteria depend on the length of time-delays.

4.2 H_∞ Filtering

4.2.1 Filter Analysis

First, let us give the following lemmas that will be used in the proofs of our main results in this chapter.

In the following theorem, the delay-dependent technique and an LMI method are used to deal with the stability analysis problem for the H_∞ filter design of the stochastic system (4.1)–(4.4), and a sufficient condition is derived that ensures the solvability of the H_∞ filtering problem.

Theorem 5 Consider the filtering error system (4.11)–(4.12), with given filter parameters. If there exist positive definite matrices $P_i > 0$, $T_i > 0$, $Q > 0$, and $R > 0$, such that the following matrix inequalities

$$\begin{bmatrix} \Omega_{1i} & 0 & P_i\bar{B}_i & \mathcal{G}_i & \mathcal{A}_i & 0 & P_iN^T & P_i\bar{A}_{di}N \\ * & \Omega_{2i} & 0 & C_{di}^TK_i & \mathcal{A}_{di} & 0 & 0 & 0 \\ * & * & -\gamma^2 I & 0 & \mathcal{B}_i & 0 & 0 & 0 \\ * & * & * & -2I & 0 & 0 & 0 & 0 \\ * & * & * & * & -\bar{\tau}R & R & 0 & 0 \\ * & * & * & * & * & -\varepsilon_{2i}I & 0 & 0 \\ * & * & * & * & * & * & -\varepsilon_{1i}I & 0 \\ * & * & * & * & * & * & * & -I \end{bmatrix} < 0, \quad (4.22)$$

$$\begin{bmatrix} T_i & P_i\bar{A}_{di} \\ \bar{A}_{di}^TP_i & R \end{bmatrix} > 0, \quad (4.23)$$

for $\forall\, i \in S$ hold, where

$$\begin{aligned} \Omega_{1i} &:= P_i(\bar{A}_i + \bar{A}_{di}N) + (\bar{A}_i + \bar{A}_{di}N)^TP_i + \Sigma_{i,j=1}^N \gamma_{ij}P_j + \bar{L}_i^T\bar{L}_i + N^TQN \\ &\quad +2(\varepsilon_{1i} + \varepsilon_{2i})(N^TM_{1i}^TM_{1i}N) + \bar{\tau}T_i,\ \mathcal{G}_i := P_i\bar{G}_i + N^TC_i^TK_i, \\ \Omega_{2i} &:= 2(\varepsilon_{1i} + \varepsilon_{2i})(M_{2i}^TM_{2i}) + N^T\bar{E}_{di}^TP_i\bar{E}_{di}N + \bar{\tau}c_eN^TN - Q, \\ \mathcal{A}_i &:= \bar{\tau}\bar{A}_i^TN^TR,\ \mathcal{A}_{di} := \bar{\tau}\bar{A}_{di}^TN^TR,\ \mathcal{B}_i := \bar{\tau}\bar{B}_i^TN^TR, \end{aligned} \quad (4.24)$$

with $c_e = \max_{i\in S}\|\bar{E}_{di}\|^2$, then the filtering error system is stochastically stable, with the disturbance attenuation level γ for $\tau \le \bar{\tau}$ ($\bar{\tau}$ is the upper bound of the time-delays).

Proof 5 *Recall the Newton–Leibniz formula and (4.18), we can write*

$$\xi_\tau = \xi(t) - \int_{t-\tau}^t d\xi(s) = \xi(t) - \int_{t-\tau}^t \eta_f(s)ds - \int_{t-\tau}^t \bar{E}_{di}N\xi_\tau d\omega(s). \quad (4.25)$$

It is easy to know from (4.25) that the following system is equivalent to (4.11)-(4.12):

$$\begin{aligned} d\xi(t) &= \Big[(\bar{A}_i + \bar{A}_{di}N)\xi(t) - \bar{A}_{di}N\int_{t-\tau}^{t}\eta_f(s)ds - \bar{A}_{di}N\int_{t-\tau}^{t}\bar{E}_{di}N\xi_\tau d\omega(s) \\ &\quad + \bar{B}_i v(t) + N^T f(x(t), x(t-\tau), i) + \bar{G}_i\psi_s(u)\Big]dt \\ &\quad + \bar{E}_{di}N\xi_\tau d\omega(t), \qquad (4.26) \\ e(t) &= \bar{L}_i\xi(t), \qquad (4.27) \\ \xi(t) &= \rho(t),\ r(t) = r(0), \quad \forall t\in[-2\tau, 0], \qquad (4.28) \end{aligned}$$

where $\rho(t) \in L^p_{\mathcal{F}_0}([-2\tau, 0]; \mathbb{R}^{2n})$ is the initial function. Hence, we only need to show that the system (4.26)-(4.28) is stochastically stable, with the disturbance attenuation level γ.

Now, let $P_i > 0$, $Q > 0$, $R > 0$, $c_e = \max_{i\in S}\|\bar{E}_{di}\|^2$, and define the following Lyapunov–Krasovskii functional candidate for the system (4.26):

$$\begin{aligned} V(x(t), t, i) &= \xi^T(t)P_i\xi(t) + \int_{t-\tau}^{t}\xi^T(s)N^TQN\xi(s)ds \\ &\quad + \int_{t-\tau}^{t}\int_{s}^{t}\eta^T(\beta)N^TRN\eta(\beta)d\beta ds \\ &\quad + \int_{t-\tau}^{t}\int_{s}^{t}c_e\xi^T(\beta)N^TN\xi(\beta)d\beta ds. \qquad (4.29) \end{aligned}$$

It can be derived by Itô's differential formula [81] that

$$dV(\xi(t), t, i) = \mathcal{L}V(\xi(t), t, i)dt + 2\xi^T(t)P_i\bar{E}_{di}N\xi_\tau d\omega(t), \qquad (4.30)$$

where

$$\begin{aligned} \mathcal{L}V(\xi(t), t, i) &= \xi^T(t)[(\bar{A}_i + \bar{A}_{di}N)^TP_i + P_i(\bar{A}_i + \bar{A}_{di}N) + \sum_{j=1}^{N}\gamma_{ij}P_j]\xi(t) \\ &\quad - 2\xi^T(t)P_i\bar{A}_{di}N\left(\int_{t-\tau}^{t}\eta_f(s)ds + \int_{t-\tau}^{t}\bar{E}_{di}N\xi_\tau d\omega(s)\right) \\ &\quad + \xi^T(t)N^TQN\xi(t) - \xi_\tau^TN^TQN\xi_\tau + 2\xi^T(t)P_i\bar{B}_iv(t) \\ &\quad + 2\xi^T(t)P_iN^Tf(x(t), x(t-\tau), i) + 2\xi^T(t)P_i\bar{G}_i\psi_s(u) \\ &\quad + \xi_\tau^TN^T\bar{E}_{di}^TP_i\bar{E}_{di}N\xi_\tau + \tau\eta^T(t)N^TRN\eta(t) \\ &\quad + \tau c_e\xi_\tau^TN^TN\xi_\tau + \tau\xi^T(t)T_i\xi(t) - \int_{t-\tau}^{t}(\xi^T(t)T_i\xi(t) \\ &\quad + c_e\xi_\tau^TN^TN\xi_\tau + \eta^T(s)N^TRN\eta(s))ds. \qquad (4.31) \end{aligned}$$

Noting (4.5) and Lemma 2.2, we have

$$
\begin{aligned}
& 2\xi(t)^T P_i N^T f(x(t), x(t-\tau), i) \\
\leq\ & \varepsilon_{1i}^{-1}\xi^T(t)P_iN^TNP_i\xi(t) + \varepsilon_{1i}f^T(x(t),x(t-\tau),i)f(x(t),x(t-\tau),i) \\
\leq\ & \varepsilon_{1i}^{-1}\xi^T(t)P_iN^TNP_i\xi(t) + \varepsilon_{1i}(|M_{1i}x(t)| + |M_{2i}x(t-\tau)|)^2 \\
\leq\ & \varepsilon_{1i}^{-1}\xi^T(t)P_iN^TNP_i\xi(t) + 2\varepsilon_{1i}(\xi^T(t)N^TM_{1i}^TM_{1i}N\xi(t) \\
& +\xi_\tau^TN^TM_{2i}^TM_{2i}N\xi_\tau).
\end{aligned} \tag{4.32}
$$

Again, from Lemma 2.2, we obtain

$$
\begin{aligned}
-2\xi^T(t)P_i\bar{A}_{di}N\int_{t-\tau}^{t} E_{di}N\xi_\tau d\omega(s) \ \leq\ & \xi^T(t)P_i\bar{A}_{di}NN^T\bar{A}_{di}^TP_i\xi(t) \\
& +|\int_{t-\tau}^{t} E_{di}N\xi_\tau d\omega(s)|^2.
\end{aligned} \tag{4.33}
$$

Note that, during deriving (4.33), we have fixed the scalar parameter ε as 1, which is to maintain the simplicity of the Lyapunov functional.

Moreover,

$$
\mathbb{E}|\int_{t-\tau}^{t} E_{di}N\xi_\tau d\omega(s)|^2 \leq \int_{t-\tau}^{t} \mathbb{E}|E_{di}N\xi_\tau|^2 ds. \tag{4.34}
$$

From (4.17) and Lemma 2.2, it is not difficult to see that

$$
\begin{aligned}
& \tau\eta^T(t)N^TRN\eta(t) \\
=\ & [\bar{A}_i\xi(t) + \bar{A}_{di}N\xi_\tau + \bar{B}_iv(t) + N^Tf(x(t),x(t-\tau),i) \\
& +\bar{G}_i\psi_s(u)]^TN^T(\tau R)N[\bar{A}_i\xi(t) + \bar{A}_{di}N\xi_\tau + \bar{B}_iv(t) \\
& +N^Tf(x(t),x(t-\tau),i) + \bar{G}_i\psi_s(u)] \\
\leq\ & [\bar{A}_i\xi(t) + \bar{A}_{di}N\xi_\tau + \bar{B}_iv(t) + \bar{G}_i\psi_s(u)]^TN^T \\
& \cdot((\tau R)^{-1} - \varepsilon_{2i}^{-1}NN^TNN^T)^{-1}N \\
& \cdot[\bar{A}_i\xi(t) + \bar{A}_{di}N\xi_\tau + \bar{B}_iv(t) + \bar{G}_i\psi_s(u)] \\
& +\varepsilon_{2i}f^T(x(t),x(t-\tau),i)f(x(t),x(t-\tau),i) \\
\leq\ & [\bar{A}_i\xi(t) + \bar{A}_{di}N\xi_\tau + \bar{B}_iv(t) + \bar{G}_i\psi_s(u)]^TN^T \\
& \cdot((\tau R)^{-1} - \varepsilon_{2i}^{-1}NN^TNN^T)^{-1}N \\
& \cdot[\bar{A}_i\xi(t) + \bar{A}_{di}N\xi_\tau + \bar{B}_iv(t) + \bar{G}_i\psi_s(u)] \\
& +2\varepsilon_{2i}(\xi^T(t)N^TM_{1i}^TM_{1i}N\xi(t) + \xi_\tau^TN^TM_{2i}^TM_{2i}N\xi_\tau).
\end{aligned} \tag{4.35}
$$

Substituting (4.32)–(4.35) into (4.31) and taking the mathematical expectation on both sides, we have

$$
\begin{aligned}
&\mathbb{E}\mathcal{L}V(\xi(t),t,i)\\
\le&\mathbb{E}\{\xi^T(t)[(\bar{A}_i+\bar{A}_{di}N)^TP_i+P_i(\bar{A}_i+\bar{A}_{di}N)+\sum_{i=1}^{N}\gamma_{ij}P_j+N^TQN\\
&+2(\varepsilon_{1i}+\varepsilon_{2i})N^TM_{1i}^TM_{1i}N+P_i\bar{A}_{di}NN^T\bar{A}_{di}^TP_i+\varepsilon_{1i}^{-1}P_iN^TNP_i+\tau T_i]\xi(t)\\
&-\xi_\tau^T[2(\varepsilon_{1i}+\varepsilon_{2i})N^TM_{2i}^TM_{2i}N+N^T\bar{E}_{di}^TP_i\bar{E}_{di}N+\tau c_eN^TN+N^TQN]\xi_\tau\\
&+2\xi^T(t)P_i\bar{B}_iv(t)+2\xi^T(t)P_i\bar{G}_i\psi_s(u)-2\psi_s^T(u)\psi_s(u)+2\psi_s(u)K_iC_iN\xi(t)\\
&+2\psi_s(u)K_iC_{di}N\xi_\tau+[\bar{A}_i\xi(t)+\bar{A}_{di}N\xi_\tau+\bar{B}_iv(t)+\bar{G}_i\psi_s(u)]^TN^T((\tau R)^{-1}\\
&-\varepsilon_{2i}^{-1}NN^TNN^T)^{-1}N[\bar{A}_i\xi(t)+\bar{A}_{di}N\xi_\tau+\bar{B}_iv(t)+\bar{G}_i\psi_s(u)]\\
&-\int_{t-\tau}^{t}(\eta_f^T(s)N^TRN\eta_f(s)+2\xi^T(t)P_i\bar{A}_{di}N\eta_f(s)+\xi^T(t)T_i\xi(t))ds\}\\
\le&\mathbb{E}\{\bar{\xi}^T(t)\Omega_i\bar{\xi}(t)\}-\int_{t-\tau}^{t}\mathbb{E}\{\bar{\xi}^T(t,s)\Gamma_i\bar{\xi}(t,s)\}ds,
\end{aligned}
\tag{4.36}
$$

where

$$\bar{\xi}(t) = [\xi^T(t)\quad \xi_\tau^TN^T\quad \psi_s^T(u)]^T,\ \bar{\xi}(t,s)=[\xi^T(t)\quad \eta^T(s)N^T]^T$$

$$
\begin{aligned}
\Omega_i: =& \begin{bmatrix} \Omega_{11i} & 0 & P_i\bar{G}_i+N^TC_i^TK_i\\ * & \Omega_{2i} & C_{di}^TK_i\\ * & * & -2I\end{bmatrix}\\
&+\begin{bmatrix}\bar{A}_i^TN^T\\ \bar{A}_{di}^TN^T\\ 0\end{bmatrix}\Phi_i^{-1}[N\bar{A}_i\ N\bar{A}_{di}\ 0]
\end{aligned}
\tag{4.37}
$$

$$\Gamma_i = \begin{bmatrix} T_i & P_i\bar{A}_{di}\\ \bar{A}_{di}^TP_i & R\end{bmatrix}, \tag{4.38}$$

where Ω_{2i} *is defined in (4.70) and*

$$\Omega_{11i} := \Omega_{1i}+\varepsilon_{1i}^{-1}P_iN^TNP_i+P_i\bar{A}_{di}NN^T\bar{A}_{di}^TP_i-\bar{L}_i^T\bar{L}_i \tag{4.39}$$

$$\Phi_i := (\tau R)^{-1}-\varepsilon_{2i}^{-1}NN^TNN^T. \tag{4.40}$$

By Schur complement, we can obtain from (4.68) and (4.69) that, for $\tau\le\bar{\tau}$,

$$\Omega_i<0,\ \Gamma_i>0,\ \forall\, i\in S. \tag{4.41}$$

Based on the inequality (4.36), the mean-square exponential stability of the system (4.26) can be proved as follows. Define

$$\lambda_P=\max_{i\in S}\lambda_{\max}(P_i),\ \lambda_p=\min_{i\in S}\lambda_{\min}(P_i),\ \lambda_\Omega=\min_{i\in S}(-\lambda_{\max}(\Omega_i)).$$

It follows from (4.36) that

$$\mathbb{E}\mathcal{L}V(\xi(t),t,i)\le-\lambda_\Omega\mathbb{E}|\bar{\xi}(t)|^2\le-\lambda_\Omega\mathbb{E}|\xi(t)|^2. \tag{4.42}$$

From the definition of $\eta(t)$ and (4.29), there exist positive scalars δ_1, δ_2, such that

$$\lambda_p \mathbb{E}|\xi(t)|^2 \leq \mathbb{E}V(\xi(t), t, i) \leq \lambda_P \mathbb{E}|\xi(t)|^2 + \int_{t-2\tau}^{t} \delta_1 \mathbb{E}|\xi(s)|^2 ds, \tag{4.43}$$

and

$$\mathbb{E}V(\xi(0), 0, r(0)) \leq \delta_2 \mathbb{E}\|\rho\|^2, \tag{4.44}$$

where ρ, $\forall\, t \in [-\tau, 0]$ is the initial function of (4.26). Let δ be a root to the inequality

$$\delta(\lambda_P + 2\tau\delta_1 e^{2\delta\tau}) \leq \lambda_\Omega. \tag{4.45}$$

To prove the mean-square exponential stability, we modify the Lyapunov functional candidate (4.29) as

$$V_1(\xi(t), t, i) = e^{\delta t} V(\xi(t), t, i), \tag{4.46}$$

and then, by Dynkin's formula [81], we obtain that for each $r(t) = i$, $i \in S$, $t > 0$

$$\begin{aligned} \mathbb{E}V_1(\xi(t), t, i) &= \mathbb{E}V_1(\xi(0), 0, r(0)) \\ &+ \mathbb{E}\int_0^t e^{\delta s}\left[\delta V(\xi(s), s, r(s)) + \mathcal{L}V(\xi(s), s, r(s))\right] ds. \end{aligned} \tag{4.47}$$

It then follows from (4.42), (4.43) that

$$\begin{aligned} \mathbb{E}V_1(\xi(t), t, i) \;\leq\;& \delta_2 \mathbb{E}\|\rho\|^2 + \mathbb{E}\int_0^t e^{\delta s}\delta\left(\lambda_P|\xi(s)|^2 + \int_{s-2\tau}^{s} \delta_1|\xi(\beta)|^2 d\beta\right) ds \\ & -\lambda_\Omega \mathbb{E}\int_0^t e^{\delta s}|\xi(s)|^2 ds. \end{aligned} \tag{4.48}$$

Noticing the definition of δ and the fact of

$$\begin{aligned} &\int_0^t e^{\delta s}\int_{s-2\tau}^{s} (\delta_1|\xi(\beta)|^2 d\beta ds \leq \int_{-2\tau}^{t} \delta_1|\xi(\beta)|^2 \int_{\beta}^{\beta+2\tau} e^{\delta s} ds d\beta \\ \leq\;& 2\tau e^{2\delta\tau}\int_{-2\tau}^{t} \delta_1|\xi(\beta)|^2 e^{\delta\beta} d\beta \\ \leq\;& 2\tau e^{2\delta\tau}\left(\int_0^t \delta_1|\xi(s)|^2 e^{\delta s} ds + \int_{-2\tau}^{0} (\delta_1|\xi(s)|^2 e^{\delta s} ds\right). \end{aligned} \tag{4.49}$$

Finally, it follows from (4.43), (4.48), and (4.49) that

$$e^{\delta t}\lambda_p \mathbb{E}|\xi(t)|^2 \leq (\delta_2 + 2\tau\delta\delta_1 e^{2\delta\tau})\mathbb{E}\|\rho\|^2,$$

or

$$\lim_{t\to\infty} \sup \frac{1}{t} \log(\mathbb{E}|\xi(t, \rho)|^2) \leq -\delta,$$

which indicates that, for $\tau \leq \bar{\tau}$, the trivial solution of (4.25) is exponentially stable in the mean-square sense.

In the sequel, we shall deal with the H_∞ performance of the the system (4.26)–(4.28). Assume zero initial condition, i.e., $\xi(t) = 0$ for $t \in [-2\tau, 0]$, and define

$$J(t) = \mathbb{E}\left\{\int_0^t [e^T(s)e(s) - \gamma^2 v^T(s)v(s)]ds\right\}. \tag{4.50}$$

It follows from Dynkin's formula [81] and fact $\xi(0) = 0$ that

$$\mathbb{E}\{V(\xi(t), t, r(t))\} = \mathbb{E}\left\{\int_0^t \mathcal{L}V(\xi(s), s, r(s))ds\right\}. \tag{4.51}$$

From (4.50) and (4.51), it is easy to see that

$$\begin{aligned} J(t) &= \mathbb{E}\left\{\int_0^t [e^T(s)e(s) - \gamma^2 v^T(s)v(s) + \mathcal{L}V(\xi(s), s, r(s))]ds\right\} \\ &\quad -\mathbb{E}\{V(\xi(t), t, r(t))\} \\ &\leq \mathbb{E}\left\{\int_0^t [e^T(s)e(s) - \gamma^2 v^T(s)v(s) + \mathcal{L}V(\xi(s), s, r(s))]ds\right\}. \end{aligned} \tag{4.52}$$

Next, let

$$\bar{\xi}(s, v) = [\xi^T(s) \quad \xi_\tau^T N^T \quad \psi_s^T(u) \quad v^T(s)]^T, \ \bar{\xi}(s, \beta) = [\xi^T(s) \quad \eta^T(\beta) N^T]^T, \tag{4.53}$$

and then, it follows from (4.36) that

$$\begin{aligned} &\mathbb{E}\{c^T(s)c(s) - \gamma^2 v^T(s)v(s) + \mathcal{L}V(x(s), s, i)\} \\ \leq\ &\mathbb{E}\{\bar{\xi}^T(s, v)\Xi_i\bar{\xi}(s, v)\} - \int_{s-\tau}^s \mathbb{E}\{\bar{\xi}^T(s, \beta)\Gamma_i\bar{\xi}(s, \beta)\}d\beta, \end{aligned} \tag{4.54}$$

where

$$\begin{aligned} \Xi_i : &= \begin{bmatrix} \Omega_{11i} + \bar{L}_i^T\bar{L}_i & 0 & P_i\bar{B}_i & P_i\bar{G}_i + N^T C_i^T K_i \\ * & \Omega_{2i} & 0 & C_{di}^T K_i \\ * & * & -\gamma^2 I & 0 \\ * & * & * & -2I \end{bmatrix} \\ &\quad + \begin{bmatrix} \bar{A}_i^T N^T \\ \bar{A}_{di}^T N^T \\ \bar{B}_i^T N^T \\ 0 \end{bmatrix} \Phi_i^{-1}[N\bar{A}_i \ N\bar{A}_{di} \ N\bar{B}_i \ 0] \end{aligned} \tag{4.55}$$

and Ω_{2i}, Γ_i, Ω_{11i}, and Φ_i are defined in (4.70), (4.38), (4.39), and (4.40), respectively.

By the Schur complement and the conditions (4.68)–(4.69), for $\tau \leq \bar{\tau}$, it follows that

$$\Xi_i < 0, \quad \Gamma_i > 0 \quad \forall\, i \in S, \tag{4.56}$$

and we can obtain from (4.52), (4.54), and (4.56) that, for all $t > 0$, $J(t) < 0$. *Therefore, we arrive at*

$$\mathbb{E}\left\{\int_0^t [e^T(s)e(s)\right\} ds \leq \gamma^2 \left\{\int_0^t v^T(s)v(s)ds\right\},$$

which implies that

$$\|e(t)\|_{E_2} < \gamma \|v(t)\|_2. \tag{4.57}$$

From Definition 6, it is concluded that the filtering error system (4.11)–(4.12) is stochastically stable, with a disturbance attenuation level $\gamma > 0$ for $\tau \leq \bar{\tau}$. The proof is now completed.

In the next subsection, our attention is focused on the design of filter parameters F_i, G_i, and H_i, for $i \in S$, by using the results in Theorem 5. The explicit expression of the expected filter parameters is obtained in term of the solution to a set of LMIs.

4.2.2 Filter Synthesis

The following theorem shows that the desired filter parameters can be derived by solving several LMIs.

Theorem 6 Consider the system (4.11)–(4.12). If there exist matrices $X_i > 0$, $\mathcal{Y}_i > 0$, $\bar{T}_{11i} > 0$, $\bar{T}_{22i} > 0$, $Q > 0$, $R > 0$, a matrix $\bar{T}_{12i}$ and scalars $\varepsilon_{1i} > 0$, $\varepsilon_{2i} > 0$, such that the following linear matrix inequalities

$$\left[\begin{array}{cccccccc}
\Pi_{1i} & \Pi_{2i} & 0 & \mathcal{Y}_i B_{1i} & C_i^T K_i & \bar{\tau} A_i^T R & 0 & \mathcal{Y}_i A_{di} \\
* & \Pi_{3i} & 0 & \Pi_{4i} & \tilde{G}_i + C_i^T K_i & \bar{\tau} A_i^T R & 0 & \Pi_{5i} \\
* & * & \Omega_{2i} & 0 & C_{di}^T K_i & \bar{\tau} A_{di} R & 0 & 0 \\
* & * & * & -\gamma^2 I & 0 & \bar{\tau} B_{1i}^T R & 0 & 0 \\
* & * & * & * & -2I & 0 & 0 & 0 \\
* & * & * & * & * & -\bar{\tau} R & \bar{\tau} R & 0 \\
* & * & * & * & * & * & -\varepsilon_{2i} I & 0 \\
* & * & * & * & * & * & * & -I \\
* & * & * & * & * & * & * & * \\
* & * & * & * & * & * & * & * \\
* & * & * & * & * & * & * & * \\
* & * & * & * & * & * & * & * \\
* & * & * & * & * & * & * & * \\
* & * & * & * & * & * & * & * \\
* & * & * & * & * & * & * & *
\end{array}\right.$$

$$
\left.\begin{array}{ccccccc}
\mathcal{Y}_i & 0 & 0 & L_i^T - H_i^T & M_{1i}^T\varepsilon_{1i} & M_{1i}^T\varepsilon_{2i} & Q \\
X_i & 0 & 0 & L_i^T & M_{1i}^T\varepsilon_{1i} & M_{1i}^T\varepsilon_{2i} & Q \\
0 & E_{di}^T\mathcal{Y}_i & E_{di}^T X_i & 0 & 0 & 0 & 0 \\
0 & 0 & 0 & 0 & 0 & 0 & 0 \\
0 & 0 & 0 & 0 & 0 & 0 & 0 \\
0 & 0 & 0 & 0 & 0 & 0 & 0 \\
0 & 0 & 0 & 0 & 0 & 0 & 0 \\
0 & 0 & 0 & 0 & 0 & 0 & 0 \\
-\varepsilon_{1i}I & 0 & 0 & 0 & 0 & 0 & 0 \\
* & -\mathcal{Y}_i & \mathcal{Y}_i & 0 & 0 & 0 & 0 \\
* & * & -X_i & 0 & 0 & 0 & 0 \\
* & * & * & -I & 0 & 0 & 0 \\
* & * & * & * & -2\varepsilon_{1i}I & 0 & 0 \\
* & * & * & * & * & -2\varepsilon_{2i}I & 0 \\
* & * & * & * & * & * & -Q
\end{array}\right] < 0, \quad (4.58)
$$

$$
\left[\begin{array}{ccc}
\bar{T}_{11i} & \bar{T}_{12i} & \mathcal{Y}_i A_{di} \\
\bar{T}_{12i}^T & \bar{T}_{22i} & \Pi_{5i} \\
A_{di}^T\mathcal{Y}_i & \Pi_{5i}^T & R
\end{array}\right] > 0, \quad (4.59)
$$

for $\forall\, i \in S$, hold, where

$$
\begin{aligned}
\Pi_{1i} &:= \mathcal{Y}_i A_i + A_i^T\mathcal{Y}_i + \mathcal{Y}_i A_{di} + A_{di}^T\mathcal{Y}_i + \Sigma_{j=1}^N \gamma_{ij}\mathcal{Y}_j + \bar{\tau}\bar{T}_{11i}, \\
\Pi_{2i} &:= \mathcal{Y}_i A_i + A_i^T X_i + \mathcal{Y}_i A_{di} + A_{di}^T X_i + \Sigma_{j=1}^N \gamma_{ij}\mathcal{Y}_j + \bar{\tau}\bar{T}_{12i}, \\
&\quad\; + C_i^T K_{1i}\tilde{G}_i^T + C_{di}^T K_{1i}\tilde{G}_i^T, \\
\Pi_{3i} &:= X_i A_i + A_i^T X_i + \bar{\tau}c_e I + \tilde{G}_i K_{1i} C_i + C_i^T K_{1i}\tilde{G}_i^T + X_i A_{di} + \tilde{G}_i K_{1i} C_{di}, \\
&\quad\; + C_{di}^T K_{1i}\tilde{G}_i^T + A_{di}^T X_i + \Sigma_{j=1}^N \gamma_{ij} X_j + \bar{\tau}c_e I + \bar{\tau}\bar{T}_{22i}, \\
\Pi_{4i} &:= X_i B_{1i} + \tilde{G}_i B_{2i}, \\
\Pi_{5i} &:= X_i A_{di} + \tilde{G}_i K_{1i} C_{di}, \qquad (4.60)
\end{aligned}
$$

then the system (4.11)–(4.12) is stochastically stable, with disturbance attenuation γ for $\tau \leq \bar{\tau}$. In this case, the parameters of the desired H_∞ filter (Σ_f) are given as follows:

$$
F_i := (\mathcal{Y}_i - X_i)^{-1}\tilde{F}_i,\ G_i := (\mathcal{Y}_i - X_i)^{-1}\tilde{G}_i,\ H_i := \tilde{H}_i. \qquad (4.61)
$$

Proof 6 *Define*

$$
P_i = \left[\begin{array}{cc} X_i & \mathcal{Y}_i - X_i \\ \mathcal{Y}_i - X_i & X_i - \mathcal{Y}_i \end{array}\right] > 0, \quad \Upsilon = \left[\begin{array}{cc} Y_i & I \\ Y_i & 0 \end{array}\right], \qquad (4.62)
$$

where $Y_i = \mathcal{Y}_i^{-1} > 0$.

From (4.59), we have

$$T := \Upsilon^{-T}\text{diag}(Y_i,\ I)\begin{pmatrix} \bar{T}_{11i} & \bar{T}_{12i} \\ \bar{T}_{12i}^T & \bar{T}_{22i} \end{pmatrix}\text{diag}(Y_i,\ I)\Upsilon > 0. \qquad (4.63)$$

Premultiplying and postmultiplying the LMIs in (4.58) by diag $(Y_i, \underbrace{I, ...I}_{8}, Y_i, \underbrace{I, ..., I}_{5})$*, and (4.59) by* $\text{diag}(Y_i, I, I)$*, we have*

$$\left[\begin{array}{ccccccccc}
\bar{\Pi}_{1i} & \bar{\Pi}_{2i} & 0 & B_{1i} & Y_iC_i^TK_i & \bar{\tau}Y_iA_i^TR & 0 & A_{di} & I \\
* & \Pi_{3i} & 0 & \Pi_{4i} & \tilde{G}_i & \bar{\tau}A_i^TR & 0 & \Pi_{5i} & X_i \\
* & * & \Omega_{2i} & 0 & C_{di}^TK_i & \bar{\tau}A_{di}R & 0 & 0 & 0 \\
* & * & * & -\gamma^2 I & 0 & \bar{\tau}B_{1i}^TR & 0 & 0 & 0 \\
* & * & * & * & -2I & 0 & 0 & 0 & 0 \\
* & * & * & * & * & -\bar{\tau}R & \bar{\tau}R & 0 & 0 \\
* & * & * & * & * & * & -\varepsilon_{2i}I & 0 & 0 \\
* & * & * & * & * & * & * & -I & 0 \\
* & * & * & * & * & * & * & * & -\varepsilon_{1i}I \\
* & * & * & * & * & * & * & * & * \\
* & * & * & * & * & * & * & * & * \\
* & * & * & * & * & * & * & * & * \\
* & * & * & * & * & * & * & * & * \\
* & * & * & * & * & * & * & * & * \\
* & * & * & * & * & * & * & * & * \\
* & * & * & * & * & * & * & * & *
\end{array}\right.$$

$$\left.\begin{array}{cccccc}
0 & 0 & \bar{L}_i & Y_iM_{1i}^T\varepsilon_{1i}Y_i & Y_iM_{1i}^T\varepsilon_{1i} & Y_iQ \\
0 & 0 & L_i^T & M_{1i}^T\varepsilon_{1i} & M_{1i}^T\varepsilon_{1i} & Q \\
E_{di}^T & E_{di}^TX_i & 0 & 0 & 0 & 0 \\
0 & 0 & 0 & 0 & 0 & 0 \\
0 & 0 & 0 & 0 & 0 & 0 \\
0 & 0 & 0 & 0 & 0 & 0 \\
0 & 0 & 0 & 0 & 0 & 0 \\
0 & 0 & 0 & 0 & 0 & 0 \\
0 & 0 & 0 & 0 & 0 & 0 \\
-Y_i & I & 0 & 0 & 0 & 0 \\
* & -X_i & 0 & 0 & 0 & 0 \\
* & * & -I & 0 & 0 & 0 \\
* & * & * & -2\varepsilon_{1i}I & 0 & 0 \\
* & * & * & * & -2\varepsilon_{2i}I & 0 \\
* & * & * & * & * & -Q
\end{array}\right] < 0, \qquad (4.64)$$

$$\begin{bmatrix} Y_i\bar{T}_{11i} & Y_i\bar{T}_{12i} & A_{di} \\ * & \bar{T}_{22i} & X_iA_{di} + \tilde{G}_iK_{1i}C_{di} \\ * & * & R \end{bmatrix} > 0, \quad \forall\, i \in S, \qquad (4.65)$$

where

$$\begin{aligned}
\bar{\Pi}_{1i} &:= A_iY_i + Y_iA_i^T + A_{di}Y_i + Y_iA_{di}^T + \Sigma_{j=1}^N \gamma_{ij}Y_iY_j^{-1}Y_i \\
&\quad +\bar{\tau}c_eY_i^2 + \bar{\tau}Y_i\bar{T}_{11i}Y_i, \\
\bar{\Pi}_{2i} &:= A_i + Y_iA_i^TX_i + A_{di} + Y_iA_{di}^TX_i + \Sigma_{j=1}^N \gamma_{ij}Y_iY_j^{-1}Y_i \\
&\quad +\bar{\tau}c_eY_i + \bar{\tau}Y_iT_{12i} \\
&\quad Y_iC_i^TK_{1i}\tilde{G}_i^T + Y_i\tilde{F}_i^T + Y_iC_{di}K_{1i}\tilde{G}_i^T, \\
\bar{L}_i &:= Y_iL_i^T - Y_iH_i^T.
\end{aligned}$$

From the definitions of P_i and Υ_i, the LMIs in (4.64)-(4.65) are equivalent to the following matrix inequalities:

$$\begin{bmatrix}
\bar{\bar{\Omega}}_{1i} & 0 & \Upsilon_i^TP_i\bar{B}_i & \Lambda_{1i} & \Lambda_{4i} & 0 & \Upsilon_i^TP_iN^T & \Lambda_{2i} & 0 \\
* & \Omega_{2i} & 0 & C_{di}^TK & \Lambda_{5i} & 0 & 0 & 0 & \Lambda_{3i} \\
* & * & -\gamma^2I & 0 & \Lambda_{6i} & 0 & 0 & 0 & 0 \\
* & * & * & -2I & 0 & 0 & 0 & 0 & 0 \\
* & * & * & * & -\bar{\tau}R & \bar{\tau}R & 0 & 0 & 0 \\
* & * & * & * & * & -\varepsilon_{2i}I & 0 & 0 & 0 \\
* & * & * & * & * & * & -\varepsilon_{1i}I & 0 & 0 \\
* & * & * & * & * & * & * & -I & 0 \\
* & * & * & * & * & * & * & * & -\bar{P}
\end{bmatrix} < 0, \tag{4.66}$$

$$\begin{bmatrix} \Upsilon_i^TT_i\Upsilon_i & \Upsilon_i^TP_i\bar{A}_{di} \\ * & R \end{bmatrix} > 0, \tag{4.67}$$

where

$$\begin{aligned}
\bar{\Omega}_{1i} &= \Omega_{1i} - N^TE_i^TP_iE_iN,\ \bar{\bar{\Omega}}_{1i} = \Upsilon_i^T\bar{\Omega}_{1i}\Upsilon_i,\ \bar{P} = \Upsilon_i^TP_i\Upsilon_i, \\
\Lambda_{1i} &= \Upsilon_i^TP_iG_i + \Upsilon_i^TN^TC_i^TK_i,\ \Lambda_{2i} = \Upsilon_i^TP_i\bar{A}_{di}N,\ \Lambda_{3i} = \Upsilon_i^TN^TE_{di}^TP_i\Upsilon_i, \\
\Lambda_{4i} &= \bar{\tau}\Upsilon_i^T\bar{A}_i^TN^TR,\ \Lambda_{5i} = \bar{\tau}\bar{A}_{di}^TN^TR,\ \Lambda_{6i} = \bar{\tau}\bar{B}_i^TN^TR.
\end{aligned}$$

Finally, premultiplying and postmultiplying (4.66) by diag $(\Upsilon_i^{-T}, \underbrace{I, ..., I}_{7}, \Upsilon_i^{-T})$ *and its transpose, (4.67) by* $\mathrm{diag}(\Upsilon_i^{-T}, I)$ *and its transpose, we can obtain from Theorem 5 and Schur complement lemma that, with the given filter parameters in (4.61), the system (4.11)-(4.12) is stochastically stable, with disturbance attenuation γ for $\tau \leq \bar{\tau}$.*

Remark 4.3 *The H_∞ filter design problem is solved in Theorem 6 for the addressed delayed stochastic jumping systems, with sensor nonlinearities and external nonlinear disturbances. LMI-based sufficient conditions are obtained for the existence of full-order filters that ensure the mean-square exponential stability of the resulting filtering error system and reduce the effect of the disturbance input on the estimated signal to a prescribed level for all admissible time-delays and nonlinearities. The feasibility of the filter design problem can* *be readily checked by the solvability of two sets of LMIs, which can be determined by using the MATLAB LMI Toolbox in a straightforward way. In the next section, an illustrative example will be provided to show the usefulness of the proposed techniques.*

4.2.3 An Illustrative Example

In this section, a simulation example is presented to illustrate the usefulness and flexibility of the filter design method developed in this chapter. We are interested in obtaining the upper bound $\bar{\tau}$ of the time-delays and designing the H_∞ filter for the stochastic jumping system with nonlinear disturbances and sensor nonlinearities.

The system data of (4.1)–(4.3) are given as follows:

$$\begin{bmatrix} \gamma_{11} & \gamma_{12} \\ \gamma_{21} & \gamma_{22} \end{bmatrix} = \begin{bmatrix} -2.5 & 2.5 \\ 0.9 & -0.9 \end{bmatrix}, \qquad \gamma = 1.8.$$

Mode 1:

$$\begin{aligned} A_1 &= \begin{bmatrix} -3.5 & 1 \\ 0 & -2.7 \end{bmatrix}, \quad A_{d1} = \begin{bmatrix} 0.15 & 0 \\ 0 & 0.21 \end{bmatrix}, \\ E_1 &= \begin{bmatrix} 0.13 & 0 \\ 0 & 0.15 \end{bmatrix}, B_{11} = \begin{bmatrix} 0.2 \\ 0.1 \end{bmatrix}, B_{21} = \begin{bmatrix} 0.13 \\ 0.02 \end{bmatrix}, \\ M_{11} &= \begin{bmatrix} 0.5 & 0 \\ 0 & 0.1 \end{bmatrix}, \quad M_{21} = \begin{bmatrix} 0.2 & 0 \\ 0 & 0.5 \end{bmatrix}, \\ C_1 &= 0.5I_2, \quad C_{d1} = 0.5I_2, \quad L_1 = [0.3 \ \ 0.7], \\ K_{11} &= \mathrm{diag}\{0.3, 0.4\}, \quad K_{21} = \mathrm{diag}\{0.6, 0.5\}. \end{aligned}$$

Mode 2:

$$\begin{aligned} A_2 &= \begin{bmatrix} -4.3 & 1 \\ 0 & -2.5 \end{bmatrix}, \quad A_{d2} = \begin{bmatrix} 0.22 & 0 \\ 0 & 0.1 \end{bmatrix}, \\ E_2 &= \begin{bmatrix} 0.12 & 0 \\ 0 & 0.31 \end{bmatrix}, B_{12} = \begin{bmatrix} 0.1 \\ 0.2 \end{bmatrix}, B_{22} = \begin{bmatrix} 0.2 \\ 0.15 \end{bmatrix}, \\ M_{12} &= \begin{bmatrix} 0.4 & 0 \\ 0 & 0.2 \end{bmatrix}, \quad M_{22} = \begin{bmatrix} 0.3 & 0 \\ 0 & 0.2 \end{bmatrix}, \\ C_2 &= 0.6I_2, \quad C_{d2} = I_2, \quad L_2 = [0.6 \ \ 0.8], \\ K_{12} &= \mathrm{diag}\{0.4, 0.6\}, \quad K_{22} = \mathrm{diag}\{0.6, 0.9\}. \end{aligned}$$

Using MATLAB LMI Control Toolbox to solve the LMIs in (4.58) and (4.59), we obtain the upper bound of time-delays as $\bar{\tau} = 2.2520$. Therefore, by Theorem 6, it can be calculated that for all $0 < \tau \leq 2.2520$, there exist the desired H_∞ filters. For demonstration purpose, let us fix $\tau = 1.5$. In this case, by the LMI toolbox, we can calculate that

$$X_1 = \begin{bmatrix} 58.5508 & -10.3863 \\ -10.3863 & 56.9008 \end{bmatrix}, \quad Y_1 = \begin{bmatrix} 34.2934 & -6.2039 \\ -6.2039 & 31.9526 \end{bmatrix},$$

$$T_1 = \begin{bmatrix} 45.1367 & -16.5489 & 13.1486 & -6.4879 \\ -16.5489 & 29.2881 & -6.8324 & 25.7375 \\ 13.1486 & -6.8324 & 87.1278 & -28.4811 \\ -6.4879 & 25.7375 & -28.4811 & 54.8590 \end{bmatrix},$$

$$X_2 = \begin{bmatrix} 48.3957 & -9.1244 \\ -9.1244 & 44.5760 \end{bmatrix}, \quad Y_2 = \begin{bmatrix} 29.6427 & -4.9378 \\ -4.9378 & 25.6274 \end{bmatrix},$$

$$T_2 = \begin{bmatrix} 35.0120 & -10.6401 & 14.8960 & -3.7476 \\ -10.6401 & 10.3716 & -2.1487 & 1.4473 \\ 14.8960 & -2.1487 & 73.7385 & -22.6410 \\ -3.7476 & 1.4473 & -22.6410 & 40.0518 \end{bmatrix},$$

$$Q = \begin{bmatrix} 29.9297 & -8.6883 \\ -8.6883 & 26.0809 \end{bmatrix}, R = \begin{bmatrix} 2.2850 & 0.0177 \\ -0.0177 & 2.5803 \end{bmatrix},$$

$$s11 = 132.4881, \quad s12 = 94.5044, \quad s21 = 50.6824, \quad s22 = 65.9064.$$

The filter parameters to be determined are as follows:

$$F_1 = \begin{bmatrix} -5.4748 & 1.8614 \\ 0.3389 & -1.7616 \end{bmatrix}, \quad G_1 = \begin{bmatrix} 0.1978 & 0.0053 \\ 0.0183 & 0.2870 \end{bmatrix},$$

$$F_2 = \begin{bmatrix} -6.4333 & 1.6877 \\ 0.1598 & -3.2487 \end{bmatrix}, \quad G_2 = \begin{bmatrix} 0.5151 & 0.0248 \\ 0.0561 & 0.2546 \end{bmatrix},$$

$$H_1 = [0.2951, \quad 0.7075], \quad H_2 = [0.6013, \quad 0.7987].$$

Figure 4.1–Figure 4.6 are the simulation results for the performance of the designed H_∞ filter, where the sensor nonlinearities are taken as

$$\psi(u) = \frac{K_{1i} + K_{2i}}{2} u + \frac{K_{2i} - K_{1i}}{2} \sin(u),$$

which satisfies (4.6). It is confirmed from the simulation results that all the expected objectives are well achieved.

4.3 H_∞ Dynamical Output Feedback Control

4.3.1 H_∞ Controller Analysis

In the following theorem, a delay-dependent technique and an LMI method are developed to deal with the stability analysis problem for the output feedback

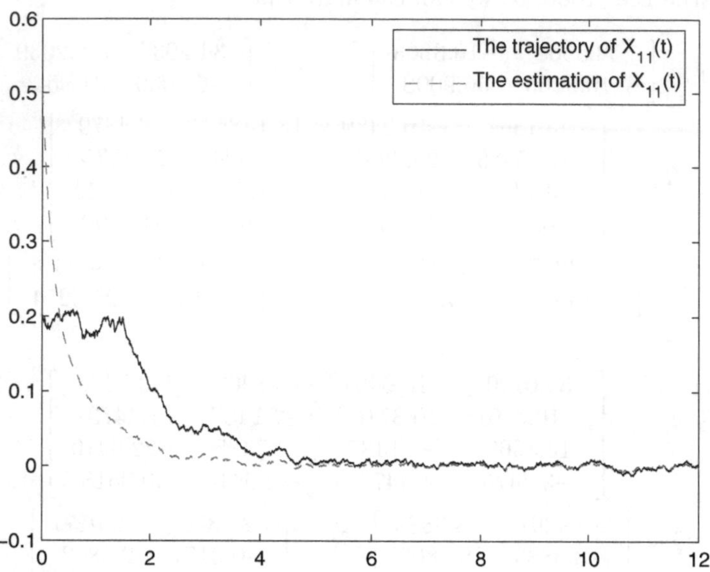

FIGURE 4.1
The state and estimation of $x_{11}(t)$.

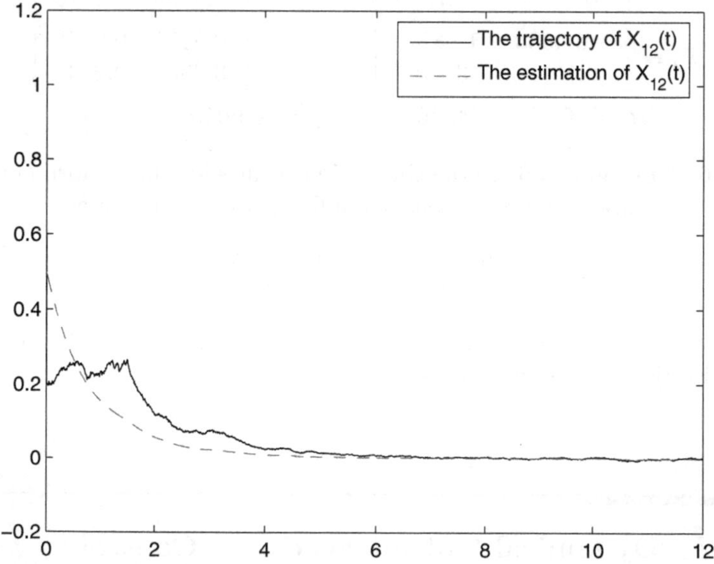

FIGURE 4.2
The state and estimation of $x_{12}(t)$.

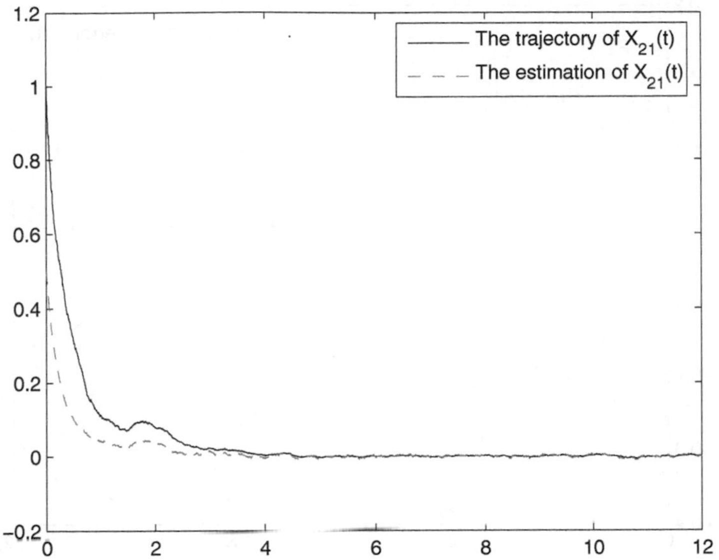

FIGURE 4.3
The state and estimation of $x_{21}(t)$.

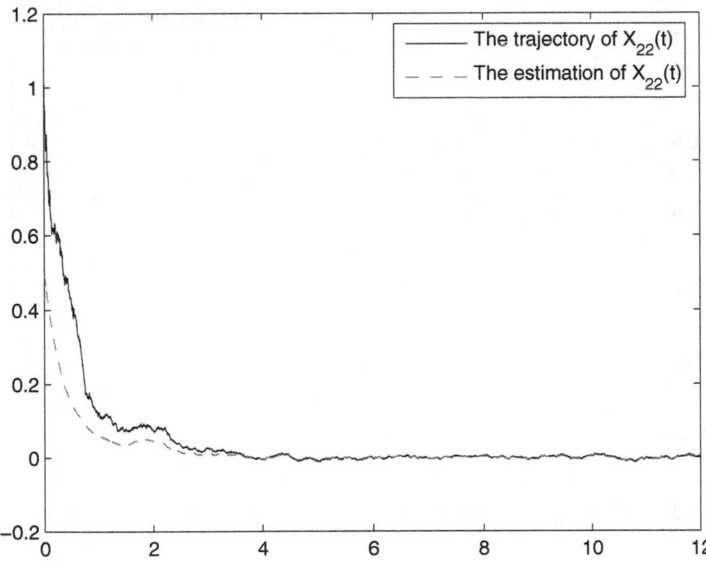

FIGURE 4.4
The state and estimation of $x_{22}(t)$.

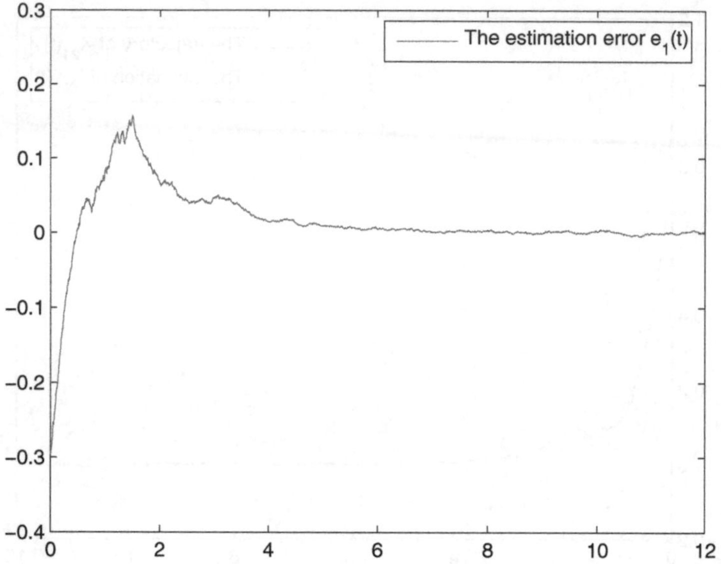

FIGURE 4.5
Estimation error $e_1(t)$.

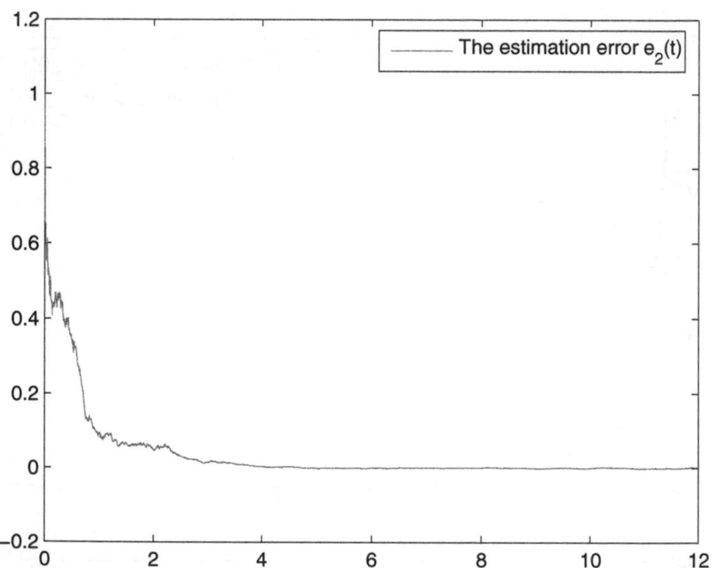

FIGURE 4.6
Estimation error $e_2(t)$.

H_∞ controller design of the stochastic system (4.1)–(4.4), and a sufficient condition is derived that ensures the solvability of the output feedback H_∞ control problem.

Theorem 7 Consider the augmented system (4.15)–(4.16), with given controller parameters. If there exist positive definite matrices $P_i > 0$, $T_i > 0$, $Q > 0$, and $R > 0$, such that the following linear matrix inequalities

$$\begin{bmatrix} \Delta_{1i} & 0 & P_i\tilde{B}_i & \mathbb{G}_i & \mathbb{A}_i & 0 & P_iN^T & P_i\tilde{A}_{di}N \\ * & \Delta_{2i} & 0 & C_{di}^TK_i & \mathbb{A}_{di} & 0 & 0 & 0 \\ * & * & -\gamma^2 I & 0 & \mathbb{B}_i & 0 & 0 & 0 \\ * & * & * & -2I & 0 & 0 & 0 & 0 \\ * & * & * & * & -\bar{\tau}R & R & 0 & 0 \\ * & * & * & * & * & -\varepsilon_{2i}I & 0 & 0 \\ * & * & * & * & * & * & -\varepsilon_{1i}I & 0 \\ * & * & * & * & * & * & * & -I \end{bmatrix} < 0, \quad (4.68)$$

$$\begin{bmatrix} T_i & P_i\tilde{A}_{di} \\ \tilde{A}_{di}^TP_i & R \end{bmatrix} > 0, \quad (4.69)$$

for $\forall\, i \in S$, hold, where

$$\begin{aligned} \Delta_{1i} &:= P_i(\tilde{A}_i + \tilde{A}_{di}N) + (\tilde{A}_i + \tilde{A}_{di}N)^TP_i + \Sigma_{i,j=1}^N \gamma_{ij}P_j + \tilde{L}_i^T\tilde{L}_i + N^TQN \\ &\quad +2(\varepsilon_{1i} + \varepsilon_{2i})(N^TM_{1i}^TM_{1i}N) + \bar{\tau}T_i,\ \mathbb{G}_i := P_i\tilde{G}_i + N^TC_i^TK_i, \\ \Delta_{2i} &:= 2(\varepsilon_{1i} + \varepsilon_{2i})(M_{2i}^TM_{2i}) + N^T\tilde{E}_{di}^TP_i\tilde{E}_{di}N + \bar{\tau}c_eN^TN - Q, \\ \mathbb{A}_i &:= \bar{\tau}\tilde{A}_i^TN^TR,\ \mathbb{A}_{di} := \bar{\tau}\tilde{A}_{di}^TN^TR,\ \mathbb{B}_i := \bar{\tau}\tilde{B}_i^TN^TR, \end{aligned} \quad (4.70)$$

with $c_e = \max_{i\in S}\|\tilde{E}_{di}\|^2$, then the augmented system (Σ_k) is stochastically stable, with the disturbance attenuation level γ for $\tau \le \bar{\tau}$ ($\bar{\tau}$ is the upper bound of the time-delays).

Proof 7 *The proof of this theorem can be obtained along the similar line of that of Theorem 5, and is therefore omitted here to avoid unnecessary duplication.*

In the next subsection, our attention is focused on the design of controller parameters A_{ci}, B_{ci}, and C_{ci}, for $i \in S$, by using the results in Theorem 7. The explicit expression of the expected controller parameters is obtained in terms of the solution to a set of LMIs.

4.3.2 Output Feedback H_∞ Controller Synthesis

The following theorem shows that the desired controller parameters can be derived by solving several LMIs.

Theorem 8 Consider the system (4.15)–(4.16). If there exist matrices $X_i > 0$, $Y_i > 0$, $\bar{T}_{11i} > 0$, $\bar{T}_{22i} > 0$, matrix $\bar{T}_{12i}$, and scalars $\epsilon_{1i} > 0$, $\epsilon_{2i} > 0$, such that the following linear matrix inequalities

$$\left[\begin{array}{cccccccccccccc}
\bar{\Delta}_{1i} & \bar{\Delta}_{2i} & 0 & B_{1i} & Y_i C_i^T K_i & \mathbb{A}_i & 0 & A_{di} & Y_i Q & I & Y_i L_i^T & Y_i H_{1i} & 0 & H_{3i} \\
* & \bar{\Delta}_{3i} & 0 & \bar{\Delta}_{4i} & \bar{\Delta}_{5i} & \bar{\tau} A_i^T R & 0 & \bar{\Delta}_{6i} & Q & X_i & L_i^T & H_{1i} & 0 & 0 \\
* & * & \bar{\Delta}_{7i} & 0 & C_{di}^T K_i & \bar{\tau} A_{di}^T R & 0 & 0 & 0 & 0 & 0 & 0 & H_{2i} & 0 \\
* & * & * & -\gamma^2 I & 0 & \bar{\tau} B_{1i}^T R & 0 & 0 & 0 & 0 & 0 & 0 & 0 & 0 \\
* & * & * & * & -2I & 0 & 0 & 0 & 0 & 0 & 0 & 0 & 0 & 0 \\
* & * & * & * & * & -\bar{\tau} R & \bar{\tau} R & 0 & 0 & 0 & 0 & 0 & 0 & 0 \\
* & * & * & * & * & * & -\epsilon_{2i} I & 0 & 0 & 0 & 0 & 0 & 0 & 0 \\
* & * & * & * & * & * & * & -I & 0 & 0 & 0 & 0 & 0 & 0 \\
* & * & * & * & * & * & * & * & -Q & 0 & 0 & 0 & 0 & 0 \\
* & * & * & * & * & * & * & * & * & -\epsilon_{1i} I & 0 & 0 & 0 & 0 \\
* & * & * & * & * & * & * & * & * & * & -I & 0 & 0 & 0 \\
* & * & * & * & * & * & * & * & * & * & * & -V_{1i} & 0 & 0 \\
* & * & * & * & * & * & * & * & * & * & * & * & -V_{2i} & 0 \\
* & * & * & * & * & * & * & * & * & * & * & * & * & -V_{3i}
\end{array}\right] < 0, \quad (4.71)$$

$$\left[\begin{array}{ccc}
\bar{T}_{11i} & \bar{T}_{12i} & A_{di} \\
\bar{T}_{12i}^T & \bar{T}_{22i} & \bar{\Delta}_{6i} \\
A_{di}^T & \bar{\Delta}_{4i}^T & R
\end{array}\right] > 0, \quad (4.72)$$

for $\forall\, i \in S$, hold for some given scalars $\epsilon_{1i} > 0$, $\epsilon_{2i} > 0$, and matrices $Q > 0$,

$R > 0$, where

$$
\begin{aligned}
\bar{\Delta}_{1i} &:= A_iY_i + Y_iA_i^T + A_{di}Y_i + Y_iA_{di}^T + D_i\tilde{C}_{ci} + \tilde{C}_{ci}^TD_i^T + \gamma_{ii}Y_i + \bar{\tau}\bar{T}_{11i},\\
\bar{\Delta}_{2i} &:= \tilde{A}_{ci}^T + \bar{\tau}\bar{T}_{12i},\\
\bar{\Delta}_{3i} &:= X_iA_i + A_i^TX_i + X_iA_{di} + A_{di}^TX_i + \tilde{B}_{ci}K_{1i}(C_i + C_{di})\\
&\quad +(C_i^T + C_{di}^T)K_{1i}\tilde{B}_{ci}^T + \Sigma_{j=1}^N\gamma_{ij}X_j + \bar{\tau}\bar{T}_{22i},\\
\bar{\Delta}_{4i} &:= X_iB_{1i} + \tilde{B}_{ci}B_{2i},\\
\bar{\Delta}_{5i} &:= \tilde{B}_{ci} + C_i^TK_i,\\
\bar{\Delta}_{6i} &:= X_iA_{di} + \tilde{B}_{ci}K_{1i}C_{di},\\
\bar{\Delta}_{7i} &:= 2(\varepsilon_{1i} + \varepsilon_{2i})(M_{2i}^TM_{2i}) + \bar{\tau}c_eN^TN - Q,\\
\mathbb{A}_i &:= \bar{\tau}Y_iA_i^TR + \bar{\tau}\tilde{C}_{ci}^TD_i^TR,\\
H_{1i} &:= [M_{1i}^T\ \ M_{1i}^T],\\
H_{2i} &:= [E_i^T\ \ E_i^TX_i],\\
H_{3i} &:= \underbrace{[Y_i^T, ..., Y_i]}_{N-1},\\
V_{2i} &:= \begin{bmatrix} Y_i & I \\ I & X_i \end{bmatrix},\\
V_{1i} &:= \mathrm{diag}\{2\epsilon_{1i}^{-1}, 2\epsilon_{2i}^{-1}\},\\
V_{3i} &:= \mathrm{diag}\{\gamma_{ij}^{-1}Y_1, ..., \gamma_{ij}^{-1}Y_{i-1}, \gamma_{ij}^{-1}Y_{i+1}, ..., \gamma_{ij}^{-1}Y_N\},
\end{aligned}
\tag{4.73}
$$

then the system (4.15)–(4.16) is stochastically stable with disturbance attenuation γ for $\tau \leq \bar{\tau}$. In this case, the parameters of the desired output feedback H_∞ controller (Σ_k) are given as follows:

$$
\begin{aligned}
A_{ci} &:= (\mathcal{Y}_i - X_i)^{-1}(\tilde{A}_{ci} - A_i^T - A_{di}^T - X_i(A_i + A_{di})Y_i - X_iB_{1i}\tilde{C}_{ci}\\
&\quad -\tilde{B}_{ci}K_{1i}(C_i + C_{di}) - \tilde{B}_{ci}B_{2i}\tilde{C}_{ci} - \Sigma_{j=1}^N\gamma_{ij}\mathcal{Y}_jY_i)\mathcal{Y}_i,\\
B_{ci} &:= (\mathcal{Y}_i - X_i)^{-1}\tilde{B}_{ci},\ C_{ci} := \tilde{C}_{ci}\mathcal{Y}_i,
\end{aligned}
\tag{4.74}
$$

where $\mathcal{Y}_i^{-1} = Y_i > 0$.

Proof 8 *Define*

$$
P_i = \begin{bmatrix} X_i & \mathcal{Y}_i - X_i \\ \mathcal{Y}_i - X_i & X_i - \mathcal{Y}_i \end{bmatrix} > 0, \quad \Upsilon = \begin{bmatrix} Y_i & I \\ Y_i & 0 \end{bmatrix}. \tag{4.75}
$$

From (4.72), we have

$$
T := \Upsilon^{-T}\begin{pmatrix} \bar{T}_{11i} & \bar{T}_{12i} \\ \bar{T}_{12i}^T & \bar{T}_{22i} \end{pmatrix}\Upsilon > 0. \tag{4.76}
$$

Finally, premultiplying and postmultiplying (4.71) by diag $(\Upsilon_i^{-T}, \underbrace{I, ..., I}_{11}, \Upsilon_i^{-T})$ *and its transpose, (4.72) by* $\mathrm{diag}(\Upsilon_i^{-T}, I)$ *and its transpose, we can obtain from Theorem 7 and Schur complement lemma that, with* *the given controller parameters in (4.74), the system (4.15)–(4.16) is stochastically stable with disturbance attenuation γ for $\tau \leq \bar{\tau}$.*

Remark 4.4 *The output feedback H_∞ controller design problem is solved in Theorem 8 for the addressed delayed stochastic jumping systems, with sensor nonlinearities and external nonlinear disturbances. LMI-based sufficient conditions are obtained for the existence of full-order controllers that ensure the mean-square exponential stability of the resulting system (Σ_k) and reduce the effect of the disturbance input to a prescribed level for all admissible time-delays and nonlinearities. The feasibility of the controller design problem can be readily checked by the solvability of two sets of LMIs, which can be determined by using the MATLAB LMI Toolbox in a straightforward way. In the next section, an illustrative example will be provided to show the usefulness of the proposed techniques.*

4.3.3 An Illustrative Example

In this section, a numerical example is presented to illustrate the usefulness and flexibility of the controller design method developed in this chapter. We are interested in obtaining the upper bound $\bar{\tau}$ of the time-delays and designing the output feedback H_∞ controller for the stochastic jumping system, with nonlinear disturbances and sensor nonlinearities.

The system data of (4.1)–(4.3) are given as follows:

$$\begin{bmatrix} \gamma_{11} & \gamma_{12} \\ \gamma_{21} & \gamma_{22} \end{bmatrix} = \begin{bmatrix} -1.5 & 1.5 \\ 0.9 & -0.9 \end{bmatrix},$$
$$\gamma = 1.5,\ Q = 0.2I,\ R = 0.1I.$$

Mode 1:

$$A_1 = \begin{bmatrix} -1.5 & 1 \\ 0 & -2 \end{bmatrix},\ A_{d1} = \begin{bmatrix} 0.15 & 0 \\ 0 & 0.21 \end{bmatrix},$$
$$E_{d1} = \begin{bmatrix} 0.13 & 0 \\ 0 & 0.15 \end{bmatrix},\ B_{11} = \begin{bmatrix} 0.03 & 0.2 \\ 0.4 & 0.02 \end{bmatrix},$$
$$B_{21} = \begin{bmatrix} 0.21 & 0.31 \\ 0.12 & 0.14 \end{bmatrix},\ D_1 = \begin{bmatrix} 0.02 \\ 0.1 \end{bmatrix},$$
$$M_{11} = \mathrm{diag}\{0.05, 0.1\},\ M_{12} = \mathrm{diag}\{0.05, 0.1\},$$
$$C_1 = 0.05I_2,\ C_{d1} = 0.05I_2,$$
$$L_1 = [0.1\ 0.11],\ K_{11} = \mathrm{diag}\{0.03, 0.04\},$$
$$K_{21} = \mathrm{diag}\{0.06, 0.05\},\ \epsilon_{11} = 0.03,\ \epsilon_{21} = 0.013,$$

Mode 2:

$$A_2 = \begin{bmatrix} -1.3 & 1 \\ 0 & -1.5 \end{bmatrix}, \ A_{d2} = \begin{bmatrix} 0.022 & 0 \\ 0 & 0.1 \end{bmatrix},$$

$$E_{d2} = \begin{bmatrix} 0.12 & 0 \\ 0 & 0.031 \end{bmatrix}, \ B_{12} = \begin{bmatrix} 0.03 & 0.2 \\ 0.04 & 0.2 \end{bmatrix},$$

$$D_2 = \begin{bmatrix} 0.1 \\ 0.2 \end{bmatrix}, \ B_{22} = \begin{bmatrix} 0.21 & 0.31 \\ 0.12 & 0.14 \end{bmatrix},$$

$$M_{12} = \begin{bmatrix} 0.04 & 0 \\ 0 & 0.1 \end{bmatrix}, \ M_{22} = \begin{bmatrix} 0.05 & 0 \\ 0 & 0.1 \end{bmatrix},$$

$$C_2 = 0.06I_2, \ C_{d2} = 0.1I_2,$$

$$L_2 = [0.2\ 0.12], \ K_{12} = \mathrm{diag}\{0.04, 0.06\},$$

$$K_{22} = \mathrm{diag}\{0.06, 0.09\}, \ \epsilon_{12} = 0.01, \ \epsilon_{22} = 0.021.$$

Using MATLAB LMI Control Toolbox to solve the LMIs in (4.71) and (4.72), we obtain the upper bound of time-delays as $\bar{\tau} = 2.3896$. Therefore, by Theorem 8, it can be calculated that for all $0 < \tau \leq 2.5001$, there exist the desired output feedback H_∞ controllers. For demonstration purpose, let us fix $\tau = \bar{\tau} = 2.5001$. In this case, by the LMI toolbox, the controller parameters can be determined as follows:

$$A_{c1} = \begin{bmatrix} 6.9240 & -2.3038 \\ -1.2898 & 3.5304 \end{bmatrix}, \ B_{c1} = \begin{bmatrix} 0.3973 & -0.1108 \\ 0.0266 & 0.3508 \end{bmatrix},$$

$$C_{c1} = \begin{bmatrix} -0.3387 & 2.7095 \\ -0.4391 & -5.6313 \end{bmatrix}, \ A_{c2} = \begin{bmatrix} 13.7578 & -46.9813 \\ -1.9732 & -4.5679 \end{bmatrix},$$

$$B_{c2} = \begin{bmatrix} 2.0381 & 0.3460 \\ -0.1325 & 0.7878 \end{bmatrix}, \ C_{c2} = \begin{bmatrix} -6.5704 & 120.8731 \\ 0.9309 & -23.5211 \end{bmatrix}.$$

4.4 Summary

In this chapter, we have developed a delay-dependent approach to deal with the stochastic H_∞ filtering and stochastic H_∞ control problem for a class of Itô type stochastic time-delay jumping systems subject to both the sensor nonlinearity and the exogenous nonlinear disturbances. The time-delays are allowed to exist in the system states, the sensor nonlinearity, as well as the external nonlinear disturbances. By using Itô's differential formula and the Lyapunov stability theory, we have proposed a linear matrix inequality method to derive sufficient conditions under which the desired controllers exist, which are dependent on the length of the time-delays. We have also characterized the expression of the filter parameters, and employed a simulation example to demonstrate the effectiveness of the proposed results. Moreover, we can extend the main results in this chapter to more complex and realistic systems, such as systems with polytopic or norm-bounded uncertainties, and systems with general nonlinearities. We will also focus on the real-time applications in network-based communications and bio-informatics. The corresponding results will appear in the next chapters.

5

H_∞ Analysis for Stochastic Differential Systems by Razumikhin Theory

CONTENTS

In this chapter, we tackle the H_∞ analysis problem for a general class of nonlinear stochastic systems with time-delays, which are described by general stochastic functional differential equations. By using the Razumikhin-type method, we first establish sufficient conditions for the time-delay stochastic systems to be internally stable, and then deal with the H_∞ analysis problem in order to quantify the disturbance rejection attenuation level of the nonlinear stochastic time-delays system. General conditions are derived under which the L_2 gain of the system is less than or equal to a given constant, and some easy-to-test criteria are subsequently given, which can be used to check the stability and determine the H_∞ performance index of the system.

The rest of the chapter is arranged as follows. Section 10.1 formulates the H_∞ analysis problem for a general class of nonlinear stochastic systems with time-delays, which are described by general stochastic functional differential equations. In Section 10.2, the Razumikhin-type method is used to establish some certain conditions to achieve the H_∞ performance. Section 10.3 gives the methodology to deal with the L_2 gain problem for a special class of stochastic differential delay systems. A numerical example is presented in Section 10.4 to show the effectiveness of the proposed algorithm. Section 10.5 draws the summary.

5.1 Problem Formulation

Consider an n-dimension nonlinear stochastic time-delay system:

$$dx(t) = f(t,x_t)dt + g_1(t,x_t)d\omega(t) + g_2(t,x_t)v(t)dt, \tag{5.1}$$

$$z(t) = h(t,x(t)), \quad t \geq 0, \tag{5.2}$$

where $x_t = \{x(t+\theta) : -r \leq \theta \leq 0\}$ is a $C([-r,0];R^n)$-valued stochastic process, and $r \geq 0$ is the upper bound for the time-delay. $x(t) \in L^2_\omega([0,T];L^2(\Omega,R^n))$, $v(t) \in L^2_\omega([0,T];L^2(\Omega,R^l))$, and $z(t) \in L^2_\omega([0,T];L^2(\Omega,R^p))$ are the system state, the external disturbance input, and the system output, respectively. $\omega(t)$ is the m-dimension Brownian motion. Moreover, the functionals $f(\cdot,\cdot)$, $g_1(\cdot,\cdot)$, $g_2(\cdot,\cdot)$, and $h(\cdot,\cdot)$ are defined as:

$$\begin{aligned} f &: R_+ \times C([-r,0];R^n) \to R^n,\ g_1 : R_+ \times C([-r,0];R^n) \to R^{n\times m}, \\ g_2 &: R_+ \times C([-r,0];R^n) \to R^{n\times l},\ h : R_+ \times C([-r,0];R^n) \to R^p. \end{aligned} \tag{5.3}$$

Assumption 5.1 . *The functionals $f(\cdot,\cdot)$, $g_1(\cdot,\cdot)$, $g_2(\cdot,\cdot)$, $h(\cdot,\cdot)$ all satisfy the local Lipschitz condition and the linear growth condition.*

It can be easily known that, under Assumption 5.1, the system defined by (5.1) and (5.2) has a unique global solution, which is denoted by $x(t;\xi)$ in this chapter. Moreover, $\mathbb{E}(\sup_{0\leq s\leq t}|x(s;\xi)|^2) < \infty$ for all $t > 0$. For the benefit of stability analysis in this chapter, we also assume that $f(t,0)=0$, $g_1(t,0)=0$, and $g_2(t,0)=0$, hence, the system defined by (5.1) and (5.2) admits a zero solution or trivial solution $x(t,0)=0$.

Let $C^{2,1}([-r,\infty)\times R^n;R_+)$ denote the family of all nonnegative functions $V(t,x)$ on $[-r,\infty)\times R^n$ that are continuously twice differentiable in x and once differentiable in t. For $V(t,x) \in C^{2,1}([-r,\infty)\times R^n;R_+)$, define an operator $\mathcal{LV}$ from $R_+ \times C([-r,0];R^n)$ to R by

$$\begin{aligned} \mathcal{LV}(t,\phi) &= V_t(t,\phi(0)) + V_x(t,\phi(0))(f(t,\phi)+g_2(t,\phi)v(t)) \\ &\quad +\frac{1}{2}tr[g_1^T(t,\phi)V_{xx}(t,\phi(0))g_1(t,\phi)], \end{aligned} \tag{5.4}$$

where

$$\begin{aligned} V_t(t,x) &= \frac{\partial V(t,x)}{\partial t}, \\ V_x(t,x) &= \left(\frac{\partial V(t,x)}{\partial x_1},\cdots,\frac{\partial V(t,x)}{\partial x_n}\right), \\ V_{xx}(t,x) &= \left(\frac{\partial^2 V(t,x)}{\partial x_i \partial x_j}\right)_{n\times n}. \end{aligned} \tag{5.5}$$

Definition 7 *The system (5.1) and (5.2) is called stochastically internally stable if there exists a constant $c > 0$, such that for any $\xi \in C^b_{\mathcal{F}_0}([-r;0];R^n)$*

$$\mathbb{E}\int_0^\infty (|x(t;\xi)|^2)dt \leq c\mathbb{E}\|\xi\|^2,\ \ t \geq 0,$$

where $x(\cdot) = x(\cdot,0,\xi)$ is the free trajectory of (5.1) and (5.2) starting at ξ(i.e., $v=0$).

Definition 8 *The system (5.1)and (5.2) is said to have an L_2 gain less than or equal to γ (or be externally stable), if its zero state response $(x(\theta) = 0, v(\theta) = 0,\ -r \le \theta \le 0)$ satisfies*

$$\|z(s)\|_{L_\omega} \le \gamma \|v(s)\|_{L_\omega},$$

for all $v(\cdot) \in L^2_\omega(R_+; L^2(\Omega, R^l))$ and $T \ge 0$.

Let us now establish a Razumikhin-type lemma on the internal stability by applying the method of [112].

Lemma 5.1 *Let f, and g_1, g_2 satisfy Assumption 5.1. Let λ, c_1, c_2, and be all positive scalars and $q > 1$. Assume that there exists a functional $V(t,x) \in C^{2,1}([-r,\infty) \times R^n; R_+)$, such that*

$$c_1|x|^2 \le V(t,x) \le |x|^2, \forall (t,x) \in [-r,\infty) \times R^n, \tag{5.6}$$

$$\mathbb{E}\mathcal{LV}(t,\phi) \le -\lambda \mathbb{E}V(t,\phi(0)),\ \ \forall\, t \ge 0, \tag{5.7}$$

for $\phi = \{\phi(0) : -r \le \theta \le 0\} \in L_{\mathcal{F}^b_0}([-r,0]; R^n)$, satisfying

$$\mathbb{E}V(t+\theta, \phi(\theta)) \le q\mathbb{E}V(t,\phi(0)),\ \ -r \le \theta \le 0. \tag{5.8}$$

Then, for all $\xi \in C_{\mathcal{F}^b_0}([-r,0]; R^n)$ and $t \ge 0$, we have

$$\mathbb{E}\int_0^\infty |x(t;\xi)|^2 dt \le c\mathbb{E}\|\xi\|^2. \tag{5.9}$$

Proof 9 *Fix the initial datum $\xi \in C_{\mathcal{F}^b_0}([-r,0]; R^n)$ arbitrarily and write $x(t,\xi) = x(t)$ for simplicity. Denote $\gamma = \min\{\lambda, \log(q)/r\}$. Let $\varepsilon \in (0,\gamma)$ be arbitrary, and set $\bar{\gamma} = \gamma - \varepsilon$. Define*

$$U(t) = \max_{-r\le\theta\le0}[e^{\bar{\gamma}(t+\theta)}\mathbb{E}V(t+\theta, x(t+\theta))],\ \ \forall t \ge 0. \tag{5.10}$$

Since $\mathbb{E}(\sup_{-r\le s\le t} x^2(s)) < \infty$ and both $x(t)$ and $V(t,x)$ are continuous, $U(t)$ is then well defined and continuous. We claim that

$$D_+U(t) := \limsup_{h\to 0_+} \frac{U(t+h) - U(t)}{h} \le 0,\ \ \forall t \ge 0. \tag{5.11}$$

To show this, for each $t \ge 0$ (fixed for the moment), define

$$\bar{\theta} = \max\{\theta \in [-r,0] : e^{\bar{\gamma}(t+\theta)}\mathbb{E}V(t+\theta, x(t+\theta)) = U(t)\}. \tag{5.12}$$

Obviously, $\bar{\theta}$ is well defined, $\bar{\theta} \in [-r,0]$ and $U(t) = \max_{-r\le\theta\le0}[e^{\bar{\gamma}(t+\theta)}\mathbb{E}V(t+\theta, x(t+\theta))]$. If $\bar{\theta} < 0$, then

$$e^{\bar{\gamma}(t+\theta)}\mathbb{E}V(t+\theta, x(t+\theta)) < e^{\bar{\gamma}(t+\bar{\theta})}\mathbb{E}V(t+\bar{\theta}, x(t+\bar{\theta})). \tag{5.13}$$

It is easy to observe that, for all sufficiently small $h > 0$*, we can obtain*

$$e^{\bar{\gamma}(t+h)}\mathbb{E}V(t+h, x(t+h)) < e^{\bar{\gamma}(t+\bar{\theta})}\mathbb{E}V(t+\bar{\theta}, x(t+\bar{\theta})), \tag{5.14}$$

hence, $U(t+h) \leq U(t)$ *and* $D_+U(t) \leq 0$*.*
If $\bar{\theta} = 0$*, then*

$$e^{\bar{\gamma}(t+\theta)}\mathbb{E}V(t+\theta, x(t+\theta)) \leq e^{\bar{\gamma}t}\mathbb{E}V(t, x(t)), \ \forall -r \leq \theta \leq 0, \tag{5.15}$$

and, subsequently,

$$\mathbb{E}V(t+\theta, x(t+\theta)) \leq e^{-\bar{\gamma}\theta}\mathbb{E}V(t, x(t)) \leq e^{-\bar{\gamma}r}\mathbb{E}V(t, x(t)), \ \forall -r \leq \theta \leq 0. \tag{5.16}$$

Note that either $\mathbb{E}V(t, x(t)) = 0$ *or* $\mathbb{E}V(t, x(t)) > 0$*. In the case of* $\mathbb{E}V(t, x(t)) = 0$*, (5.6) and (5.16) yield that* $x(t+h) = 0$*, a.s.* $\forall -r \leq \theta \leq 0$*. Recalling the fact that* $f(t, 0) = 0$ *and* $g(t, 0) = 0$*, we can see that* $x(t+h) = 0$*, a.s.* $\forall h > 0$*, hence,* $U(t+h) = 0$ *and* $D_+U(t) = 0$*. On the other hand, in the case* $\mathbb{E}V(t, x(t)) > 0$*, (5.16) implies*

$$\mathbb{E}V(t+\theta, x(t+\theta)) \leq q\mathbb{E}V(t, x(t)), \ \forall -r \leq \theta \leq 0.$$

Since $e^{\bar{\gamma}r} < q$*, by condition (5.7) , we have* $\mathbb{E}\mathcal{L}\mathcal{V}(t, x_t))$*. Moreover, by Itô's formula, one can derive that for all* $h > 0$*,*

$$\begin{aligned} &e^{\bar{\gamma}(t+h)}\mathbb{E}V(t+h, x(t+h)) - e^{\bar{\gamma}t}\mathbb{E}V(t, x(t)) \\ &= \int_t^{t+h} e^{\bar{\gamma}s}[\bar{\gamma}\mathbb{E}V(s, x(s)) + \mathbb{E}\mathcal{L}\mathcal{V}(s, x_s)]ds. \end{aligned}$$

Noting that

$$\bar{\gamma}\mathbb{E}V(t, x(t)) + \mathbb{E}\mathcal{L}\mathcal{V}(t, x_t) \leq -(\lambda - \bar{\gamma})\mathbb{E}V(t, x(t)) < 0,$$

we can see from the continuity of V *that for all sufficiently small* $h > 0$*,*

$$\bar{\gamma}\mathbb{E}V(s, x(s)) + \mathbb{E}\mathcal{L}\mathcal{V}(s, x_s) \leq 0, \quad \text{if} t \leq s \leq t+h,$$

and, consequently,

$$e^{\bar{\gamma}(t+h)}\mathbb{E}V(t+h, x(t+h)) \leq e^{\bar{\gamma}t}\mathbb{E}V(t, x(t)).$$

Therefore, $U(t+h) = U(t)$ *holds for all sufficiently small* $h > 0$*, hence,* $D_+U(t) = 0$*. To this end, it is proven that (5.11) is true.*
It now follows immediately from (5.11) that

$$U(t) \leq U(0), \quad \forall t \geq 0.$$

By the definition of $U(t)$ *and condition (5.6), one obtains*

$$\mathbb{E}|x(t)|^2 \leq \frac{c_1}{c_2}\mathbb{E}\|\xi\|^2 e^{-(\gamma-\varepsilon)t},$$

and

$$\lim_{T\to\infty} \mathbb{E}\int_0^T |x(t)|^2 ds \leq \frac{c_1}{c_2} \lim_{T\to\infty} \mathbb{E}\int_0^T \|\xi\|^2 e^{-(\gamma-\varepsilon)s} ds.$$

Since $\varepsilon > 0$ is arbitrary, the conclusion of Lemma 5.1 follows immediately by straightforward calculation. The proof is completed.

5.2 Main Results

In the following theorem, the Razumikhin-type method is used to establish the conditions for the system (5.1) and (5.2) to achieve the H_∞-performance, that is, the L_2 gain of the system is not more than a certain constant $c > 0$. Our aim is to evaluate the influence from the external disturbance inputs $v(t)$ to the overall system outputs $y(t)$.

Theorem 9 *Assume that there exists a functional $V(t,x) \in C^{2,1}([-r,\infty) \times R^n; R_+)(V(t,0) = 0)$, positive scalars $\gamma > 0$, $\lambda > 0$, and $q > 1$, such that the following conditions are satisfied for every trajectory $x(t)$ of the system (5.1) and (5.2):*

(C1) for $t \geq 0$, if

$$\mathbb{E}V(t+\theta, x(t+\theta)), \quad \theta \in [-r, 0], \tag{5.17}$$

then,

$$\mathbb{E}\mathcal{L}\mathcal{V}(t, x_t) \leq \gamma^2 \mathbb{E}\|v(t)\|^2 - \mathbb{E}\|z(t)\|^2. \tag{5.18}$$

(C2) for $t \geq 0$,

$$\mathbb{E}\|z(t)\|^2 \leq \lambda \mathbb{E}V(t,x) + \lambda^2 \mathbb{E}\|v(t)\|^2. \tag{5.19}$$

(C3) The delay bound satisfies

$$r < \frac{q-1}{q\lambda}. \tag{5.20}$$

Then, the system (5.1) and (5.2) has an L_2 gain less than or equal to λ.

Proof 10 *Let $x(t)$ be the trajectory of system (5.1) and (5.2) with its initial condition satisfying $x(\theta) = 0$ and $v(\theta) = 0$ for all $-r \leq \theta \leq 0$. Define*

$$l(t) = \int_0^t \left[\gamma^2 \mathbb{E}\|v(s)\|^2 - \mathbb{E}\|z(s)\|^2\right] ds - \mathbb{E}V(t,x), \tag{5.21}$$

and

$$T = \{t | l(t) < 0, \, t \in [-r, \infty)\}\}. \tag{5.22}$$

In the following, we shall prove that $T = \phi$ by contradiction, where ϕ is an empty set. Assume that $T \neq \phi$. Since $l(t) = 0$ for all $t \in [-r, 0]$, we have $[-r, 0] \not\subset T$, and obtain that $\bar{t} = \inf(T) \geq 0$. So, $l(\bar{t}) = 0$ and $l(t) \geq 0$ for all $t \in [-r, \bar{t}]$. Therefore, for $\theta \in [-r, 0]$, it can be shown that

$$\begin{aligned} \mathbb{E}V(\bar{t}+\theta, x(\bar{t}+\theta)) &\leq \int_0^{\bar{t}+\theta} \left[\gamma^2\mathbb{E}\|v(s)\|^2 - \mathbb{E}\|z(s)\|^2\right] ds \leq \mathbb{E}V(\bar{t}, x(\bar{t})) \\ &\quad - \int_{\bar{t}+\theta}^{\bar{t}} \left[\gamma^2\mathbb{E}\|v(s)\|^2 - \mathbb{E}\|z(s)\|^2\right] ds \\ &\leq \mathbb{E}V(\bar{t}, x(\bar{t})) + \lambda \int_{\bar{t}-r}^{\bar{t}} \mathbb{E}V(s, x(s)) ds \\ &\leq \mathbb{E}V(\bar{t}, x(\bar{t})) + \lambda r \sup_{-r\leq\theta\leq 0} |\mathbb{E}V(\bar{t}+\theta, x(\bar{t}+\theta))|, \end{aligned}$$

and, subsequently,

$$\begin{aligned} \mathbb{E}V(\bar{t}+\theta, x(\bar{t}+\theta)) &\leq \sup_{-r\leq\bar{\theta}\leq 0} \mathbb{E}V(\bar{t}+\bar{\theta}, x(\bar{t}+\bar{\theta})) \leq \frac{1}{1-\lambda r}\mathbb{E}V(\bar{t}, x(\bar{t})) \\ &< q\mathbb{E}V(\bar{t}, x(\bar{t})), \ \forall -r \leq \theta \leq 0. \end{aligned}$$

Following the properties of continuous functions, it can be concluded that there exists a positive number $d > 0$, such that for $t \in [\bar{t}, \bar{t}+\delta)$,

$$\mathbb{E}V(\bar{t}+\theta, x(\bar{t}+\theta)) \leq q\mathbb{E}V(t, x(t)), \ \forall -r \leq \theta \leq 0. \tag{5.23}$$

Next, from the condition (C1), we can show that, for all $t \in [\bar{t}, \bar{t}+\delta)$,

$$\begin{aligned} \mathbb{E}V(t, x(t)) - \mathbb{E}V(\bar{t}, x(\bar{t})) &\leq \int_{\bar{t}}^{t} \mathbb{E}\mathcal{L}\mathcal{V}(s, x_s) ds \qquad (5.24) \\ &\leq \int_{\bar{t}}^{t} \left[\gamma^2\mathbb{E}\|v(s)\|^2 - \mathbb{E}\|z(s)\|^2\right] ds. \qquad (5.25) \end{aligned}$$

Since $l(\bar{t}) = 0$, we can obtain from (5.24) that $l(t) \geq 0$, $\forall t \in [\bar{t}, \bar{t}+\delta)$. This contradicts the definition of $\bar{t}$, and, hence, $T = \phi$. The proof is then completed according to the fact that $V(t, x) \geq 0$.

Theorem 10 *Given $\gamma > 0$. For the system (5.1) and (5.2), if there exists a smooth solution $V(t, x) \in C^{2,1}([-r, \infty) \times R^n; R_+)(V(t, 0) = 0)$ to the Hamilton–Jacobi inequality*

$$\mathbb{E}\left[V_x f + \frac{1}{2}tr(g_1^T V_{xx} g_1) + \frac{1}{2\gamma^2} V_x g_2 g_2^T V_x^T\right] + \frac{1}{2}\mathbb{E}[h^T h] \leq 0, \tag{5.26}$$

then we have $\mathbb{E}\mathcal{L}\mathcal{V}(t, x_t) \leq \gamma^2\mathbb{E}\|v(t)\|^2 - \mathbb{E}\|z(t)\|^2$.

Proof 11 *By "completing the squares", we can derive that*

$$\begin{aligned}\mathbb{E}\mathcal{L}\mathcal{V}(t,x_t) &= \frac{1}{2}\gamma^2\mathbb{E}\|v-\frac{1}{\gamma^2}g_2^TV_x^T\|^2 \\ &+\mathbb{E}\left[V_xf+\frac{1}{2}tr(g_1^TV_{xx}g_1)+\frac{1}{2\gamma^2}V_xg_2g_2^TV_x^T\right] \\ &+\frac{1}{2}\gamma^2\mathbb{E}\|v(s)\|^2,\end{aligned}$$

and

$$\mathbb{E}\mathcal{L}\mathcal{V}(t,x_t)=\frac{1}{2}\gamma^2\mathbb{E}\|v\|^2-\frac{1}{2}\mathbb{E}\|y\|^2-2\frac{1}{2}\gamma^2\mathbb{E}\|v-\frac{1}{\gamma^2}g_2^TV_x^T\|^2\leq\gamma^2\mathbb{E}\|v\|^2-\mathbb{E}\|y\|^2.$$

The proof follows directly.

Corollary 5.1 *Assume that there exists a functional* $V(t,x)\in C^{2,1}([-r,\infty)\times R^n;R_+)(V(t,0)=0)$ *and scalars* $\gamma>0$, $\lambda>0$, *and* $q>1$, *such that the following conditions are satisfied for every trajectory* $x(t)$ *of system (5.2) and (5.2):*

(D1) For $t\geq 0$, *if*

$$\mathbb{E}V(t+\theta,x(t+\theta))\leq q\mathbb{E}V(t,x(t)),\quad \theta\in[-r,0], \tag{5.27}$$

then

$$\mathbb{E}\left[V_xf+\frac{1}{2}tr(g_1^TV_{xx}g_1)+\frac{1}{2\gamma^2}V_xg_2g_2^TV_x^T\right]+\frac{1}{2}\mathbb{E}[h^Th]\leq 0. \tag{5.28}$$

(D2) For $t\geq 0$,

$$\mathbb{E}\|z(t)\|^2\leq\lambda\mathbb{E}V(t,x)+\gamma^2\mathbb{E}\|v(t)\|^2. \tag{5.29}$$

(D3) The delay bound satisfies

$$r<\frac{q-1}{q\lambda}. \tag{5.30}$$

Then, the system (5.2) and (5.2) has an L_2 *gain less than or equal to* γ.

Proof 12 *It follows from Theorem 9 and Theorem 10 immediately.*

The following corollary provides another simple criterion for testing the stability and H_∞ performance. The proof is straightforward, and is, hence, omitted.

Corollary 5.2 *Assume that there exists a functional* $V(t,x)\in C^{2,1}([-r,\infty)\times R^n;R_+)(V(t,0)=0)$ *and scalars* $\gamma>0$, $\lambda>0$, *and* $q>1$, *such that the following conditions are satisfied for every trajectory* $x(t)$ *of system (5.1) and (5.2):*

(E1) $c_1|x|^2 \le V(t,x) \le c_2|x|^2$.

(E2) For $t \ge 0$, *if* $\mathbb{E}V(t+\theta, x(t+\theta)) \le q\mathbb{E}V(t,x(t))$, $\theta \in [-r, 0]$, *then* $\mathbb{E}\mathcal{L}\mathcal{V}(t,x(t)) + \lambda q\mathbb{E}V(t,x) \le 0$.

(E3) For $t \ge 0$, $\mathbb{E}\|z(t)\|^2 \le \lambda\mathbb{E}V(t,x) + \gamma^2\mathbb{E}\|v(t)\|^2$.

(E4) The delay bound satisfies $r < \frac{q-1}{q\lambda}$. *Then, the system (5.1) and (5.2) has an* L_2 *gain less than or equal to* γ.

5.3 The L_2 Gain Problem for Nonlinear Stochastic Delay System

In the previous section, we have established a framework for testing the stability and evaluating the H_∞ performance for a general class of stochastic functional differential systems with time-delay. In this section, we shall apply the result obtained to deal with the L_2 gain problem for a special class of stochastic differential delay systems. Consider a nonlinear stochastic system with multiple time-delays as follows:

$$\begin{aligned} dx(t) &= F(t, x(t), x(t-\delta_1)), \cdots, x(t-\delta_k)))dt \\ &\quad + G_1(t, x(t), x(t-\delta_1)), \cdots, x(t-\delta_k)))d\omega(t) \\ &\quad + G_2(t, x(t), x(t-\delta_1)), \cdots, x(t-\delta_k)))v(t)dt \qquad (5.31) \\ z(t) &= h(t, x(t)), \qquad (5.32) \end{aligned}$$

where $t \ge 0$, the initial data $x_0 = \xi C^b_{\mathcal{F}_0}([-r,0]; R^n)$, $\delta_i : R_+ \to [0, r](1 \le i \le k)$ are Borel measurable, and the functionals F, G_1, G_2, and h are defined on

$$\begin{aligned} F: &\quad R_+ \times R^n \times R^{n\times k} \to R^n; \\ G_1: &\quad R_+ \times R^n \times R^{n\times k} \to R^{n\times m}; \\ G_2: &\quad R_+ \times R^n \times R^{n\times k} \to R^{n\times k}; \\ h: &\quad R_+ \times R^n \to R^p. \qquad (5.33) \end{aligned}$$

The variables $x(t)$, $v(t)$, and $x(t)$ have the same meaning as in the system (5.1) and (5.2). The following assumption is needed.

Assumption 5.2 *The functionals* $F(\cdot,\cdot,\cdot)$, $G_1(\cdot,\cdot,\cdot)$, *and* $G_2(\cdot,\cdot,\cdot)$ *satisfy the local Lipschitz condition and the linear growth condition.*

Obviously, under Assumption 5.2, the system (5.31) and (5.33) has a unique global solution that is again denoted by $x(t,n)$. Besides, we also assume that

$$F(t, 0, \cdots, 0) \equiv 0, \quad G_1(t, 0, \cdots, 0) \equiv 0, \quad G_2(t, 0, \cdots, 0) \equiv 0.$$

Theorem 11 *Let λ, λ_1, $\cdots$, λ_k, γ, γ_1, and γ_2 be all positive scalars, and $q>1$. Assume that there exists a functional $V(t,x)\in C^{2,1}([-r,\infty)\times R^n;R_+)(V(t,0)=0)$, such that the following conditions are satisfied for every trajectory $x(t)$ of the system (5.31) and (5.33):*

(F1) $c_1|x|^2\le V(t,x)\le c_2|x|^2,\ \ \forall(t,x)\in[-r,\infty)\times R^n.$

(F2) $\forall t\ge 0,$

$$\mathbb{E}\mathcal{L}V(t,x_t)=\le-\gamma^2\mathbb{E}V(t,x_t)+\sum_{i=1}^{k}\lambda_i\mathbb{E}V(t-\delta_i(t),y_i). \tag{5.34}$$

(F3) $\forall t\ge 0,$

$$\mathbb{E}\|z(t)\|^2\ge\beta\mathbb{E}V(t,x)+\gamma^2\mathbb{E}\|v(t)\|^2,\ \ \beta=\lambda-q\sum_{i=1}^{k}\lambda_i. \tag{5.35}$$

(F4) The delay bound satisfies $r<\frac{q-1}{q\lambda}$. Then, the system (5.31) and (5.33) is internally stable and has an L_2 gain less than or equal to γ.

Proof 13 *For $t\ge 0$ and $\phi\in C([-r,0];R^n)$, we define*

$$\begin{aligned}
f(t,\phi) &:= F(t,\phi(0),\phi(-\delta_1(t)),\cdots,\phi(-\delta_k(t)))\\
g_1(t,\phi) &:= G_1(t,\phi(0),\phi(-\delta_1(t)),\cdots,\phi(-\delta_k(t)))\\
g_2(t,\phi) &:= G_2(t,\phi(0),\phi(-\delta_1(t)),\cdots,\phi(-\delta_k(t)))\\
g_3(t,\phi) &:= G_3(t,\phi(0),\phi(-\delta_1(t)),\cdots,\phi(-\delta_k(t))),
\end{aligned}$$

and then the system (5.31) and (5.33) is in the similar form of (5.1) and (5.2). Moreover, the operator $\mathcal{L}V$ can be written as

$$\begin{aligned}
\mathbb{E}\mathcal{L}V(t,\phi(0)) =\ & V_t(t,\phi(0))+V_x(t,\phi(0))[F(t,\phi(0),\phi(-\delta_1(t)),\cdots,\phi(-\delta_k(t)))\\
&+G_2(t,\phi(0),\phi(-\delta_1(t)),\cdots,\phi(-\delta_k(t)))\\
&+\frac{1}{2}tr[G_1^T(t,\phi(0),\phi(-\delta_1(t)),\cdots,\phi(-\delta_k(t)))V_{xx}(t,\phi(0))\\
&\cdot G_1(t,\phi(0),\phi(-\delta_1(t)),\cdots,\phi(-\delta_k(t)))].
\end{aligned} \tag{5.36}$$

If $t\ge 0$ and $\Phi\in L_{\mathcal{F}_t^2}([-r,0];R^n)$ satisfies

$$\mathbb{E}V(t+\theta,\Phi(\theta))\le q\mathbb{E}V(t,\Phi(0)),\ \ \forall\ -r\le\theta\le 0, \tag{5.37}$$

then by condition (5.34), we have

$$\begin{aligned}
&\mathbb{E}[V_t+V_x(F(t,x,y_1,\cdots,y_k)+G_2(t,x,y_1,\cdots,y_k)v(t)\\
&+\frac{1}{2}tr(G_1^T(t,x,y_1,\cdots,y_k)V_{xx}G_1(t,x,y_1,\cdots,y_k))]\\
\le\ & -\lambda\mathbb{E}V(t,x)+\sum_{i=1}^{k}\lambda_i\mathbb{E}V(t-\delta_i(t),y_i)\le-\left(\lambda-q\sum_{i=1}^{k}\lambda_i\right)\mathbb{E}V(t,x).
\end{aligned}$$

Therefore, we can know from Corollary 5.2 that the system (5.31) and (5.33) is internally stable and has an L_2 gain less than or equal to γ.

Corollary 5.3 *Let $\beta > 0$, $\lambda > 0$, $q > 1$. Assume that there are nonnegative scalars a_i, b_i, $c_i (0 \le i \le k)$ such that the following conditions are all met for every trajectory $x(t)$ of system (5.31) and (5.33):*

(F1) $\forall\ (t,x) \in [-r,\infty) \times R^n$,

$$x^T F(t,x,y_1,\cdots,y_k) \le -\lambda |x|^2. \tag{5.38}$$

(F2) For $t \ge 0,\ x, \bar{x}, y_1, \cdots, y_k \in R^n$,

$$|F(t,x,0,\cdots,0) - F(t,\bar{x},y_1,\cdots,y_k)|^2 \le a_0|x-\bar{x}| + \sum_{i=1}^{k} a_i|y_i|, \tag{5.39}$$

$$tr[G^T(t,\bar{x},y_1,\cdots,y_k)G(t,\bar{x},y_1,\cdots,y_k)] \le 2b_0|x|^2 + \sum 2b_i|y_i|^2, \tag{5.40}$$

$$x^T G_2(t,\bar{x},y_1,\cdots,y_k)v(t) \le c_0|x|^2 + \sum_{i=1}^{k} c_i|y_i|^2. \tag{5.41}$$

(F3) $\forall t \ge 0,\ \mathbb{E}\|z(t)\|^2 \le \lambda|x|^2$,

$$\beta = \lambda - \left(b_0 + c_0 + \sum_{i=1}^{k}((1+q+s_i+qb_i+qc_i) \right) > 0. \tag{5.42}$$

(F4) The delay bound satisfies $r < \frac{q-1}{q\lambda}$. Then, the system (5.31) and (5.33) has an L_2 gain less than or equal to γ.

Proof 14 *Let $V(t,x) = |x|^2$. Along the trajectory of the system (5.31) and (5.33), we obtain*

$$\begin{aligned}
& \mathbb{E}\mathcal{L}\mathcal{V}(t,\phi(0)) \\
= \;& x^T F(t,x,0,\cdots,0) + 2x^T[F(t,x,y_1,\cdots,y_k) - F(t,x,0,\cdots,0)] \\
& + \frac{1}{2} tr[G_1^T(t,x,y_1,\cdots,y_k)V_{xx}(t,\phi(0))G_1(t,x,y_1,\cdots,y_k) \\
& + V_x G_2(t,x,y_1,\cdots,y_k)]v(t) \\
\le \;& -\lambda|x|^2 + 2\sum_{i=1}^{k} a_i|x||y_i| + b_0|x|^2 + \sum_{i=1}^{k} b_i|y_i|^2 \\
& + c_0|x|^2 + \sum_{i=1}^{k} c_i|y_i|^2.
\end{aligned} \tag{5.43}$$

Utilizing the elementary inequality $u^{\frac{1}{2}}v^{\frac{1}{2}} \le \frac{1}{2}u + \frac{1}{2}v$, we have

$$|x||y_i| = \frac{1}{2}(|x|^2)^{\frac{1}{2}} + \frac{1}{2}(|y_i|^2)^{\frac{1}{2}}.$$

Substituting the above inequality into (5.44) gives

$$\mathbb{E}\mathcal{L}\mathcal{V}(t,\phi(0)) \quad \leq \quad -\left(\lambda - b_0 - c_0\sum_{i=1}^{k} a_i\right)|x|^2 + \sum_{i=1}^{k}(a_i + b_i + c_i)|y_i|^2.$$

The rest of the proof follows from Theorem 11 immediately.

5.4 Examples

In this section two examples are presented in order to illustrate our theory. It is always assumed that the initial datum is in $C_{\mathcal{F}_0^b}([-r,0];R^n)$.

Example 1 *In this example, we consider a linear stochastic differential delay system*

$$\begin{aligned} dx(t) &= -[Ax(t) + Bx(t-\delta(t))]dt + Cx(t-\delta(t))d\omega(t) \\ &\quad +Dx(t)v(t)dt, \qquad (5.44)\\ z(t) &= Hx(t), \qquad (5.45) \end{aligned}$$

where A*,* B*,* C*,* D*, and* H *are all constant matrices with appropriate dimensions,* $\omega(t)$ *is a one-dimensional Brownian motion, and* $\delta : R_+ \to [-\tau, 0]$ *is Borel measurable.*

Assume that $A + A^T$ *is positive definite, and its smallest eigenvalue is denoted by* $\lambda(A + A^T)$*. In this case, one can easily check by Corollary 5.3 that, if*

$$\frac{1}{2}\lambda_{\min}(A + A^T) - (q\|B\| + q\|C\| + \|D\|) > 0, \qquad (5.46)$$

then the zero solution of (5.44) and (5.45) is internally stable, and the system (5.44) and (5.45) has an L_2 *gain less than or equal to* γ*.*

Example 2 *Consider a semilinear stochastic functional differential delay system*

$$\begin{aligned} dx(t) &= [Ax(t) + F(x(t_t)]dt + G_1(x_t)d\omega(t) + G_2(x_t)v(t)dt \qquad (5.47)\\ z(t) &= H(t, x(t)), \qquad (5.48) \end{aligned}$$

where

$$A = \begin{bmatrix} 0 & 1 \\ -2 & -3 \end{bmatrix}, \; F(\phi) = \begin{bmatrix} 0 \\ \sigma_1(\phi) \end{bmatrix}, \; G_1(\phi) = \begin{bmatrix} 0 \\ \sigma_2(\phi) \end{bmatrix},$$

$$G_2(\phi) = \begin{bmatrix} 0 \\ \sigma_1(\phi)/C \end{bmatrix}, \; H(\phi) = \begin{bmatrix} 0 \\ 2\sigma_4(\phi) \end{bmatrix}.$$

We assume that all the functionals σ_1, σ_2, σ_3, and σ_4, defined on $C([-\tau,0];R^2)\to R$, are locally Lipschitz continuous and satisfy

$$|\sigma_1(\phi)|\vee|\sigma_2(\phi)|\vee|\sigma_3(\phi)|\vee|\sigma_4(\phi)|\le\int_{-\tau}^{0}|\phi(\theta)|d\theta,\ \Phi\in C([-\tau,0];R^2).$$

We claim that if

$$\tau<\frac{\sqrt{28}-\sqrt{14}}{14}, \tag{5.49}$$

then the zero solution of (5.47) and (5.48) is internally stable, and the system (5.47) and (5.48) has an L_2 gain less than or equal to γ. It is easy to find $H=\begin{bmatrix}1&1\\-1&-2\end{bmatrix}$, such that $H^{-1}AH=\begin{bmatrix}-1&0\\0&-2\end{bmatrix}$. Setting $Q=(H^{-1})^TH^{-1}=\begin{bmatrix}5&3\\3&2\end{bmatrix}$, and, defining $V(t,x)=x^TQx$ for $x\in R^2$, we can compute that

$$\begin{aligned}
&V_xf+\frac{1}{2}tr(g_1^TV_{xx}g_1)+V_xg_2v(t)-\gamma^2|v(t)|^2+\frac{1}{2}h^Th\\
=\ &2\phi^T(0)Q[A\phi(0)+F(\phi)]+G_1^T(\phi)QG_1(\phi)+2\phi(0)^TQG_2(\phi)C\\
&-\gamma^2C^2+\frac{1}{2}H^T(\phi)H(\phi)\\
\le\ &-2V(\phi(0))+2|\phi^T(0)(H^{-1})^T||H^{-1}F(\phi)|+2|\sigma_2(\phi)|^2\\
&+2|\phi^T(0)(H^{-1})^T||H^{-1}G_2(\phi)C|2|\sigma_4(\phi)|^2\\
\le\ &-2V(\phi(0))+2\sqrt{14}V(\phi(0))+\frac{2}{\sqrt{14\tau}}(|\sigma_1(\phi)|^2\\
&+2|\sigma_4(\phi)|^2+|\sigma_2(\phi)|^2+2|\sigma_3(\phi)|^2\\
\le\ &-(2-2\sqrt{14\tau}V(\phi(0))+(2\sqrt{14}+28\tau)\int_{-\tau}^{0}V|\phi\theta|d\theta.
\end{aligned}$$

By condition (5.49), one can find a scalar $q>1$, such that

$$2-2\sqrt{14}(1+q)\tau-28q\tau^2>0.$$

Therefore, for any $\Phi\in L^2_{\mathcal{F}_t}([-\tau,0];R^2)$ satisfying $\mathbb{E}V(\Phi(\theta))\le q\mathbb{E}V(\Phi(0))$, $\theta\in[-\tau,0]$, we have

$$\mathbb{E}\mathcal{L}\mathcal{V}(t,x_t)-\gamma^2\mathbb{E}\|v(t)\|^2+\mathbb{E}\|z(t)\|^2\le 0.$$

On the other hand, by choosing suitable $\lambda>0$, the conditions (C1) and (C2) in Theorem 9 can always be satisfied. Hence, our claim is confirmed from Theorem 9.

5.5 Summary

The H_∞ analysis problems for a general class of nonlinear stochastic systems with time-delay are studied in this paper. The Razumikhin-type method and Hamilton–Jacobi–Bellman (HJB) inequality theory are employed to obtain some easy-to-test criteria, which are used to determine whether the stochastic system under investigation is internally stochastic stable and whether a prescribed H_∞ performance level is satisfied. Illustrating examples are given to show the effectiveness and convenience of the proposed method.

6

Robust Filtering with Nonlinearities and Multiple Missing Measurements

CONTENTS

In this chapter, the filtering problems are addressed for a class of discrete-time stochastic nonlinear time-delay systems with sensors information dropout and stochastic disturbances. The sensor measurement missing is assumed to be random and different for individual sensors, which is modeled by individual random variable satisfying a certain probabilistic distribution on the interval [0 1]. Such a probabilistic distribution could be any commonly used discrete distributions. The multiplicative stochastic disturbances are in the form of a scalar Gaussian white noise with unit variance. We are interested in designing a filter, such that the overall filtering error is exponentially mean-square stable. By using the linear matrix inequality (LMI) method, sufficient conditions are derived to ensure the existence of the desired filters that are then characterized by the solution to a set of LMIs. An illustrative example is exploited to show the effectiveness of the proposed design procedures.

The rest of the chapter is organized as follows. In Section 6.1, the filtering problem is addressed for a class of discrete-time stochastic nonlinear time-delay systems with sensors information dropout and stochastic disturbances. Section 6.2 addresses filter analysis and synthesis problems. A simulation example is presented to illustrate the usefulness and flexibility of the filter design method developed in Section 6.3. Section 6.4 provides the summary of the chapter.

6.1 Problem Formulation

Consider the following discrete-time uncertain stochastic system with state-delay and stochastic nonlinearities:

$$\begin{cases} x(k+1) &= (A+\Delta A(k))x(k)+(A_d+\Delta A_d(k))x(k-d) \\ &\quad +f(k,x(k),x(k-d))+E_1x(k)\omega(k) \\ x(k) &= \mu(k), \quad k=-d,-d+1,\ldots,0, \end{cases} \tag{6.1}$$

where $x(k)\in\mathbb{R}^n$ is the state; $d\in\mathbb{Z}^+$ is a known constant time-delay with $d\geq 1$; $\omega(k)$ is a one-dimensional Gaussian white noise sequence satisfying $\mathbb{E}\{\omega(k)\}=0$ and $\mathbb{E}\{\omega^2(k)\}=1$; $\mu(k)$ is the initial state of the system; and $\Delta A(k)$ and $\Delta A_d(k)$ are unknown matrices representing parameter uncertainties in the following form

$$[\Delta A(k)\ \ \Delta A_d(k)] = MF(k)[N\ \ N_d], \tag{6.2}$$

with $F(k)$ satisfying $F^T(k)F(k)\leq I$.

The measurement with sensor data missing is described by

$$\begin{aligned} y(k) &= \Xi Cx(k)+g(k,x(k),x(k-d))+E_2x(k)\omega(k) \\ &= \sum_{i=1}^{m}\xi_iC_ix(k)+g(k,x(k),x(k-d))+E_2x(k)\omega(k), \end{aligned} \tag{6.3}$$

where $y(k)\in\mathbb{R}^m$ is the measured output vector, $\Xi=\text{diag}\{\xi_1,\cdots,\xi_m\}$, with ξ_i $(i=1,\ldots,m)$ being m unrelated random variables, which are also independent of $\omega(k)$. It is assumed that ξ_i has the probabilistic density function $p_i(s)$ $(i=1,\ldots,m)$ on the interval $[0\ 1]$, with mathematical expectation μ_i and variance σ_i^2. $C_i:=\text{diag}\{\underbrace{0,\cdots,0}_{i-1},1,\underbrace{0,\cdots,0}_{m-i}\}C$ $(i=1,\ldots,m)$, and A, A_d, C, E_1, E_2, M, N, and N_d are known constant matrices.

Remark 6.1 *In real systems, the measurement data may be transferred through multiple sensors rather than a single sensors. For different sensor, the data missing probability may be different. In other words, it makes more sense to assume that the missing law for each individual sensor satisfies individual probabilistic distribution. (6.3) describes a measurement equation with multiple sensors, in which the diagonal matrix Ξ represents the whole missing states and the random variable ξ_i corresponds to the ith sensor $(i=1,\ldots,m)$. Note that the measurement missing phenomenon has been extensively considered and several models have been introduced. One popular model is the arguable Bernoulli distributed model in which 0 is used to stand for an entire missing of signals and 1 denotes the intactness; see, e.g., [186, 187]. However, in practice, the proportion of the data missing at one moment is usually a fraction rather than 0 or 1. In (6.3), ξ_i can take value on the interval $[0\ 1]$, and the* *probability for ξ_i to take different values may be different. ξ_i could satisfy any discrete probabilistic distributions on the interval [0 1], and therefore includes the Bernoulli distribution as a special case.*

Let $f(k,d) := f(k,x(k),x(k-d))$ and $g(k,d) := g(k,x(k),x(k-d))$ describe the so-called stochastic nonlinear functions of the states and delayed states, which are bounded in a statistical sense as follows:

$$\mathbb{E}\left\{\begin{bmatrix} f(k,d) \\ g(k,d) \end{bmatrix} |x(k)\right\} = 0, \tag{6.4}$$

$$\mathbb{E}\left\{\begin{bmatrix} f(k,d)) \\ g(k,d)) \end{bmatrix} [f^T(l,d)) \;\; g^T(l,d))]|x(l)\right\} = 0, \quad k \neq l, \tag{6.5}$$

$$\begin{aligned}&\mathbb{E}\left\{\begin{bmatrix} f(k,d)) \\ g(k,d)) \end{bmatrix} [f^T(k,d))^T \;\; g^T(k,d))]|x(k)\right\} \\ &\leq \sum_{i=1}^{q} \begin{bmatrix} \pi_{1i} \\ \pi_{2i} \end{bmatrix} \begin{bmatrix} \pi_{1i} \\ \pi_{2i} \end{bmatrix}^T \left[x^T(k)\Phi_i x(k) + x^T(k-d)\Psi_i x(k-d)\right],\end{aligned} \tag{6.6}$$

where $\pi_{1i} \in \mathbb{R}^{n\times 1}$ and $\pi_{2i} \in \mathbb{R}^{m\times 1}$ $(i = 1,\dots,q)$ have compatible dimensions with $f(k,x(k),x(k-d))$ and $g(k,x(k),x(k-d))$, and Φ_i and Ψ_i $(i = 1,\dots,q)$ are positive-definite matrices with appropriate dimensions. For convenience, we assume that $f(k,d) := f(k,x(k),x(k-d))$ and $g(k,d) := g(k,x(k),x(k-d))$ are unrelated to ξ_i $(i = 1,\dots,m)$ and $\omega(k)$.

Remark 6.2 *The so-called stochastic nonlinear functions described in (6.4)–(6.6) have been extensively considered in some references; see, e.g., [211, 219], because such a description contains several well-studied nonlinear functions as special cases. It should be noted that, in this chapter, we assume that such nonlinear functions involve delayed states, and are therefore more general that those in [211, 219].*

In this chapter, we are interested in designing a linear filter of the following structure:

$$x_f(k+1) = A_f x_f(k) + Ky(k), \tag{6.7}$$

where $x_f(k) \in \mathbb{R}^n$ is the state estimate, and A_f and K are filter parameters to be determined.

By augmenting the state variables

$$\tilde{x}(k) := \begin{bmatrix} x(k) \\ x_f(k) \end{bmatrix},$$

$$\tilde{x}_d(k) := \begin{bmatrix} x(k-d) \\ x_f(k-d) \end{bmatrix},$$

$$h(k) := \begin{bmatrix} f(k,x(k),x(k-d)) \\ g(k,x(k),x(k-d)) \end{bmatrix},$$

and combining (6.1) and (6.7), we obtain the filtering error dynamics as follows:

$$\tilde{x}(k+1) = \mathcal{A}\tilde{x}(k) + \mathcal{A}_d Z\tilde{x}_d(k) + \mathcal{C}\tilde{x}(k) + \mathcal{B}h(k) + \mathcal{E}Z\tilde{x}(k)\omega(k), \tag{6.8}$$

where

$$\begin{aligned}
\mathcal{A} &= \begin{bmatrix} A+\Delta A(k) & 0 \\ K\bar{\Xi}C & A_f \end{bmatrix}, \; Z = \begin{bmatrix} I & 0 \end{bmatrix}, \\
\mathcal{A}_d &= \begin{bmatrix} A_d+\Delta A_d(k) \\ 0 \end{bmatrix}, \; \mathcal{C} = \begin{bmatrix} 0 & 0 \\ K(\Xi-\bar{\Xi})C & 0 \end{bmatrix}, \\
\mathcal{B} &= \begin{bmatrix} I & 0 \\ 0 & K \end{bmatrix}, \; \mathcal{E} = \begin{bmatrix} E_1 \\ KE_2 \end{bmatrix}, \; \bar{\Xi} = \mathbb{E}\{\Xi\}.
\end{aligned}$$

Observe the system (6.8), and let $\tilde{x}(k;\nu)$ denote the state trajectory from the initial data $\tilde{x}(s) = \nu(s)$ on $-d \le s \le 0$. Obviously, $\tilde{x}(k,0) \equiv 0$ is the trivial solution of system (6.8) corresponding to the initial datum $\nu = 0$.

Before formulating the problem to be investigated, we first introduce the following stability concept for the augmented system (6.8).

Definition 9 For the system (6.8) and every initial condition ν, the trivial solution is said to be *exponentially mean-square stable* if there exist scalars α ($\alpha > 0$) and β ($0 < \beta < 1$), such that

$$\mathbb{E}|\tilde{x}(k,\nu)|^2 \le \alpha\beta^k \sup_{-d\le s\le 0} \mathbb{E}|\nu(s)|^2. \tag{6.9}$$

In this chapter, we aim to design an exponential filter of the form (6.7) for the system (6.1), such that, for all admissible time-delays, uncertainties, sensors data missing, stochastic nonlinearities, and exogenous stochastic disturbances, the filtering error system (6.8) is exponentially mean-square stable.

6.2 Main Results

6.2.1 Filter Analysis

First, let us give the following lemmas that will be used in the proofs of our main results in this chapter.

Lemma 6.1 *(S-procedure) [13]. Let $\Upsilon = \Upsilon^T$, $\mathcal{M}$ and $\mathcal{N}$ be real matrices of appropriate dimensions, with F satisfying $F^TF \le I$, then*

$$\Upsilon + \mathcal{M}F\mathcal{N} + \mathcal{N}^TF^T\mathcal{M}^T < 0,$$

if and only if there exists a positive scalar $\delta > 0$, such that

$$\Upsilon + \frac{1}{\delta}\mathcal{M}\mathcal{M}^T + \delta\mathcal{N}^T\mathcal{N} < 0,$$

or, equivalently

$$\begin{bmatrix} \Upsilon & \mathcal{M} & \delta\mathcal{N}^T \\ \mathcal{M}^T & -\delta I & 0 \\ \delta\mathcal{N} & 0 & -\delta I \end{bmatrix} < 0.$$

For convenience of presentation, we first deal with the nominal system of (6.1) (i.e. without parameter uncertainties) and will eventually extend our main results to the ones including the robustness. In the following theorem, Lyapunov stability theorem and an LMI-based method are combined together to deal with the stability analysis problem for the filter design of the discrete-time stochastic nonlinear system with time-delays and stochastic disturbance. A sufficient condition is derived that guarantees the solvability of the exponential filtering problem.

Theorem 12 Consider the augmented filtering system (6.8) with *given* filter parameters. If there exist positive definite matrices $P > 0$, $Q > 0$, and positive scalars $\varepsilon_i > 0$ $(i = 1, \ldots, q)$, such that the following matrix inequalities

$$\begin{bmatrix} -P + Z^TQZ & * & * & * & * & * & * \\ 0 & -Q & * & * & * & * & * \\ P\mathcal{A} & P\mathcal{A}_d & -P & * & * & * & * \\ \bar{\mathcal{C}} & 0 & 0 & -\mathcal{P} & * & * & * \\ P\mathcal{E}Z & 0 & 0 & 0 & -P & * & * \\ \hat{\Phi} & 0 & 0 & 0 & 0 & -\Lambda & * \\ 0 & \hat{\Psi} & 0 & 0 & 0 & 0 & -\Lambda \end{bmatrix} < 0, \tag{6.10}$$

$$\begin{bmatrix} -\varepsilon_i & * \\ P\mathcal{B}\bar{\Pi}_i & -P \end{bmatrix} < 0, \qquad i = 1, \ldots, q, \tag{6.11}$$

hold, where

$$\begin{aligned} \bar{\mathcal{C}} &:= [\sigma_1\bar{\mathcal{C}}_1^TP, \cdots, \sigma_m\bar{\mathcal{C}}_m^TP]^T, \ \bar{\mathcal{C}}_i = \begin{bmatrix} 0 & 0 \\ KC_i & 0 \end{bmatrix}, \\ \hat{\Phi} &:= [\varepsilon_1\bar{\Phi}_1^{\frac{1}{2}}, \cdots, \varepsilon_q\bar{\Phi}_q^{\frac{1}{2}}]^T, \ \hat{\Psi} := [\varepsilon_1\Psi_1^{\frac{1}{2}}, \cdots, \varepsilon_q\Psi_q^{\frac{1}{2}}]^T, \\ \bar{\Phi}_i &:= \begin{bmatrix} \Phi_i & 0 \\ 0 & 0 \end{bmatrix}, \ \bar{\Pi}_i := \begin{bmatrix} \pi_{1i} \\ \pi_{2i} \end{bmatrix}, \\ \Lambda &:= \mathrm{diag}\{\varepsilon_1 I, \cdots, \varepsilon_q I\}, \ \mathcal{P} := \mathrm{diag}\{P, \cdots, P\}, \end{aligned} \tag{6.12}$$

then the filtering error system (6.8) is exponentially mean-square stable.

Proof 15 *Define the following Lyapunov functional candidate for the system (6.8):*

$$V(\tilde{x}(k),k) = \tilde{x}^T(k)P\tilde{x}(k) + \sum_{s=k-d}^{k-1} \tilde{x}^T(s)Z^TQZ\tilde{x}(s). \tag{6.13}$$

Calculating the difference of the Lyapunov functional (6.13) according to (6.8) gives

$$\begin{aligned}
& \mathbb{E}\{\Delta V(\tilde{x}(k),k)\} \\
= \; & \mathbb{E}\{V(\tilde{x}(k+1),k+1)|\tilde{x}(k)\} - V(\tilde{x}(k),k) \\
= \; & \mathbb{E}\{[\mathcal{A}\tilde{x}(k) + \mathcal{A}_d Z\tilde{x}_d(k) + \mathcal{C}\tilde{x}(k) + \mathcal{B}h(k) + \mathcal{E}Z\tilde{x}(k)\omega(k)]^T \\
& P[\mathcal{A}\tilde{x}(k) + \mathcal{A}_d Z\tilde{x}_d(k) + \mathcal{C}\tilde{x}(k) + \mathcal{B}h(k) + \mathcal{E}Z\tilde{x}(k)\omega(k)]\} \\
& -\tilde{x}^T(k)P\tilde{x}(k) + \tilde{x}^T(k)Z^TQZ\tilde{x}(k) \\
& -\tilde{x}^T(k-d)Z^TQZ\tilde{x}(k-d) \\
= \; & [\mathcal{A}\tilde{x}(k) + \mathcal{A}_d Z\tilde{x}_d(k)]^T P[\mathcal{A}\tilde{x}(k) + \mathcal{A}_d Z\tilde{x}_d(k)] \\
& +\tilde{x}^T(k)Z^TQZ\tilde{x}(k) - \tilde{x}^T(k)P\tilde{x}(k) \\
& +\mathbb{E}\{\tilde{x}^T(k)\mathcal{C}^TP\mathcal{C}\tilde{x}(k)\} + \mathbb{E}\{h^T(k)\mathcal{B}^TP\mathcal{B}h(k)\} \\
& +\mathbb{E}\{\omega^T(k)\tilde{x}^T(k)Z\mathcal{E}^TP\mathcal{E}Z\tilde{x}(k)\omega(k)\} \\
& -\tilde{x}^T(k-d)Z^TQZ\tilde{x}(k-d).
\end{aligned} \tag{6.14}$$

We can obtain from the definition of (6.3) that

$$\mathbb{E}\{\xi_i - \mu_i\}\{\xi_l - \mu_l\} = \begin{cases} \sigma_i^2 & i = l \\ 0 & i \neq l, \end{cases} \quad (i,l = 1,\ldots,m), \tag{6.15}$$

and then derive that

$$\begin{aligned}
& \mathbb{E}\{\mathcal{C}^TP\mathcal{C}\} \\
= \; & \mathbb{E}\left\{\begin{bmatrix} 0 & 0 \\ \sum_{i=1}^m(\xi_i - \mu_i)KC_i & 0 \end{bmatrix}^T P \begin{bmatrix} 0 & 0 \\ \sum_{i=1}^m(\xi_i - \mu_i)KC_i & 0 \end{bmatrix}\right\} \\
= \; & \sum_{i=1}^m \mathbb{E}(\xi_i - \mu_i)^2 \begin{bmatrix} 0 & 0 \\ KC_i & 0 \end{bmatrix}^T P \begin{bmatrix} 0 & 0 \\ KC_i & 0 \end{bmatrix} \\
= \; & \sum_{i=1}^m \sigma_i^2 \begin{bmatrix} 0 & 0 \\ KC_i & 0 \end{bmatrix}^T P \begin{bmatrix} 0 & 0 \\ KC_i & 0 \end{bmatrix} \\
= \; & \sum_{i=1}^m \sigma_i^2 \bar{\mathcal{C}}_i^T P \bar{\mathcal{C}}_i.
\end{aligned} \tag{6.16}$$

Again, we can have from (6.4)–(6.6) that

$$\begin{aligned}
&\mathbb{E}\left\{h^T(k)\mathcal{B}^T P\mathcal{B}h(k)\right\} \\
\leq\ & \sum_{i=1}^{q}\left[\tilde{x}^T(k)\bar{\Phi}_i\tilde{x}(k)+\tilde{x}^T(k-d)Z^T\Psi_i Z\tilde{x}(k-d)\right] \\
&\cdot \mathrm{tr}(\mathcal{B}\Pi_i\mathcal{B}^T P),
\end{aligned} \tag{6.17}$$

where $\Pi_i := \bar{\Pi}_i\bar{\Pi}_i^T$, *with* $\bar{\Pi}_i$ *and* $\bar{\Phi}_i$ $(i=1,\ldots,q)$ *being defined in (6.12).*
From (6.14)–(6.17), we have

$$\begin{aligned}
&\mathbb{E}\{\Delta V(\tilde{x}(k),k)\} \\
\leq\ & \mathbb{E}\left\{[\mathcal{A}\tilde{x}(k)+\mathcal{A}_d Z\tilde{x}_d(k)]^T P[\mathcal{A}\tilde{x}(k)+\mathcal{A}_d Z\tilde{x}_d(k)]\right. \\
&+\sum_{i=1}^{q}\left[\tilde{x}^T(k)\bar{\Phi}_i\tilde{x}(k)+\tilde{x}^T(k-d)Z^T\Psi_i Z\tilde{x}(k-d)\right]\mathrm{tr}(\mathcal{B}\Pi_i\mathcal{B}^T P) \\
&+\tilde{x}^T(k)[\sum_{i=1}^{m}\sigma_i^2\bar{\mathcal{C}}_i^T P\bar{\mathcal{C}}_i+Z^T\mathcal{E}^T P\mathcal{E}Z+Z^T QZ-P]\tilde{x}(k) \\
&\left.-\tilde{x}^T(k-d)Z^T QZ\tilde{x}(k-d)\right\} \\
=\ & \mathbb{E}\{\eta^T(k)\Omega\eta(k)\},
\end{aligned} \tag{6.18}$$

where $\eta(k) = \left[\tilde{x}^T(k)\ \ Z\tilde{x}^T(k-d)\right]$ *and*

$$\Omega := \begin{bmatrix} \Omega_1 & \mathcal{A}^T P\mathcal{A}_d \\ \mathcal{A}_d^T P\mathcal{A} & -Q+\mathcal{A}_d^T P\mathcal{A}_d+\sum_{i=1}^{q}\Psi_i\mathrm{tr}(\mathcal{B}\Pi_i\mathcal{B}^T P) \end{bmatrix}, \tag{6.19}$$

with

$$\begin{aligned}
\Omega_1 \ :=\ & -P+\sum_{i=1}^{m}\sigma_i^2\bar{\mathcal{C}}_i^T P\bar{\mathcal{C}}_i+Z^T\mathcal{E}^T P\mathcal{E}Z+Z^T QZ \\
&+\mathcal{A}^T P\mathcal{A}+\sum_{i=1}^{q}\bar{\Phi}_i\mathrm{tr}(\mathcal{B}\Pi_i\mathcal{B}^T P).
\end{aligned} \tag{6.20}$$

By Schur complement, (6.11) holds if and only if $\mathrm{tr}(\mathcal{B}\Pi_i\mathcal{B}^T P) < \varepsilon_i$ $(i = 1,\ldots,q)$. *Again, by Schur complement, we can obtain from (6.10) and (6.11) that* $\Omega < 0$ *and, subsequently,*

$$\mathbb{E}\{\Delta V(\tilde{x}(k))\} < -\lambda_{\min}(\Omega)|\tilde{x}(k)|^2. \tag{6.21}$$

Finally, we can confirm from Lemma 1 of [186] that the augmented filtering system (6.8) is exponentially mean-square stable.

In the next subsection, our attention is focused on the design of filter parameters A_f and K by using the results in Theorem 12. The explicit expression of the expected filter parameters will be obtained in term of the solution to a set of LMIs.

6.2.2 Filter Synthesis

The following theorem shows that the desired filter parameters can be derived by solving several LMIs.

Theorem 13 Consider the system (6.8). If there exist positive-definite matrices $S > 0$, $R > 0$, and $Q > 0$, and positive scalars $\varepsilon_i > 0$, $(i = 1, 2, \ldots, q)$, such that the following linear matrix inequalities

$$\Upsilon := \begin{bmatrix} -\bar{S} & * & * & * & * & * & * & * & * & * \\ -\bar{S} & -\bar{R} & * & * & * & * & * & * & * & * \\ 0 & 0 & -Q & * & * & * & * & * & * & * \\ SA & SA & SA_d & -S & * & * & * & * & * & * \\ \Pi_1 & \Pi_3 & RA_d & -S & -R & * & * & * & * & * \\ \tilde{\mathcal{C}} & \tilde{\mathcal{C}} & 0 & 0 & 0 & -\bar{\mathcal{P}} & * & * & * & * \\ SE_1 & SE_1 & 0 & 0 & 0 & 0 & -S & * & * & * \\ \Pi_2 & \Pi_4 & 0 & 0 & 0 & 0 & -S & -R & * & * \\ \tilde{\Phi} & \tilde{\Phi} & 0 & 0 & 0 & 0 & 0 & 0 & -\Lambda & * \\ 0 & 0 & \hat{\Psi} & 0 & 0 & 0 & 0 & 0 & 0 & -\Lambda \end{bmatrix} < 0, \tag{6.22}$$

$$\begin{bmatrix} -\varepsilon_i & * & * \\ S\pi_{1i} & -S & * \\ R\pi_{1i} + \tilde{K}\pi_{2i} & -S & -R \end{bmatrix} < 0, \quad i = 1, \ldots, q, \tag{6.23}$$

hold, where

$$\begin{aligned} \tilde{\mathcal{C}} &:= \left[\begin{bmatrix} 0 \\ \sigma_1 \tilde{K} C_1 \end{bmatrix}^T, \ldots, \begin{bmatrix} 0 \\ \sigma_m \tilde{K} C_m \end{bmatrix}^T \right]^T, \\ \bar{\mathcal{P}} &:= \operatorname{diag}_m \left\{ \begin{bmatrix} -S & 0 \\ -S & -R \end{bmatrix} \right\}, \\ \tilde{\Phi} &:= \left[\begin{bmatrix} \varepsilon_1 \Phi_1^{\frac{1}{2}} \\ 0 \end{bmatrix}^T, \ldots, \begin{bmatrix} \varepsilon_q \Phi_q^{\frac{1}{2}} \\ 0 \end{bmatrix}^T \right]^T, \\ \Pi_1 &:= RA + \tilde{K}\bar{\Xi}C + \tilde{A}_f, \ \Pi_2 := RE_1 + \tilde{K}E_2, \ \Pi_3 := RA + \tilde{K}\Xi, \\ \Pi_4 &:= RE_1 + \tilde{K}E_2, \ \bar{R} := R - Q, \ \bar{S} := S - Q, \end{aligned}$$

then the system (6.8) is exponentially mean-square stable. In this case, the parameters of the desired filter (6.7) are given as follows:

$$K := X_{12}^{-1}\tilde{K}, \ A_f := X_{12}^{-1}\tilde{A}_f S^{-1} Y_{12}^{-1}. \tag{6.24}$$

Proof 16 *Partition P and P^{-1} as*

$$P = \begin{bmatrix} R & X_{12} \\ X_{12}^T & X_{22} \end{bmatrix} > 0, \quad P^{-1} = \begin{bmatrix} S^{-1} & Y_{12} \\ Y_{12}^T & Y_{22} \end{bmatrix}, \tag{6.25}$$

and construct

$$T_1 = \begin{bmatrix} S^{-1} & I \\ Y_{12}^T & 0 \end{bmatrix} > 0, \quad T_2 = \begin{bmatrix} I & R \\ 0 & X_{12}^T \end{bmatrix}, \tag{6.26}$$

which imply that $PT_1 = T_2$ *and* $T_1^T P T_1 = T_1^T T_2$.

Premultiplying and postmultiplying (6.10) by

$$\mathrm{diag}\{T_1^T, I, T_1^T, \mathrm{diag}_m\{T_1^T\}, T_1^T, \mathrm{diag}_q\{I\}, \mathrm{diag}_q\{I\}\}$$

and its transpose, (6.11) by $\mathrm{diag}\{1,\ T_1^T\}$ *and its transpose, we can have*

$$\begin{bmatrix} -\bar{\bar{S}} & * & * & * & * & * & * & * & * & * \\ -\bar{I} & -\bar{\bar{R}} & * & * & * & * & * & * & * & * \\ 0 & 0 & -Q & * & * & * & * & * & * & * \\ AS^{-1} & A & A_d & -S^{-1} & * & * & * & * & * & * \\ \Gamma & \bar{\bar{A}} & RA_d & -I & -R & * & * & * & * & * \\ \check{\mathcal{C}} & \tilde{\mathcal{C}} & 0 & 0 & 0 & -\tilde{\mathcal{P}} & * & * & * & * \\ E_1S^{-1} & E_1 & 0 & 0 & 0 & 0 & -S^{-1} & * & * & * \\ \bar{\bar{E}} & \bar{\bar{E}}_1 & 0 & 0 & 0 & 0 & -I & -R & * & * \\ \check{\Phi} & \tilde{\Phi} & 0 & 0 & 0 & 0 & 0 & 0 & -\Lambda & * \\ 0 & 0 & \hat{\Psi} & 0 & 0 & 0 & 0 & 0 & 0 & -\Lambda \end{bmatrix} < 0, \tag{6.27}$$

$$\begin{bmatrix} -\varepsilon_i & * & * \\ \pi_{1i} & -S^{-1} & * \\ R\pi_{1i} + \tilde{K}\pi_{2i} & -I & -R \end{bmatrix} < 0, \qquad i = 1, \ldots, q, \tag{6.28}$$

where $\tilde{\mathcal{C}}$, $\tilde{\Phi}$, $\hat{\Psi}$, Λ *have been defined previously, and*

$$\begin{aligned} \Gamma &:= RAS^{-1} + \tilde{K}\Xi CS^{-1} + X_{12}A_fY_{12}, \\ \check{\mathcal{C}} &:= \left[\begin{bmatrix} 0 \\ \sigma_1\tilde{K}C_1S^{-1} \end{bmatrix}^T, \ldots, \begin{bmatrix} 0 \\ \sigma_m\tilde{K}C_mS^{-1} \end{bmatrix}^T\right]^T, \\ \tilde{\mathcal{P}} &:= \mathrm{diag}_m\left\{\begin{bmatrix} -S^{-1} & 0 \\ -I & -R \end{bmatrix}\right\}, \\ \check{\Phi} &:= \left[\begin{bmatrix} \varepsilon_1\Phi_1^{\frac{1}{2}}S^{-1} \\ 0 \end{bmatrix}^T, \ldots, \begin{bmatrix} \varepsilon_q\Phi_q^{\frac{1}{2}}S^{-1} \\ 0 \end{bmatrix}^T\right]^T. \\ \bar{\bar{S}} &:= S^{-1} - S^{-1}QS^{-1}, \bar{I} := I - QS^{-1},\ \bar{\bar{A}} := RA + \tilde{K}\Xi C, \\ \bar{\bar{E}} &:= RE_1S^{-1} + \tilde{K}E_2S^{-1},\ \bar{\bar{R}} := R - Q,\ \bar{\bar{E}}_1 := RE_1 + \tilde{K}E_2. \end{aligned}$$

Furthermore, Premultiplying and postmultiplying (6.27) by

$$\mathrm{diag}\{S,\ I,\ I,\ S,\ I,\ \mathrm{diag}_m\{\mathrm{diag}\{S,\ I\}\},\ S,\ I,\ \mathrm{diag}_q\{I\},\ \mathrm{diag}_q\{I\}\},$$

and (6.28) by $\mathrm{diag}\{1,\ S,\ I\}$, *we can obtain (6.22) and (6.23), which completes the proof.*

Remark 6.3 *The desired filter design problem is solved in Theorem 13 for the addressed nominal discrete-time stochastic nonlinear systems with multiple sensor data missing, external stochastic disturbances. LMI-based sufficient conditions are obtained for the existence of full-order filters that ensure the exponential mean-square stability of the resulting filtering error system for all admissible time-delays and nonlinearities. The feasibility of the filter design problem can be readily checked by the solvability of two sets of LMIs, which can be determined by using the MATLAB LMI Toolbox in a straightforward way.*

6.2.3 The Solution

In the following theorem, we extend the main results in Theorem 13 to the parameter uncertain cases. The proof of this theorem can be obtained along the similar line of that of Theorem 13, and is therefore omitted here to avoid unnecessary duplication.

Theorem 14 Consider the system (6.8). If there exist positive-definite matrices $S > 0$, $R > 0$, and $Q > 0$, and positive scalars $\delta > 0$ and $\varepsilon_i > 0$, $(i = 1, 2, \ldots, q)$, such that the following linear matrix inequalities

$$\left[\begin{array}{cccccccccccc}
-S+Q & * & * & * & * & * & * & * & * & * & * & * \\
-S+Q & -R+Q & * & * & * & * & * & * & * & * & * & * \\
0 & 0 & -Q & * & * & * & * & * & * & * & * & * \\
SA & SA & SA_d & -S & * & * & * & * & * & * & * & * \\
\bar{\Pi}_1 & \bar{\Pi}_3 & RA_d & -S & -R & * & * & * & * & * & * & * \\
\tilde{\mathcal{C}} & \tilde{\mathcal{C}} & 0 & 0 & 0 & -\bar{\mathcal{P}} & * & * & * & * & * & * \\
SE_1 & SE_1 & 0 & 0 & 0 & 0 & -S & * & * & * & * & * \\
\bar{\Pi}_2 & \bar{\Pi}_4 & 0 & 0 & 0 & 0 & -S & -R & * & * & * & * \\
\tilde{\Phi} & \tilde{\Phi} & 0 & 0 & 0 & 0 & 0 & 0 & -\Lambda & * & * & * \\
0 & 0 & \hat{\Psi} & 0 & 0 & 0 & 0 & 0 & 0 & -\Lambda & * & * \\
0 & 0 & 0 & M^T S & M^T R & 0 & 0 & 0 & 0 & 0 & -\delta I & * \\
\delta N & \delta N & \delta N_d & 0 & 0 & 0 & 0 & 0 & 0 & 0 & 0 & -\delta I
\end{array}\right] < 0, \quad (6.29)$$

$$\begin{bmatrix} -\varepsilon_i & * & * \\ S\pi_{1i} & -S & * \\ R\pi_{1i} + \tilde{K}\pi_{2i} & -S & -R \end{bmatrix} < 0, \ i = 1, \ldots, q, \tag{6.30}$$

where $\bar{\Pi}_1 = RA + \tilde{K}\bar{\Xi}C + \tilde{A}_f, \bar{\Pi}_2 = RE_1 + \tilde{K}E_2, \bar{\Pi}_3 = RA + \tilde{K}\bar{\Xi}C, \bar{\Pi}_4 = RE_1 + \tilde{K}E_2$, hold, then the system (6.8) is exponentially mean-square stable.

Proof 17 *In (6.22), replace A and A_d with $A+\Delta A(k)$ and $A_d+\Delta A_d(k)$, and then rewrite (6.22) in the form of the inequality $\Upsilon + \mathcal{M}F\mathcal{N} + \mathcal{N}^T F^T \mathcal{M}^T < 0$, where Υ has been defined in (6.22) and*

$$\mathcal{M} = [0 \ 0 \ 0 \ M^T S \ M^T R \ 0 \ 0 \ 0 \ 0 \ 0]^T, \ \mathcal{N} = [N \ N \ N_d \ 0 \ 0 \ 0 \ 0 \ 0 \ 0 \ 0].$$

From Schur complement lemma and Lemma 6.1, we can easily obtain the results of this theorem.

6.3 An Illustrative Example

In this section, a simulation example is presented to illustrate the usefulness and flexibility of the filter design method developed in this chapter. We are interested in designing the filter for the discrete-time stochastic nonlinear uncertain system with stochastic disturbances and sensor data missing.

The system data of (6.1)–(6.3) are given as follows:

$$m = n = q = 2, \ d = 2,$$
$$A = \begin{bmatrix} 0.7 & -0.4 \\ 0 & 0.1 \end{bmatrix}, A_d = \begin{bmatrix} 0.1 & 0 \\ 0 & 0.1 \end{bmatrix},$$
$$E_1 = \begin{bmatrix} 0.1 & 0 \\ 0 & 0.5 \end{bmatrix}, E_2 = \begin{bmatrix} 0.1 & 0 \\ -0.1 & 0.2 \end{bmatrix},$$
$$C = \begin{bmatrix} 0.5 & 0 \\ 0 & 0.5 \end{bmatrix}, M = \begin{bmatrix} 0.02 \\ 0.02 \end{bmatrix},$$
$$N = [0.01 \ 0.02], N_d = [0.01 \ 0.02].$$

Using MATLAB LMI Control Toolbox to solve the LMIs in (6.22) and (6.23), we can calculate the filter parameters as follows:

$$K = \begin{bmatrix} -0.0338 & 0.0213 \\ -0.0077 & -0.0705 \end{bmatrix}, \quad A_f = \begin{bmatrix} 0.5668 & -0.1363 \\ -0.2064 & -0.2217 \end{bmatrix}.$$

Figures 6.1–6.2 display the simulation results for the performance of the

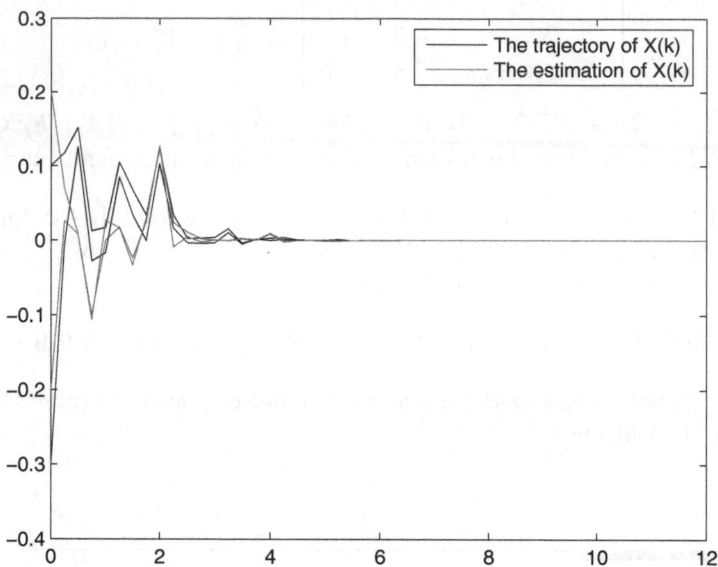

FIGURE 6.1
The trajectory and estimation of $x(k)$.

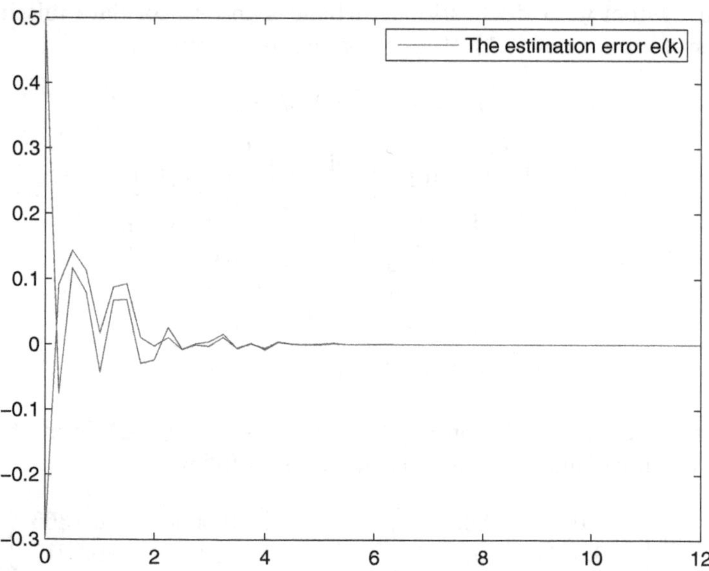

FIGURE 6.2
Estimation error $e(k)$.

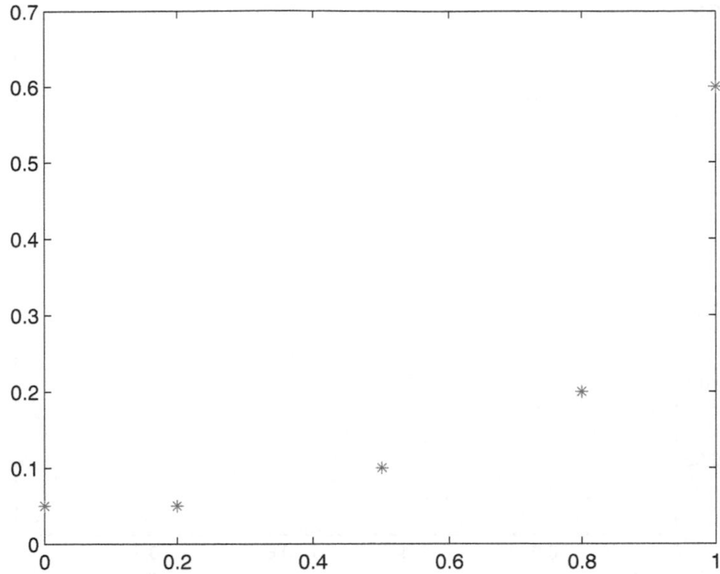

FIGURE 6.3
The probabilistic density function of ξ_1.

designed filter, where the stochastic nonlinear functions are taken as

$$\mathbb{E}\left\{\left[\begin{array}{c} f \\ g \end{array}\right]|x(k)\right\} = 0,$$

$$\begin{aligned}
\mathbb{E}\left\{ff^T|x(k)\right\} = & \left[\begin{array}{c} 0.21 \\ 0.13 \end{array}\right]\left[\begin{array}{c} 0.21 \\ 0.13 \end{array}\right]^T\left\{x^T(k)\left[\begin{array}{cc} 0.16 & 0 \\ 0 & 0.04 \end{array}\right]x(k)\right. \\
& \left. +x^T(k-d)\left[\begin{array}{cc} 0.09 & 0 \\ 0 & 0.04 \end{array}\right]x(k-d)\right\}, \\
\mathbb{E}\left\{gg^T|x(k)\right\} = & \left[\begin{array}{c} 0.11 \\ 0.1 \end{array}\right]\left[\begin{array}{c} 0.11 \\ 0.1 \end{array}\right]^T\left\{x^T(k)\left[\begin{array}{cc} 0.04 & 0 \\ 0 & 0.04 \end{array}\right]x(k)\right. \\
& \left. +x^T(k-d)\left[\begin{array}{cc} 0.09 & 0 \\ 0 & 0.09 \end{array}\right]x(k-d)\right\}.
\end{aligned}$$

Figures 6.3–6.4 show the probabilistic density functions of ξ_1 and ξ_2 in

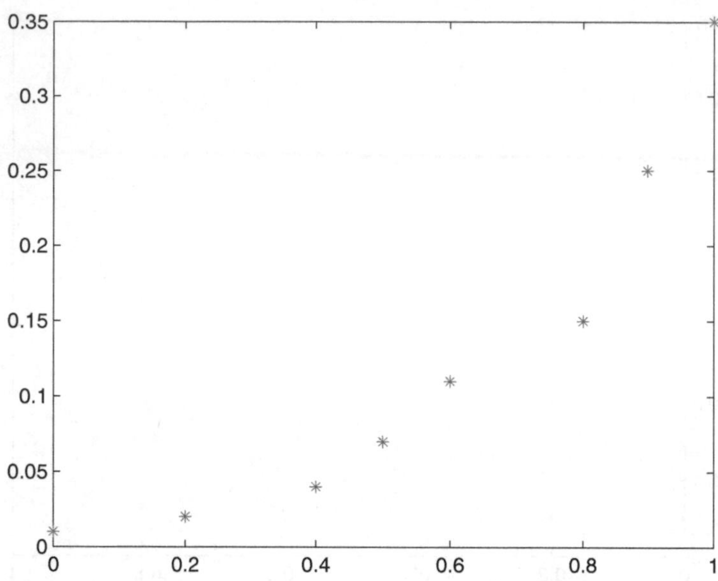

FIGURE 6.4
The probabilistic density function of ξ_2.

[0 1] described by

$$p_1(s_1) = \begin{cases} 0.05 & s_1 = 0 \\ 0.05 & s_1 = 0.2 \\ 0.1 & s_1 = 0.5 \\ 0.2 & s_1 = 0.8 \\ 0.6 & s_1 = 1 \end{cases} \text{ and}$$

$$p_2(s) = \begin{cases} 0.01 & s_2 = 0 \\ 0.02 & s_2 = 0.2 \\ 0.04 & s_2 = 0.4 \\ 0.07 & s_2 = 0.5 \\ 0.11 & s_2 = 0.6 \\ 0.15 & s_2 = 0.8 \\ 0.25 & s_2 = 0.9 \\ 0.35 & s_2 = 1 \end{cases}, \tag{6.31}$$

from which the expectations and variances can be easily calculated as $\mu_1 = 0.82$, $\mu_2 = 0.816$, $\sigma_1^2 = 0.0826$, and $\sigma_2^2 = 0.0469$.

6.4 Summary

In this chapter, the filtering problem has been studied for a class of discrete-time uncertain stochastic nonlinear time-delay systems, with both the probabilistic missing measurements and external stochastic disturbances. The measurement missing phenomenon is assumed to occur in a random way, and the missing probability for each sensor is governed by an individual random variable satisfying a certain probabilistic distribution in the interval [0 1]. We have designed a filter such that, for the admissible random measurement missing, stochastic disturbances, norm-bounded uncertainties as well as stochastic nonlinearities, the error dynamics of the filtering process is exponentially mean-square stable. By using the LMI method, sufficient conditions have been established that ensure the exponential mean-square stability of the filtering error, and then the filter parameters have been characterized by the solution to a set of LMIs. An illustrative example has been exploited to show the effectiveness of the proposed design procedures.

7

Probability-Dependent Control with Randomly Occurring Nonlinearities

CONTENTS

In this chapter, the gain-scheduled control problem is addressed by using probability-dependent Lyapunov functions for a class of discrete-time stochastic delayed systems with randomly occurring sector-nonlinearities. The occurrence of the sector-nonlinearities is assumed to be a time-varying Bernoulli distribution with measurable probability in real time. The multiplicative noises are given by means of a scalar Gaussian white noise sequence with known variances. The aim of the addressed gain-scheduled control problem is to design a controller with scheduled gains, such that, for the admissible randomly occurring nonlinearities (RONs), time-delays, and external noise disturbances, the closed-loop system is exponentially mean-square stable. Note that the designed gain-scheduled controller is based on the measured time-varying probability and is therefore less conservative than the conventional controller with constant gains. It is shown that the time-varying controller gains can be derived in terms of the measurable probability by solving a convex optimization problem via the semidefinite program method. A simulation example is exploited to illustrate the effectiveness of the proposed design procedures.

In this chapter, we will aim to consider the gain-scheduled state-feedback control problem for the discrete-time stochastic delayed systems with RONs. The main contributions of this chapter can be summarized as follows: *1) a new state-feedback control problem is investigated by a gain-scheduling approach for a class of discrete-time stochastic delayed systems with RONs; 2) the nonlinearity disturbances are assumed to occur in a random way, and a stochastic variable sequence satisfying time-varying Bernoulli distributions is introduced to describe the time-varying feature of the RONs; 3) the time-varying controller gains are affected by not only the constant parameters but also the time-varying probability; 4) an easily implementable algorithm is developed to design the controller with both constant and time-varying parameters.* The desired controller is designed by employing the gain-scheduling method, which leads to less conservatism than the traditional one with constant gains only. In the simultaneous presence of RONs, time-delays, and external noise disturbances, the closed-loop system is guaranteed to be exponentially mean-square stable. A simulation example is exploited to illustrate the effectiveness of the proposed design procedures.

The remainder of this chapter is organized as follows: Section 7.1 formulates the the gain-scheduled state-feedback control problem for the discrete-time stochastic delayed systems with RONs. In section 7.2, controller analysis and synthesis problems are addressed. Section 7.3 illustrates the usefulness and flexibility of the filter design method developed. Section 7.4 is the summary.

7.1 Problem Formulation

Consider the following discrete-time stochastic delayed systems with RONs:

$$\begin{aligned} x(k+1) &= Ax(k) + A_d x(k-d) + Bu(k) + \xi(k)Cf(z(k) \\ &\quad +Dx(k)\omega(k) \qquad (7.1) \\ x(0) &= \rho, \qquad (7.2) \end{aligned}$$

where $x(k) \in \mathbb{R}^n$ is the state, ρ is the initial state of the system, d is a constant delay, and $z(k) := Gx(k) + G_d x(k-d)$. $\omega(k)$ is a one-dimensional Gaussian white noise sequence satisfying $\mathbb{E}\{\omega(k)\} = 0$ and $\mathbb{E}\{\omega^2(k)\} = \sigma^2$. A, A_d, B, C, D, G, and G_d are constant matrices with appropriate dimensions.

The function $f(\cdot)$ with ($f(0) = 0$) represents nonlinear disturbances satisfying the following sector-bounded condition:

$$[f(z(k)) - F_1 z(k)]^T [f(z(k)) - F_2 z(k)] \leq 0, \qquad (7.3)$$

where F_1 and F_2 are constant real matrices of appropriate dimensions with $F_2 - F_1 > 0$. In this case, $f(\cdot)$ is said to belong to the sector $[F_1,\ F_2]$.

For technical convenience, the nonlinear function $f(z(k))$ is decomposed into a linear part and a nonlinear part as

$$f(z(k)) = F_1 z(k) + f_s(z(k)), \qquad (7.4)$$

and then it follows from (7.3) that

$$f_s^T(z(k))(f_s(z(k)) - Fz(k)) \leq 0, \qquad (7.5)$$

where $F = F_2 - F_1 > 0$.

Remark 7.1 *The sector nonlinearity is more general than the usual Lipschitz condition and has been widely employed to model some nonlinear phenomena in reality, for example, the saturation nonlinearities and nonlinear activation functions in neural networks and gene regulatory networks. In the past few years, the control, filtering and model reduction problems for the systems with sector nonlinearities have been intensively studied; see, e.g., [54,77,83,97,191].*

In (7.1), $\xi(k)$ is a random variable sequence to account for the RONs, satisfying the following Bernoulli distribution,

$$\begin{aligned} \text{Prob}\{\xi(k)=1\} &= \mathbb{E}\{\xi(k)\} = p(k), \\ \text{Prob}\{\xi(k)=0\} &= 1-\mathbb{E}\{\xi(k)\} = 1-p(k), \end{aligned} \tag{7.6}$$

where $p(k)$ is a time-varying positive scalar sequence that takes values on the interval $[p_1\ p_2] \subseteq [0\ 1]$, with p_1 and p_2 being the lower and upper bounds of $p(k)$, respectively. In this chapter, for simplicity, we assume that $\xi(k)$, $\omega(k)$, and ρ are uncorrelated.

Remark 7.2 *The nonlinear disturbances may occur randomly due to some environmental reasons, and a Bernoulli distribution model has been introduced in [169, 185] to describe such randomly occurring phenomenon. However, the Bernoulli distribution employed in the literature has been assumed to be time-invariant. This is certainly conservative to deal with the time-varying systems with RONs according to time-varying probabilities. As such, we are going to utilize a random variable sequence in (7.6) that satisfies time-varying Bernoulli distributions in order to better describe the randomly intermittent nature of the concerned nonlinear disturbances.*

In this chapter, we are interested in designing the following gain-scheduled controller

$$u(k) = K(p)x(k), \tag{7.7}$$

where $K(p)$ is the controller gain sequence to be designed that is of the following structure

$$K(p) = K_0 + p(k)K_u, \tag{7.8}$$

where, for every time step k, $p(k)$ is the time-varying parameter of the controller gain taking value in $[p_1,\ p_2]$, and K_0, K_u are the constant parameters of the controller gain to be designed.

Remark 7.3 *The controller gain given in (7.8) includes not only the constant parameters K_0, K_u, but also the time-varying parameter $p(k)$. Such controller will be scheduled with the time-varying parameter (the time-varying probability in this chapter), which would certainly give rise to less conservatism than the conventional controller with constant gains only. Note that gain-scheduled control/filtering problems have recently stirred considerable research attentions [3, 16, 61, 136, 154, 166].*

Remark 7.4 *Because of the special structure of the controller gain in (7.8), the implementation of the gain-scheduled controller deserves particular attention. In fact, the implementation consists of the following steps: 1) compute the constant gains K_0 and K_u of the controller by using the main results developed in this chapter; 2) estimate/measure the time-varying probability $p(k)$ by statistical tests in practice; and 3) obtain the controller gain from (7.8).*

The closed-loop system of (7.1), with the state-feedback gain-scheduled controller (7.7), is given as follows:

$$\begin{aligned} x(k+1) &= Ax(k) + A_d x(k-d) + \xi(k)Cf(z(k)) + BK(p)x(k) \\ &\quad + Dx(k)\omega(k). \end{aligned} \tag{7.9}$$

In this chapter, our purpose is to design a probability-dependent gain-scheduled controller of the form (7.7) for the system (7.1) by exploiting a probability-dependent Lyapunov function and LMI method, such that, for all admissible time-delays, RONs and exogenous stochastic noises, the closed-loop system (7.9) is exponentially mean-square stable.

7.2 Gain-Scheduled Controller Design

In the following theorem, a probability-dependent gain-scheduled control problem is dealt with for a class of discrete-time stochastic delayed systems (7.1) with RONs by exploiting Lyapunov theory and convex programming method. A sufficient condition is derived to guarantee the solvability of the desired gain-scheduled control problem and, meanwhile, the parameters of the gain-scheduled controller can be obtained by solving the convex optimization problem via the semidefinite programme method in terms of the measured time-varying probability.

Theorem 15 *Consider the discrete-time stochastic delayed systems (7.9). Assume that there exist positive-definite matrices $\mathcal{Q}(p(k)) > 0$ and $\mathcal{Q}_d > 0$, slack matrix S, and matrices $\mathcal{K}_0$, $\mathcal{K}_u$, such that the following LMIs hold:*

$$\begin{bmatrix} \mathcal{Q}_d - \mathcal{Q}(p(k)) & * & * & * & * & * \\ 0 & -\mathcal{Q}_d & * & * & * & * \\ FGS & FG_dS & -2I & * & * & * \\ \breve{\Omega}_{4,1} & \breve{\Omega}_{4,2} & \breve{\Omega}_{4,3} & -\Delta_{k+1} & * & * \\ \breve{\Omega}_{5,1} & 0 & 0 & 0 & -\sigma^2\Delta_{k+1} & * \\ \breve{\Omega}_{6,1} & \breve{\Omega}_{6,2} & \breve{\Omega}_{6,3} & 0 & 0 & -\Delta_p(k)\Delta_{k+1} \end{bmatrix} < 0, \tag{7.10}$$

where

$$\begin{aligned}
\breve{\Omega}_{4,1} &= \bar{A}S + B\mathcal{K}(p),\ \breve{\Omega}_{4,2} = \bar{A}_d S,\ \breve{\Omega}_{4,3} = p(k)C,\ \breve{\Omega}_{5,1} = \sigma^2 DS, \\
\breve{\Omega}_{6,1} &= \Delta_p(k)CF_1GS,\ \breve{\Omega}_{6,2} = \Delta_p(k)CF_1G_dS,\ \breve{\Omega}_{6,3} = \Delta_p(k)C, \\
\bar{A} &= A + p(k)CF_1G,\ \mathcal{K}(p) = \mathcal{K}_0 + p(k)\mathcal{K}_u, \Delta_p(k) = p(k)(1-p(k)), \\
\bar{A}_d &= A_d + p(k)CF_1G_d, \Delta_{k+1} = -\mathcal{Q}(p(k+1)) + S + S^T. \qquad (7.11)
\end{aligned}$$

In this case, the constant gains of the desired controller can be obtained as follows:

$$K_0 = \mathcal{K}_0 S^{-1},\ K_u = \mathcal{K}_u S^{-1}, \qquad (7.12)$$

and the closed-loops system (7.9) is then exponentially mean-square stable for all $p(k) \in [p_1\ \ p_2]$.

Proof 18 *Let* $Q(p(k)) = S^{-T}\mathcal{Q}(p(k))S^{-1}$, $Q_d = S^{-T}\mathcal{Q}_d S^{-1}$ *and define the following probability-dependent Lyapunov function:*

$$V(k) := x^T(k)Q(p(k))x(k) + \sum_{s=k-d}^{k-1} x^T(s)Q_d x(s). \qquad (7.13)$$

Then, noting $\mathbb{E}\{\xi(k) - p(k)\} = 0$ *and* $\mathbb{E}\{\omega(k) = 0\}$, *we have from (7.9) that*

$$\begin{aligned}
&\mathbb{E}\{\Delta V(k)\} \\
=\ &\mathbb{E}\big\{x^T(k+1)Q(p(k+1))x(k+1) \\
&-x^T(k)(Q(p(k)) - Q_d)x(k) \\
&-x^T(k-d)Q_d x(k-d)\big\} \\
=\ &\mathbb{E}\{[Ax(k) + A_d x(k-d) + p(k)Cf(z(k)) \\
&+(\xi(k) - p(k))Cf(z(k)) + BK(p)x(k) + Dx(k)\omega(k)]^T \\
&\times Q(p(k+1))[Ax(k) + A_d x(k-d) + p(k)Cf(z(k)) \\
&+(\xi(k) - p(k))Cf(z(k)) + BK(p)x(k) + Dx(k)\omega(k)] \\
&-x^T(k)(Q(p(k)) - Q_d)x(k) - x^T(k-d)Q_d x(k-d)\big\} \\
=\ &\mathbb{E}\big\{[Ax(k) + A_d x(k-d) + p(k)Cf(z(k)) + BK(p)x(k)]^T \\
&\times Q(p(k+1))[Ax(k) + A_d x(k-d) + p(k)Cf(z(k)) \\
&+BK(p)x(k)] - x^T(k)(Q(p(k)) - Q_d)x(k) \\
&-x^T(k-d)Q_d x(k-d) + \sigma^2 x^T(k)D^T Q(p(k+1))Dx(k) \\
&+p(k)(1-p(k))f^T(z(k))C^T Q(p(k+1))Cf(z(k))\big\}. \qquad (7.14)
\end{aligned}$$

From (7.3)–(7.5) and (7.14), it can be seen that

$$
\begin{aligned}
& \mathbb{E}\{\Delta V(k)\} \\
\leq\ & \mathbb{E}\{[(A+BK(p)+p(k)CF_1G)x(k)+(A_d+p(k)CF_1G_d)x(k-d) \\
& +p(k)Cf_s(z(k))]^T Q(p(k+1))[(A+BK(p)+p(k)CF_1G)x(k) \\
& +(A_d+p(k)CF_1G_d)x(k-d)+p(k)Cf_s(z(k))] \\
& +p(k)(1-p(k))[f_s(z(k))+F_1Gx(k)+F_1G_dx(k-d))]^T C^T \\
& \times Q(p(k+1))C[f_s(z(k))+F_1Gx(k)+F_1G_dx(k-d))] \\
& -x^T(k)(Q(p(k))-Q_d)x(k)-x^T(k-d)Q_dx(k-d) \\
& -2f_s^T(z(k))f_s(z(k))+\sigma^2x^T(k)D^TQ(p(k+1))Dx(k) \\
& +2f_s^T(z(k))FGx(k)+2f_s^T(z(k))FG_dx(k-d))\} \\
=\ & \mathbb{E}\{\bar{x}^T(k)\Omega\bar{x}(k)\},
\end{aligned}
\tag{7.15}
$$

where $\bar{x}(k) = [x^T(k)\ \ x^T(k-d)\ \ f_s^T(z(k))]^T$ *and*

$$
\Omega = \begin{bmatrix} \Omega_1 & * & * \\ \Omega_2 & \Omega_3 & * \\ \Omega_4 & \Omega_5 & \Omega_6 \end{bmatrix},
\tag{7.16}
$$

with

$$
\begin{aligned}
\Omega_1 &= [A+BK(p)+p(k)CF_1G]^TQ(p(k+1))[A+BK(p)+p(k)CF_1G] \\
&\quad -Q(p(k))+p(k)(1-p(k))G^TF_1^TC^TQ(p(k+1))CF_1G \\
&\quad +\sigma^2D^TQ(p(k+1))D+Q_d, \\
\Omega_2 &= [A_d+p(k)CF_1G_d]^TQ(p(k+1))[A+BK(p)+p(k)CF_1G] \\
&\quad +p(k)(1-p(k))G_d^TF_1^TC^TQ(p(k+1))CF_1G, \\
\Omega_3 &= [A_d+p(k)CF_1G_d]^TQ(p(k+1))[A_d+p(k)CF_1G_d] \\
&\quad +p(k)(1-p(k))G_d^TF_1^TC^TQ(p(k+1))CF_1G_d-Q_d, \\
\Omega_4 &= p(k)C^TQ(p(k+1))[A+BK(p)+p(k)CF_1G] \\
&\quad +p(k)(1-p(k))C^TQ(p(k+1))CF_1G+FG, \\
\Omega_5 &= p(k)C^TQ(p(k+1))[A_d+p(k)CF_1G_d] \\
&\quad +p(k)(1-p(k))C^TQ(p(k+1))CF_1G_d+FG_d, \\
\Omega_6 &= -2I+p(k)C^TQ(p(k+1))C.
\end{aligned}
\tag{7.17}
$$

In the following, we are going to show that $\Omega < 0$ *can be concluded from (7.10). To do this, we Premultiply and postmultiply the LMIs in (7.10) by* $\mathrm{diag}\{S^{-T}, S^{-T}, I, S^{-T},$

$\sigma^{-2}S^{-T}$, $\Delta_p^{-1}(k)S^{-T}\}$ *and its transpose, and we can obtain*

$$\begin{bmatrix} Q_d - Q(p(k)) & * & * & * & * & * \\ 0 & -Q_d & * & * & * & * \\ FG & FG_d & -2I & * & * & * \\ \bar{\Omega}_{4,1} & \bar{\Omega}_{4,2} & \bar{\Omega}_{4,3} & -\bar{\Delta}_{k+1} & * & * \\ \bar{\Omega}_{5,1} & 0 & 0 & 0 & -\sigma^{-2}\bar{\Delta}_{k+1} & * \\ \bar{\Omega}_{6,1} & \bar{\Omega}_{6,2} & \bar{\Omega}_{6,3} & 0 & 0 & -\Delta_p^{-1}(k)\bar{\Delta}_{k+1} \end{bmatrix} < 0, \tag{7.18}$$

with $\bar{\Omega}_{4,1} = S^{-T}\bar{A} + S^{-T}BK(p)$, $\bar{\Omega}_{4,2} = S^{-T}\bar{A}_d$, $\bar{\Omega}_{4,3} = p(k)S^{-T}C$, $\bar{\Omega}_{5,1} = S^{-T}D$, $\bar{\Omega}_{6,1} = S^{-T}CF_1G$, $\bar{\Omega}_{6,2} = S^{-T}CF_1G_d$, $\bar{\Omega}_{6,3} = S^{-T}C$, *and* $\bar{\Delta}_{k+1} = -Q(p(k+1)) + S^{-1} + S^{-T}$.

Let us now show that, if (7.18) holds, then the LMIs

$$\begin{bmatrix} Q_d - Q(p(k)) & * & * & * & * & * \\ 0 & -Q_d & * & * & * & * \\ FG & FG_d & -2I & * & * & * \\ \tilde{\Omega}_{4,1} & \bar{A}_d & \tilde{\Omega}_{4,3} & -\Pi_{k+1} & * & * \\ D & 0 & 0 & 0 & -\sigma^{-2}\Pi_{k+1} & * \\ \tilde{\Omega}_{6,1} & \tilde{\Omega}_{6,2} & C & 0 & 0 & -\Delta_p^{-1}(k)\Pi_{k+1} \end{bmatrix} < 0, \tag{7.19}$$

where $\tilde{\Omega}_{4,1} = \bar{A} + BK(p)$, $\tilde{\Omega}_{4,3} = p(k)C$, $\tilde{\Omega}_{6,1} = CF_1G$, $\tilde{\Omega}_{6,2} = CF_1G_d$, *and* $\Pi_{k+1} = Q^{-1}(p(k+1))$, *are true.*

If (7.18) holds, we can know from $-Q(p(k+1)) + S^{-1} + S^{-T} > 0$ *that* S^{-1} *is a nonsingular matrix. By performing the congruence transformation* $\text{diag}\{I, I, I, S, S, S\}$ *to (7.18), we can have*

$$\begin{bmatrix} Q_d - Q(p(k)) & * & * & * & * & * \\ 0 & -Q_d & * & * & * & * \\ FG & FG_d & -2I & * & * & * \\ \tilde{\Omega}_{4,1} & \bar{A}_d & \tilde{\Omega}_{4,3} & -\Gamma_{k+1} & * & * \\ D & 0 & 0 & 0 & -\sigma^{-2}\Gamma_{k+1} & * \\ \tilde{\Omega}_{6,1} & \tilde{\Omega}_{6,2} & C & 0 & 0 & -\Delta_p^{-1}(k)\Gamma_{k+1} \end{bmatrix} < 0, \tag{7.20}$$

with $\Gamma_{k+1} = -S^TQ(p(k+1))S + S + S^T$. *Then, (7.19) can be inferred from the inequality* $Q^{-1}(p(k+1)) \geq -S^TQ(p(k+1))S + S + S^T$. *Furthermore, by Lemma 2.2, we can know from (7.19) that* $\Omega < 0$ *and, subsequently,*

$$\mathbb{E}\{\Delta V(k)\} < -\lambda_{\min}(-\Omega)\mathbb{E}|\bar{x}(k)|^2, \tag{7.21}$$

where $\lambda_{\min}(-\Omega)$ *is the minimum eigenvalue of* $-\Omega$. *Finally, it can be confirmed from Lemma 1 of [186] that the closed-loop system (7.9) is exponentially mean-square stable. The proof of this theorem is now completed.*

Remark 7.5 *In Theorem 15, in order to obtain desired controller gains by making use of the time-varying probability, probability-dependent Lyapunov functions have been exploited to reduce the conservatism of controller design.*

Note that, in the past few years, parameter-dependent Lyapunov functions have been often applied to uncertain systems and time-varying parameter systems [43, 154, 166]. Also, a slack variable S has been introduced in Theorem 15 to decouple the product terms between Lyapunov matrices and system matrices so that we can easily deal with the resulting controller design problem. Such a technique has been extensively utilized to design controllers/filters [5, 16, 43, 46, 61, 229].

Remark 7.6 *In Theorem 15, the constant parameters of the controller gain K_0 and K_u can be obtained by solving the LMIs in (7.10) and, therefore, the gain-scheduled controller gain can be derived by using the measured time-varying probability in real time. However, LMIs in (7.10) are dependent on $p(k) \in [p_1 \; p_2]$, so the number of LMIs in (7.10) is actually infinite. For this case, the desired controller cannot be obtained directly from Theorem 15 due to the infinite number of LMIs. To handle such a problem, in the next theorem, we attempt to convert the problem into a computationally accessible one by changing the description of $p(k)$.*

7.3 Computationally Accessible Solution

To deal with the computational issue, we will assign a specific form to $p(k)$. Let us set $Q(p(k)) = Q_0 + p(k)Q_p$, and we can easily know that $\mathcal{Q}(p(k)) = \mathcal{Q}_0 + p(k)\mathcal{Q}_p$.

Theorem 16 *Consider the discrete-time nonlinear stochastic systems (7.1). If there exist positive-definite matrices $\mathcal{Q}_0 > 0$, $\mathcal{Q}_p > 0$, and $\mathcal{Q}_d > 0$, and matrices S, $\mathcal{K}_0$, and $\mathcal{K}_u$, such that the following LMIs hold:*

$$\mathbb{M}^{ijl} = \begin{bmatrix} \hat{\Omega}^{ijl}_{1,1} & * & * & * & * & * & * \\ 0 & -\mathcal{Q}_d & * & * & * & * & * \\ FGS & FG_dS & -2I & * & * & * & * \\ \hat{\Omega}^{ijl}_{4,1} & \hat{\Omega}^{ijl}_{4,2} & p_iC & -\Delta^{ijl} & * & * & * \\ \sigma^2 DS & 0 & 0 & 0 & 0 & -\sigma^2\Delta^{ijl} & * \\ \hat{\Omega}^{ijl}_{6,1} & \hat{\Omega}^{ijl}_{6,2} & \Lambda^{ijl}C & 0 & 0 & 0 & -\Lambda^{ijl}\Delta^{ijl} \end{bmatrix} < 0, \quad i,j,l = 1,2, \tag{7.22}$$

where

$$\begin{aligned} \hat{\Omega}^{ijl}_{1,1} &= \mathcal{Q}_d - \mathcal{Q}_0 - p_i\mathcal{Q}_p, \\ \hat{\Omega}^{ijl}_{4,1} &= AS + B\mathcal{K}_0 + p_iB\mathcal{K}_u + p_iCF_1GS, \\ \hat{\Omega}^{ijl}_{4,2} &= A_dS + p_iCF_1G_dS, \; \Lambda^{ijl} = p_i(1-p_j), \\ \hat{\Omega}^{ijl}_{6,1} &= \Lambda^{ijl}CF_1GS, \; \hat{\Omega}^{ijl}_{6,2} = \Lambda^{ijl}CF_1G_dS \\ \Delta^{ijl} &= -\mathcal{Q}_0 - p_l\mathcal{Q}_p + S + S^T, \end{aligned} \tag{7.23}$$

and $\mathcal{K}_0$, $\mathcal{K}_u$ have been defined in (7.12), then there exists a controller in the form of (7.7), such that the closed loop system (7.9) is exponentially mean-square stable.

Proof 19 *First, letting*

$$\alpha_1(k) = \frac{p_2 - p(k)}{p_2 - p_1}, \quad \alpha_2(k) = \frac{p(k) - p_1}{p_2 - p_1}, \tag{7.24}$$

we have

$$p(k) = \alpha_1(k)p_1 + \alpha_2(k)p_2, \tag{7.25}$$

with $\alpha_i(k) \geq 0$ $(i = 1, 2)$ and $\alpha_1(k) + \alpha_2(k) = 1$.

Similarly, letting

$$\beta_1(k) = \frac{p_2 - p(k+1)}{p_2 - p_1}, \quad \beta_2(k) = \frac{p(k+1) - p_1}{p_2 - p_1}, \tag{7.26}$$

it can be obtained that

$$p(k+1) = \beta_1(k)p_1 + \beta_2(k)p_2, \tag{7.27}$$

where $\beta_i(k) \geq 0$ $(i = 1, 2)$, $\beta_1(k) + \beta_2(k) = 1$.

By means of the transformations defined above, it can be easily derived that

$$\begin{aligned} \mathcal{Q}(p(k)) &= \mathcal{Q}_0 + \sum_{i=1}^{2} \alpha_i(k)p_i\mathcal{Q}_p, \quad \mathcal{K}(p) = \mathcal{K}_0 + \sum_{i=1}^{2} \alpha_i(k)p_i\mathcal{K}_u, \\ \mathcal{Q}(p(k+1)) &= \mathcal{Q}_0 + \sum_{l=1}^{2} \beta_l(k)p_l\mathcal{Q}_p. \end{aligned} \tag{7.28}$$

On the other hand, it is easy to deduce from (7.22) that

$$\sum_{i,j,l=1}^{2} \alpha_i(k)\alpha_j(k)\beta_l(k)\mathbb{M}^{ijl} < 0. \tag{7.29}$$

From (7.24)–(7.27), it can be concluded that (7.10) is true. The proof is now completed.

Remark 7.7 *In Theorem 16, the infinite LMIs that are dependent on the time-varying probability in Theorem 15 have been changed to finite ones that are dependent on the upper and the lower bounds of $p(k)$. By Theorem 16, we can easily obtain the constant parameters of controller by solving a set of LMIs via MATLAB LMI Toolbox.*

In the following, we outline the design procedure given in Theorem 16 as a computationally appealing gain-scheduled control algorithm.

Algorithm 7.1 *Discrete-time gain-scheduled controller design algorithm.*

Step 1: Given the initial values for the positive integer N, the initial state ρ, time-delays d, p_1, and p_2, the matrices A, A_d, B, C, D, F_1, F_2, G, and G_d. Set $k = 0$.

Step 2: Solve the linear matrix inequalities in (7.22) for $i, j, l = 1, 2$ to obtain the positive-definite matrices $\mathcal{Q}_0$, $\mathcal{Q}_p$, $\mathcal{Q}_d$, matrices S, $\mathcal{K}_0$, $\mathcal{K}_u$, and then the constant parameters K_0, K_u by (7.12).

Step 3: From measured time-varying probability $p(k)$ in real time and (7.8), derive gain-scheduled controller gain $K(p)$ and then set $k = k + 1$.

Step 4: If $k < N$, then go to Step 3; otherwise, go to Step 5.

Step 5: Stop.

7.4 An Illustrative Example

In this section, the gain-scheduled controller is designed for the discrete-time stochastic delayed systems with randomly occurring nonlinearities.

The system parameters are given as follows:

$$A = \begin{bmatrix} 0.87 & 0.2 \\ 0 & 0.955 \end{bmatrix}, \; A_d = \begin{bmatrix} 0.23 & 0.2 \\ 0 & 0.33 \end{bmatrix}, \; B = \begin{bmatrix} 0.16 & 0 \\ 0 & 0.46 \end{bmatrix},$$
$$C = \begin{bmatrix} 0.1 & 0 \\ 0 & 0.2 \end{bmatrix}, \; D = \begin{bmatrix} 0.23 & 0 \\ 0.15 & 0.18 \end{bmatrix}, \; F_1 = \begin{bmatrix} 0.06 & 0 \\ 0 & 0.07 \end{bmatrix},$$
$$F_2 = \begin{bmatrix} 4.81 & 0 \\ 0 & 2.95 \end{bmatrix}, \; G = \begin{bmatrix} 0.51 & 0 \\ 0 & 0.621 \end{bmatrix}, \; G_d = \begin{bmatrix} 0.31 & 0 \\ 0 & 0.22 \end{bmatrix},$$
$$p_1 = 0.39, \; p_2 = 0.91, \; \sigma^2 = 1.$$

Set the time-varying Bernoulli distribution sequences as $p(k) = p_1 + (p_2 - p_1)|\sin(k)|$ and the sector nonlinear function $f(u)$ as

$$f(u) = \frac{F_1 + F_2}{2} u + \frac{F_2 - F_1}{2} \sin(u),$$

which satisfies (7.3). Also, select the initial state as $\rho = [2 \;\; -2]^T$.

According to Theorem 16 and Algorithm 7.1, the constant controller parameters K_0, K_u can be obtained as follows:

$$K_0 = \begin{bmatrix} -3.7662 & -0.9641 \\ 0.1053 & -2.2874 \end{bmatrix}, \; K_u = \begin{bmatrix} -0.6804 & 0.0041 \\ 0.0287 & -0.4942 \end{bmatrix},$$
$$\mathcal{Q}_0 = \begin{bmatrix} 10.1578 & 0.4002 \\ 0.4002 & 4.5863 \end{bmatrix}, \; \mathcal{Q}_p = \begin{bmatrix} 0.0791 & -0.0086 \\ -0.0086 & 0.0698 \end{bmatrix},$$
$$\mathcal{Q}_d = \begin{bmatrix} 4.1365 & 0.4313 \\ 0.4313 & 2.0879 \end{bmatrix}.$$

Then, the gain-scheduled controller gain $K(p)$ and parameter-dependent Lyapunov matrix can be calculated at every time step k, as in Table 7.1.

Figure 7.1 gives the response curves of state $x(k)$ of uncontrolled systems. Figure 7.2 depicts the simulation results of state $x(k)$ of the controlled systems. Figure 7.3 shows the time-varying probability parameters $p(k)$. The simulation results have illustrated our theoretical analysis.

TABLE 7.1
Computing Results

k	$p(k)$	$Q(p(k))$	$K(p)$
0	0.8276	$\begin{bmatrix} 10.2233 & 0.3931 \\ 0.3931 & 4.6441 \end{bmatrix}$	$\begin{bmatrix} -4.3293 & -0.9607 \\ 0.1290 & -2.6964 \end{bmatrix}$
1	0.8628	$\begin{bmatrix} 10.2261 & 0.3928 \\ 0.3928 & 4.6466 \end{bmatrix}$	$\begin{bmatrix} -4.3533 & -0.9605 \\ 0.1300 & -2.7138 \end{bmatrix}$
2	0.4634	$\begin{bmatrix} 10.1945 & 0.3962 \\ 0.3962 & 4.6186 \end{bmatrix}$	$\begin{bmatrix} -4.0815 & -0.9622 \\ 0.1186 & -2.5164 \end{bmatrix}$
3	0.7835	$\begin{bmatrix} 10.2198 & 0.3935 \\ 0.3935 & 4.6410 \end{bmatrix}$	$\begin{bmatrix} -4.2994 & -0.9609 \\ 0.1278 & -2.6746 \end{bmatrix}$
$\vdots$	$\vdots$	$\vdots$	$\vdots$

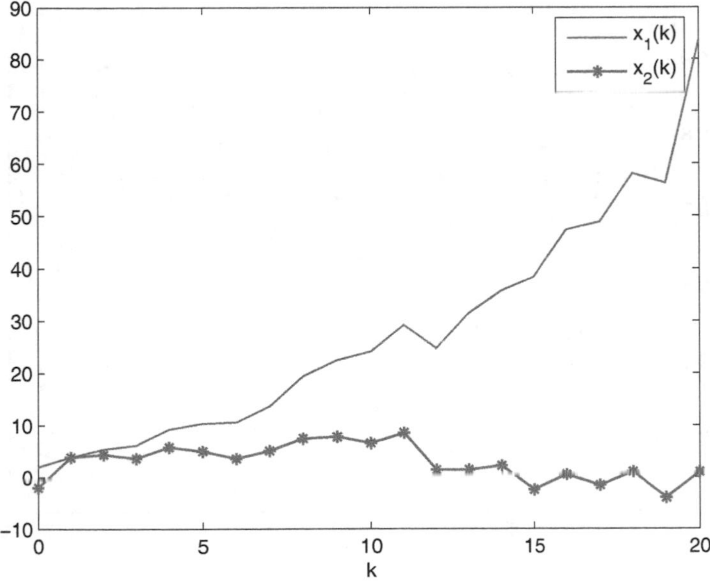

FIGURE 7.1
The state evolution $x(k)$ of uncontrolled systems.

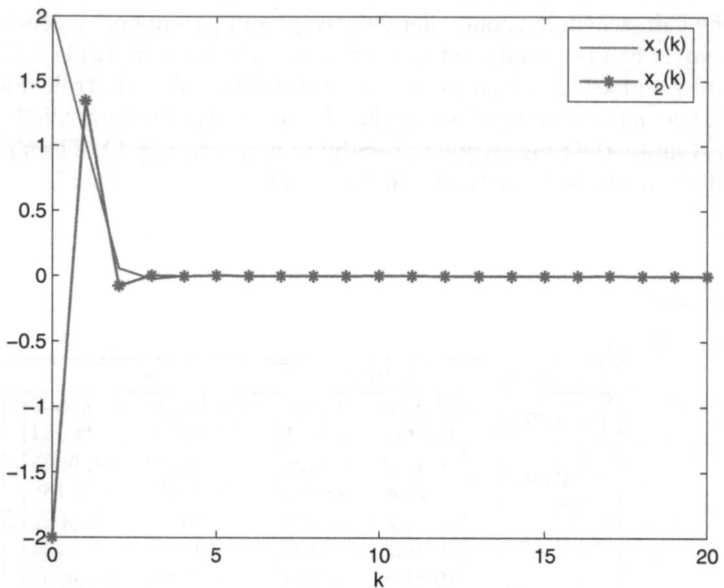

FIGURE 7.2
The state evolution $x(k)$ of controlled systems.

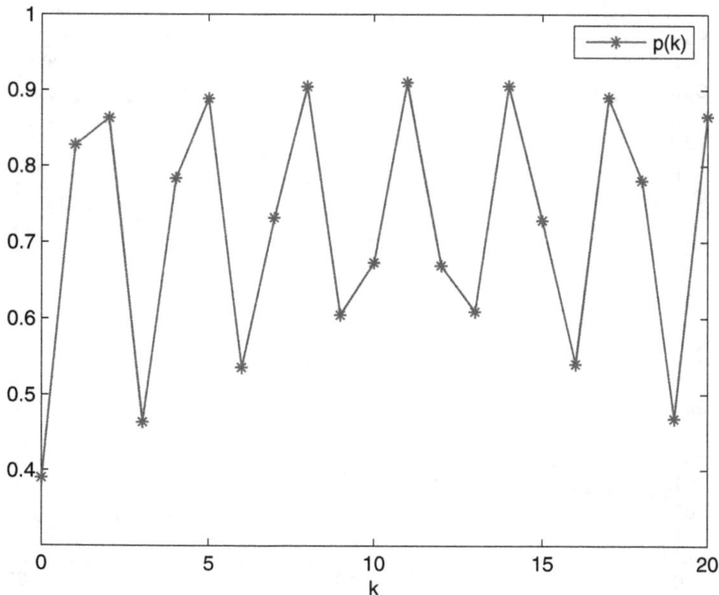

FIGURE 7.3
Time-varying probability $p(k)$.

7.5 Summary

This chapter has dealt with the probability-dependent gain-scheduled state feedback control problem for a class of discrete-time stochastic delayed systems with randomly occurring nonlinearities. We assume the nonlinear disturbances to be randomly occurring, and the occurring way is modeled by a stochastic variable sequence satisfying time-varying Bernoulli distributions. By employing probability-dependent Lyapunov functions, we have designed a gain-scheduled controller with the gain, including both constant parameters and time-varying parameters, such that, for the admissible RONs, time-delays and external noise disturbances, the closed-loop system is exponentially mean-square stable. The effectiveness of the proposed design procedure has been illustrated via a numerical example. Moreover, we can extend the main results in this chapter to more complex and realistic systems, such as systems with polytopic or norm-bounded uncertainties. On the other hand, we will also consider the corresponding filter design problem for stochastic systems with randomly occurring nonlinearities as well as the real-time applications in network-based communications and bio-informatics.

8

Probability-Dependent Filtering with Missing Measurements

CONTENTS

This chapter addresses the gain-scheduled filtering problem for a class of discrete-time systems with missing measurements, nonlinear disturbances, and external stochastic noises. The measurement missing phenomenon is assumed to occur in a random way, and the missing probability is time-varying with securable upper and low bounds that can be measured in real time. The multiplicative noise is a state-dependent scalar Gaussian white noise sequence with known variance. The aim of the addressed gain-scheduled filtering problem is to design a filter, such that, for the admissible random measurement missing, nonlinear parameters and external noise disturbances, the error dynamics is exponentially mean-square stable. The desired filter is equipped with time-varying gains based primarily on the time-varying missing probability and is therefore less conservative than the traditional filter with fixed gains. It is shown that the filter parameters can be derived in terms of the measurable probability via the semidefinite program method.

The main contributions of this chapter are summarized as follows: *1) a new filtering problem is addressed for a class of discrete-time nonlinear stochastic systems with missing measurements via a gain-scheduling approach; 2) a sequence of stochastic variables satisfying Bernoulli distributions is exploited to reflect the time-varying features of the missing measurements in sensors; 3) a time-varying Lyapunov function dependent on the missing probability is proposed and then applied to improve the performance of the gain-scheduled filters; and 4) the filter parameters can be updated online according to the missing probabilities estimated through statistical tests.*

The rest of the chapter is organized as follows. Section 8.1 formulates the gain-scheduled filtering problem for a class of discrete-time systems with missing measurements, nonlinear disturbances, and external stochastic noise. The stability analysis and gain-scheduled filter design problems are dealt with in Section 8.2 by using the parameter-dependent Lyapunov function and the convex optimization method. In Section 8.3, a numerical example is presented to show the effectiveness of the proposed algorithm. Section 8.4 gives the summary.

8.1 Problem Formulation

Consider the following class of discrete-time nonlinear stochastic systems:

$$\begin{aligned} x(k+1) &= Ax(k) + Bf(z(k)) + Dx(k)\omega(k) && (8.1)\\ y_0(k) &= Cx(k), && (8.2) \end{aligned}$$

where $x(k) \in \mathbb{R}^n$ is the state, $y_0(k) \in \mathbb{R}^m$ is the ideal measurement output (without data missing), and $z(k) := Zx(k)$. $\omega(k)$ is a one-dimensional Gaussian white noise sequence satisfying $\mathbb{E}\{\omega(k)\} = 0$ and $\mathbb{E}\{\omega^2(k)\} = \sigma^2$. A, B, C, D, and Z are constant matrices with appropriate dimensions. $x(0) = \rho$ is the initial state, and the output matrix C is assumed to be of full row rank.

The nonlinear vector-valued function $f(\cdot)$ represents the nonlinear disturbance satisfying the following sector-bounded condition, with $f(0) = 0$:

$$[f(z(k)) - F_1 z(k)]^T [f(z(k)) - F_2 z(k)] \leq 0, \tag{8.3}$$

where F_1 and F_2 are constant real matrices of appropriate dimensions, and $F = F_2 - F_1$ is a symmetric positive definite matrix. It is customary that such nonlinear function $f(\cdot)$ belongs to the sector $[F_1,\ F_2]$ [83]. In this case, the nonlinear function $f(z(k))$ can be decomposed into a linear part and a nonlinear part as

$$f(z(k)) = F_1 z(k) + f_s(z(k)), \tag{8.4}$$

and it follows from (8.3) that

$$f_s^T(z(k))(f_s(z(k)) - Fz(k)) \leq 0. \tag{8.5}$$

The measurement output with sensor data missing is described by

$$y(k) = \xi(k)y_0(k) = \xi(k)Cx(k), \tag{8.6}$$

where $\xi(k) \in \mathbb{R}$ is a random white sequence characterizing the probabilistic sensor data missing phenomenon, which obeys the following time-varying

Bernoulli distribution:

$$\begin{aligned}\text{Prob}\{\xi(k)=1\} &= \mathbb{E}\{\xi(k)\}=p(k),\\ \text{Prob}\{\xi(k)=0\} &= 1-\mathbb{E}\{\xi(k)\}=1-p(k),\end{aligned} \tag{8.7}$$

where $p(k)$ is a time-varying positive scalar sequence that belongs to $[p_1\ \ p_2] \subseteq [0\ \ 1]$, with the constant p_1 and p_2 being the lower and upper bounds of $p(k)$. In this chapter, we assume that $\xi(k)$, $\omega(k)$ and ρ are uncorrelated. Furthermore, the kind of measurements missing that obey the probability distribution law (8.7) is said to be admissible.

Remark 8.1 *In (8.6), a random white sequence satisfying the time-varying Bernoulli distribution is introduced to reflect the missing measurement phenomenon that has attracted considerable attention in the past few years; see, e.g., [186]. However, the missing probability in most relevant literature has always been assumed to be a constant. Such an assumption, unfortunately, tends to be conservative in handling time-varying missing measurements. In this chapter, the missing probability is allowed to be time-varying with known lower and upper bounds, which will then be used to schedule filter gains, thereby reducing the possible conservatism.*

In this chapter, we aim to construct the following probability-dependent gain-scheduled filter for (8.1) and (8.6):

$$x_f(k+1) \;=\; G(p(k))x_f(k)+H(p(k))y(k), \tag{8.8}$$

where $x_f(k) \in \mathbb{R}^n$ is the state estimate, and $p(k)$ is the time-varying scheduling parameter taking value in $[p_1\ \ p_2]$. $G(p(k))$ and $H(p(k))$ are the scheduled filter gains of the following structure:

$$G(p(k)) = G_0 + p(k)G_f, \quad H(p(k)) = H_0 + p(k)H_f, \tag{8.9}$$

where G_0, G_f, H_0, and H_f are the constant filter parameters to be designed, and $p(k)$ is the time-varying missing probability that can be estimated/measured via statistical tests in real time.

Remark 8.2 *Different from the conventional filters, the above gain-scheduled filter structure comprises two kinds of filter gains: the constant (fixed) parameters G_0, G_f, H_0, and H_f, and the time-varying parameter $p(k)$. Here, $p(k)$ takes value in the interval $[p_1\ \ p_2]$ and can be measured in real time. In certain applications, such as the reliability analysis for sensors, if the data missing probability for a particular sensor is greater than* 0.5 *through statistical tests, then such a sensor would be replaced or at least repaired. In other words, the upper bound for the data missing probability is* 0.5*. In this chapter, the interval constraint $[p_1\ \ p_2]$ is added to reflect such an engineering practice and also facilitate the later analysis. Obviously, with this type of gain-scheduled filters, the conservatism can be reduced, since more information about the missing* *measurement phenomenon is utilized. Note that this kind of gain-scheduling technique has been extensively applied to deal with robust control and filtering problems for uncertain systems with time-varying parameters; see, e.g., [61].*

Letting $\bar{x}(k) = [x^T(k)\ x_f^T(k)]^T$, the error dynamics of the filtering process is derived from (8.1), (8.6), and (8.8) as follows:

$$\begin{aligned}\bar{x}(k+1) =& \bar{A}(p(k))\bar{x}(k) + \bar{B}f(z(k)) + (\xi(k) - p(k))\bar{C}(p(k)) \\ & \times N\bar{x}(k) + \bar{D}N\bar{x}(k)\omega(k),\end{aligned} \tag{8.10}$$

where

$$\begin{aligned}\bar{A}(p(k)) &= \begin{bmatrix} A & 0 \\ p(k)H(p(k))C & G(p(k)) \end{bmatrix}, \bar{D} = \begin{bmatrix} D \\ 0 \end{bmatrix}, \\ \bar{C}(p(k)) &= \begin{bmatrix} 0 \\ H(p(k))C \end{bmatrix}, \bar{B} = \begin{bmatrix} B \\ 0 \end{bmatrix}, N = [I\ \ 0].\end{aligned} \tag{8.11}$$

Definition 10 The filtering error system (8.10) is said to be *exponentially mean-square stable* if, with $\omega(k) = 0$, there exist constants $\alpha > 0$ and $\tau \in (0, 1)$ such that

$$\mathbb{E}\{\|\bar{x}(k)\|^2\} \le \alpha\tau^k \mathbb{E}\{\|\bar{x}(0)\|^2\}, \quad k \in \mathbb{I}^+.$$

The purpose of this chapter is to design a desired filter of the form (8.8) for the discrete nonlinear stochastic system with time-varying parameters in (8.1) and (8.6), such that, for all admissible nonlinearities, missing measurements, and stochastic disturbances, the augmented system (8.10) is exponentially mean-square stable.

8.2 Main Results

8.2.1 Stability Analysis

In the following theorem, the parameter-dependent Lyapunov function and the convex optimization are used to deal with the stability analysis problem for the gain-scheduled filter design of the discrete-time stochastic nonlinear systems (8.1) and (8.6) with missing measurements.

Theorem 17 *Consider the augmented filtering error system (8.10) with given filter gains. If there exist positive-definite matrix sequence $Q(p(k)) > 0$ and matrix S, such that the following matrix inequalities*

$$\begin{bmatrix} -Q(p(k)) & * & * & * & * \\ FZN & -2I & * & * & * \\ \Omega_1(k) & S^T\bar{B} & -\Lambda(k) & * & * \\ \Omega_2(k) & 0 & 0 & -\Theta(k)\Lambda(k) & * \\ \sigma^2 S^T\bar{D}N & 0 & 0 & 0 & -\sigma^2\Lambda(k) \end{bmatrix} < 0, \tag{8.12}$$

hold, where $\Lambda(k) = -Q(p(k+1)) + S + S^T$ *and*

$$\begin{aligned}\Theta(k) &= p(k)(1-p(k)), \Omega_2(k) = \Theta(k) S^T \bar{C}(p(k)) N \\ \Omega_1(k) &= S^T[\bar{A}(p(k)) + \bar{B} F_1 Z N],\end{aligned} \quad (8.13)$$

then (8.10) is exponentially mean-square stable.

Proof 20 *Define the Lyapunov function* $V(k) := \bar{x}^T(k) Q(p(k)) \bar{x}(k)$, *where* $Q(p(k))$ *is a time-varying positive definite matrix sequence dependent on the missing probability* $p(k)$. *By noting* $\mathbb{E}\{\xi(k) - p(k)\} = 0$ *and* $\mathbb{E}\{\omega(k)\} = 0$, *it can be obtained from (8.10) that*

$$\begin{aligned}&\mathbb{E}\{\Delta V(k)\} \\ =&\mathbb{E}\{[\bar{A}(p(k))\bar{x}(k) + \bar{B} f(z(k))]^T Q(p(k+1))[\bar{A}(p(k))\bar{x}(k) \\ &+\bar{B} f(z(k))] + p(k)(1-p(k))\bar{x}^T(k) N^T \bar{C}^T(p(k)) \\ &\times Q(p(k+1))\bar{C}(p(k)) N \bar{x}(k) + \sigma^2 \bar{x}^T(k) N^T \bar{D}^T \\ &\times Q(p(k+1))\bar{D} N \bar{x}(k) - \bar{x}^T(k) Q(p(k))\bar{x}(k)\}.\end{aligned} \quad (8.14)$$

From (8.3) and (8.5), we can obtain

$$\begin{aligned}&\mathbb{E}\{\Delta V(k)\} \\ \leq\ &\mathbb{E}\{[(\bar{A}(p(k)) + \bar{B} F_1 Z N)\bar{x}(k) + \bar{B} f_s(z(k))]^T \\ &\times Q(p(k+1))[(\bar{A}(p(k)) + \bar{B} F_1 Z N)\bar{x}(k) \\ &+\bar{B} f_s(z(k))] + p(k)(1-p(k))\bar{x}^T(k) N^T \bar{C}^T(p(k)) \\ &\times Q(p(k+1))\bar{C}(p(k)) N \bar{x}(k) + \sigma^2 \bar{x}^T(k) N^T \bar{D}^T \\ &\times Q(p(k+1))\bar{D} N \bar{x}(k) - \bar{x}^T(k) Q(p(k))\bar{x}(k) \\ &-2 f_s^T(z(k))[f_s(z(k)) + F Z N \bar{x}(k)]\}.\end{aligned} \quad (8.15)$$

From (8.14) and (8.15), it follows that

$$\mathbb{E}\{\Delta V(k)\} \leq \mathbb{E}\{\tilde{x}^T(k) \Pi \tilde{x}(k)\}, \quad (8.16)$$

where $\tilde{x}(k) = [\bar{x}^T(k) \ \ f_s^T(z(k))]^T$ *and*

$$\Pi = \begin{bmatrix} \Pi_1 & * \\ \Pi_2 & -2I + \bar{B}^T Q(p(k+1))\bar{B} \end{bmatrix}, \quad (8.17)$$

with

$$\begin{aligned}\Pi_1 =&(\bar{A}(p(k)) + \bar{B} F_1 Z N)^T Q(p(k+1))(\bar{A}(p(k)) + \bar{B} F_1 \\ &\times Z N) + \sigma^2 N^T \bar{D}^T Q(p(k+1))\bar{D} N + p(k)(1-p(k)) \\ &\times N^T \bar{C}^T(p(k)) Q(p(k+1))\bar{C}(p(k)) N - Q(p(k)) \\ \Pi_2 =&\bar{B}^T Q(p(k+1))(\bar{A}(p(k)) + \bar{B} F_1 Z N) + F Z N.\end{aligned} \quad (8.18)$$

In the following, we will conclude from (8.12) that $\Pi < 0$. *From the relation* $-Q(p(k+1)) + S + S^T > 0$ *in (8.12), we can see that* S *is nonsingular. Performing congruence transformation* $\text{diag}\{I, I, S^{-1}, \Theta^{-1}S^{-1}, \sigma^{-2}S^{-1}\}$ *to (8.12), we have*

$$\begin{bmatrix} -Q(p(k)) & * & * & * & * \\ FZN & -2I & * & * & * \\ \bar{\Omega}_1(k) & \bar{B} & -\bar{\Lambda}(k) & * & * \\ \bar{C}(p(k))N & 0 & 0 & -\bar{\Theta}\bar{\Lambda}(k) & * \\ \bar{D}N & 0 & 0 & 0 & -\bar{\sigma}\bar{\Lambda}(k) \end{bmatrix} < 0, \tag{8.19}$$

with $\bar{\Lambda}(k) = -S^{-T}Q(p(k+1))S^{-1} + S^{-1} + S^{-T}$, $\bar{\Omega}_1(k) = \bar{A}(p(k)) + \bar{B}F_1ZN$, $\bar{\Theta} = \Theta^{-1}(k)$ *and* $\bar{\sigma} = \sigma^{-2}$. *Then, it follows from inequality* $S^{-T}Q(p(k+1))S^{-1} - S^{-1} - S^{-T} \geq -Q^{-1}(p(k+1))$ *that*

$$\begin{bmatrix} -Q(p(k)) & * & * & * & * \\ FZN & -2I & * & * & * \\ \bar{\Omega}_1(k) & \bar{B} & -\tilde{\Lambda}(k) & * & * \\ \bar{C}(p(k))N & 0 & 0 & -\bar{\Theta}\tilde{\Lambda}(k) & * \\ \bar{D}N & 0 & 0 & 0 & -\bar{\sigma}\tilde{\Lambda}(k) \end{bmatrix} < 0, \tag{8.20}$$

with $\tilde{\Lambda}(k) = Q^{-1}(p(k+1))$. *To this end, by Schur complement lemma, we can see that* $\Pi < 0$. *Subsequently, we have*

$$\mathbb{E}\{\Delta V(k)\} < -\lambda_{\min}(-\Pi)\mathbb{E}|\bar{x}(k)|^2, \tag{8.21}$$

where $\lambda_{\min}(-\Pi)$ *is the minimum eigenvalue of* $-\Pi$ *and* $|\cdot|$ *is the usual vector norm. Finally, we can confirm from Lemma 1 of [186] that the augmented filtering system (8.10) is exponentially mean-square stable, and the proof of this theorem is thus completed.*

Remark 8.3 *In Theorem 17, to improve the performance of the filter to be designed, a time-varying Lyapunov function dependent on the missing probability has been proposed. Note that, in the past few years, parameter-dependent Lyapunov functions have been intensively employed for tackling uncertain systems and time-varying parameter systems aiming to reduce the conservatism; see, e.g., [42].*

Remark 8.4 *In the controller and filter design, the product terms between Lyapunov matrices and the system matrices usually have to be decoupled to bypass the difficulty encountered in the design. In this case, it is often an effective strategy to add slack variables; see, e.g., [42, 61]. Along this line, in Theorem 17, we have introduced a slack variable* S *to facilitate the resulting filter design problem.*

8.2.2 Gain-Scheduled Filter Design

The following theorem focuses on the design of gain-scheduled filter parameters $G(p(k))$ and $H(p(k))$ according to the results in Theorem 17.

Theorem 18 *Consider the discrete-time nonlinear stochastic system (8.1). Assume that there exist positive-definite matrix sequence $\bar{Q}(p(k)) > 0$, matrix sequences $\bar{H}(p(k))$ and $\bar{G}(p(k))$, nonsingular matrices S_{11}, R_2, and matrix R_1, such that the following parameter-dependent LMIs hold:*

$$\begin{bmatrix} -\bar{Q}(p(k)) & * & * & * & * \\ FZN & -2I & * & * & * \\ \Gamma_1(k) & \tilde{B} & \Gamma_2(k) & * & * \\ \Gamma_3(k) & 0 & 0 & \Theta(k)\Gamma_2(k) & * \\ \Gamma_4(k) & 0 & 0 & 0 & \sigma^2\Gamma_2(k) \end{bmatrix} < 0,$$

where

$$\begin{aligned} \Gamma_1(k) &= \begin{bmatrix} S_{11}^T A + p(k)\bar{H}(p(k))C + S_{11}^T BF_1 Z & \bar{G}(p(k)) \\ R_1^T A + p(k)\bar{H}(p(k))C + R_1^T BF_1 Z & \bar{G}(p(k)) \end{bmatrix}, \\ \Gamma_2(k) &= \bar{Q}(p(k+1)) - \Delta, \\ \Gamma_3(k) &= \Theta(k) \begin{bmatrix} \bar{H}(p(k))C \\ \bar{H}(p(k))C \end{bmatrix} N, \ \Gamma_4(k) = \sigma^2 \begin{bmatrix} S_{11}^T D \\ R_1^T D \end{bmatrix} N, \\ \Delta &= \begin{bmatrix} S_{11} + S_{11}^T & R_1 + R_2^T \\ R_2 + R_1^T & R_2 + R_2^T \end{bmatrix}, \ \tilde{B} = \begin{bmatrix} S_{11}^T B \\ R_1^T B \end{bmatrix}. \end{aligned} \tag{8.22}$$

In this case, there exist nonsingular matrices S_{21} and S_{22}, such that $R_2 = S_{21}^T S_{22}^{-T} S_{21}$, and then the gains of the desired filter can be obtained as follows:

$$G(p(k)) = S_{21}^{-T}\bar{G}(p(k))S_{21}^{-1}S_{22}, \ H(p(k)) = S_{21}^{-T}\bar{H}(p(k)).$$

Then, there exists a desired gain-scheduled filter in the form of (8.8), such that the filtering error dynamics (8.10) is exponentially mean-square stable.

Proof 21 *Let the nonsingular matrix variable S in (8.12) be partitioned as $S = [S_{ij}]_{2\times 2}$, where S_{11}, S_{21}, and S_{22} are nonsingular matrices. Introduce matrices*

$$\begin{aligned} \mathcal{T} = \begin{bmatrix} I & 0 \\ 0 & S_{22}^{-1}S_{21} \end{bmatrix}, \ \bar{Q}(p(k)) = \mathcal{T}^T Q(p(k))\mathcal{T}, \\ R_1 = S_{12}S_{22}^{-1}S_{21}, R_2 = S_{21}^T S_{22}^{-T} S_{21}. \end{aligned} \tag{8.23}$$

By congruence transformation with $\mathrm{diag}\{\mathcal{T}^{-1}, I, \mathcal{T}^{-1}, \mathcal{T}^{-1}, \mathcal{T}^{-1}\}$, *we can see that (8.22) is equivalent to (8.12), and it then follows from Theorem 17 that (8.10) is exponentially mean-square stable.*

8.2.3 Solvable Condition for Filter Design

Apparently, the number of LMIs in Theorem 18 is actually infinite due to the time-varying parameter $p(k)$ and, therefore, it is nearly impossible to solve the LMIs directly. In the following, we will convert the LMIs into finite ones.

Theorem 19 *For system (8.1), assume that there exist positive positive-definite matrices $\bar{Q}_0 > 0$ and $\bar{Q}_p > 0$, nonsingular matrices S_{11}, R_2, and matrices R_1, $\bar{G}_0$, $\bar{G}_f$, $\bar{H}_0$, and $\bar{H}_f$, such that the following LMIs hold:*

$$\mathbb{M}^{ijrl} = \begin{bmatrix} -\bar{Q}^i & * & * & * & * \\ FBN & -2I & * & * & * \\ \Gamma_1^{ij} & \tilde{B} & \Gamma_2^l & * & * \\ \Gamma_3^{ijr} & 0 & 0 & \bar{\Theta}^{jr}\Gamma_2^l & * \\ \Gamma_4 & 0 & 0 & 0 & \sigma^2\Gamma_2^l \end{bmatrix} < 0, \tag{8.24}$$

for $i, j, r, l = 1, 2$, where Γ_5 and $\tilde{B}$ have been defined in (8.22)

$$\begin{aligned} \Gamma_1^{ij} &= \begin{bmatrix} S_{11}^T A + p_j(\bar{H}_0 + p_i\bar{H}_f)C + S_{11}^T BF_1Z & \bar{G}^i \\ R_1^T A + p_j(\bar{H}_0 + p_i\bar{H}_f)C + R_1^T BF_1Z & \bar{G}^i \end{bmatrix}, \\ \Delta &= \begin{bmatrix} S_{11} + S_{11}^T & R_1 + R_2^T \\ R_2 + R_1^T & R_2 + R_2^T \end{bmatrix}, \Gamma_3^{ijr} = \bar{\Theta}^{jr} \begin{bmatrix} \bar{H}^i C \\ \bar{H}^i C \end{bmatrix} N, \\ \bar{Q}^i &= \bar{Q}_0 - p_i\bar{Q}_p, \Gamma_2^l = \bar{Q}^l - \Delta, \bar{\Theta}^{jr} = p_j(1-p_r), \\ \bar{H}^i &= \bar{H}_0 + p_i\bar{H}_f,\ \bar{G}^i = \bar{G}_0 + p_i\bar{G}_f. \end{aligned} \tag{8.25}$$

In this case, there exist nonsingular matrices S_{21} and S_{22}, such that $R_2 = S_{21}^T S_{22}^{-T} S_{21}$, and, therefore, the constant filter gains are obtained as follows:

$$\begin{aligned} G_0 &= S_{21}^{-T}\bar{G}_0 S_{21}^{-1} S_{22},\ G_f = S_{21}^{-T}\bar{G}_f S_{21}^{-1} S_{22}, \\ H_0 &= S_{21}^{-T}\bar{H}_0,\ H_f = S_{21}^{-T}\bar{H}_f. \end{aligned} \tag{8.26}$$

Then, a gain-scheduled filter can be obtained in the form of (8.8), such that the filtering error dynamics (8.10) is exponentially mean-square stable.

Proof 22 *First, choose the probability-dependent Lyapunov matrices as*

$$Q(p(k)) = Q_0 + p(k)Q_p \tag{8.27}$$

where $Q_0 > 0$ and $Q_p > 0$. It is easily seen that $\bar{Q}(p(k)) = \bar{Q}_0 + p(k)\bar{Q}_p$, with $\bar{Q}_0 = \mathcal{T}^T Q_0 \mathcal{T}$ and $\bar{Q}_p = \mathcal{T}^T Q_p \mathcal{T}$.

Setting

$$\alpha_1(k) = \frac{p_2 - p(k)}{p_2 - p_1}, \quad \alpha_2(k) = \frac{p(k) - p_1}{p_2 - p_1}, \tag{8.28}$$

we have

$$\begin{cases} \alpha_1(k) + \alpha_2(k) = 1, \ \alpha_i(k) \geq 0 \, (i = 1, 2) \\ p(k) = \alpha_1(k)p_1 + \alpha_2(k)p_2. \end{cases} \tag{8.29}$$

Similarly, letting

$$\beta_1(k) = \frac{p_2 - p(k+1)}{p_2 - p_1}, \quad \beta_2(k) = \frac{p(k+1) - p_1}{p_2 - p_1}, \tag{8.30}$$

we have

$$\begin{cases} \beta_1(k) + \beta_2(k) = 1, \ \beta_l(k) \geq 0 \, (l = 1, 2) \\ p(k+1) = \beta_1(k)p_1 + \beta_2(k)p_2. \end{cases} \tag{8.31}$$

From the above transformations, it is easily derived that

$$\begin{aligned} \bar{Q}(p(k)) &= \sum_{i=1}^{2} \alpha_i(k)\bar{Q}^i, \quad \bar{Q}(p(k+1)) = \sum_{l=1}^{2} \beta_l(k)\bar{Q}^l, \\ \bar{G}(p(k)) &= \sum_{i=1}^{2} \alpha_i(k)\bar{G}^i, \quad \bar{H}(p(k)) = \sum_{i=1}^{2} \alpha_i(k)\bar{H}^i. \end{aligned} \tag{8.32}$$

Furthermore, it follows from (8.24) that

$$\sum_{i,j,r,l=1}^{2} \alpha_i(k)\alpha_j(k)\alpha_r(k)\beta_l(k)\mathbb{M}^{ijrl} < 0. \tag{8.33}$$

Also, it follows from (8.29) and (8.31)–(8.33) that (8.22) holds. The proof is now completed.

In Theorem 19, we convert infinite LMIs in Theorem 18 to finite ones by turning the time-varying parameter $p(k)$ into the polytopic form. By such a transformation, the constant gains of the desired gain-scheduled filter can be easily derived in terms of the available LMI Toolbox by using the computationally appealing gain-scheduled filter design algorithm listed as follows.

Algorithm 8.1 *The gain-scheduled filter design algorithm.*

Step 1*: Given the initial values for the positive integer N, the initial state ρ, the constants p_1 and p_2, the matrices A, B, C, D, F_1, F_2, and Z, choose appropriate initial state estimate ρ_f and set $k = 0$.*

Step 2*: Solve the LMI in (8.24) to obtain the positive-definite matrices $\bar{Q}_0$ and $\bar{Q}_p$, matrices R_2, $\bar{G}_0$, $\bar{G}_f$, $\bar{H}_0$, and $\bar{H}_f$. Choose appropriate nonsingular matrices S_{21} and S_{22} to derive the constant filter parameters G_0, G_f, H_0, and H_f by (8.26).*

Step 3*: Based on the measured time-varying parameter $p(k)$, compute the filter gains $G(p(k))$ and $H(p(k))$ by (8.9) and the state estimate $x_f(k+1)$ by (8.8). Then, set $k = k + 1$.*

Step 4*: If $k < N$, go to Step* 3*; otherwise, go to Step* 5.

Step 5*: Stop.*

Remark 8.5 *Our main results are based on the LMI conditions. While the interior-point LMI solvers are significantly faster than classical convex optimization algorithms, it should be kept in mind that the complexity of LMI computations remains higher than that of solving, say, a Riccati equation. For instance, problems with a thousand design variables typically take more than an hour on today's workstations. However, research on LMI optimization is a very active area in the applied math, optimization, and the operations research community, and substantial speedups can be expected in the future.*

8.3 An Illustrative Example

The system parameters of (8.1) and (8.6) are given as follows:

$$A=\begin{bmatrix}0.43 & 0\\ 0.15 & 0.36\end{bmatrix}, B=\begin{bmatrix}0.1 & 0.04\\ 0 & 0.08\end{bmatrix}, p_1=0.4, p_2=0.8,$$
$$D=\begin{bmatrix}0.3 & 0.03\\ 0.05 & 0.38\end{bmatrix}, Z=\begin{bmatrix}0.51 & 0\\ 0 & 0.621\end{bmatrix}, C=[0.38\ 0.46],$$
$$F_1=\mathrm{diag}\{0.46\ 0.37\}, F_2=\mathrm{diag}\{2.81\ 2.95\}, \sigma^2=1.$$

Assume that the measurable missing probability sequence satisfies $p(k) = p_1 + (p_2 - p_1)|\sin(k)|$. According to Theorem 19 and Algorithm 8.1, the constant filter parameters G_0, G_f, H_0, and H_f can be obtained as follows:

$$G_0=\begin{bmatrix}0.0242 & -0.0007\\ 0.0086 & 0.0140\end{bmatrix},\ H_0=\begin{bmatrix}-1.3778\\ -1.5091\end{bmatrix},$$
$$G_f=\begin{bmatrix}0.0030 & 0.0084\\ 0.0070 & 0.0059\end{bmatrix},\ H_f=\begin{bmatrix}0.9806\\ 1.0835\end{bmatrix}.$$

With the available missing probability $p(k)$, the gain-scheduled filter gains $G(p(k))$ and $H(p(k))$ and the state estimate $x_f(k)$ can be obtained. Figure 8.1 and Figure 8.2 show the estimate errors $e_1(k) = x_1(k) - x_{f1}(k)$ and $e_2(k) = x_2(k) - x_{f2}(k)$, respectively. The simulation results have illustrated our theoretical analysis.

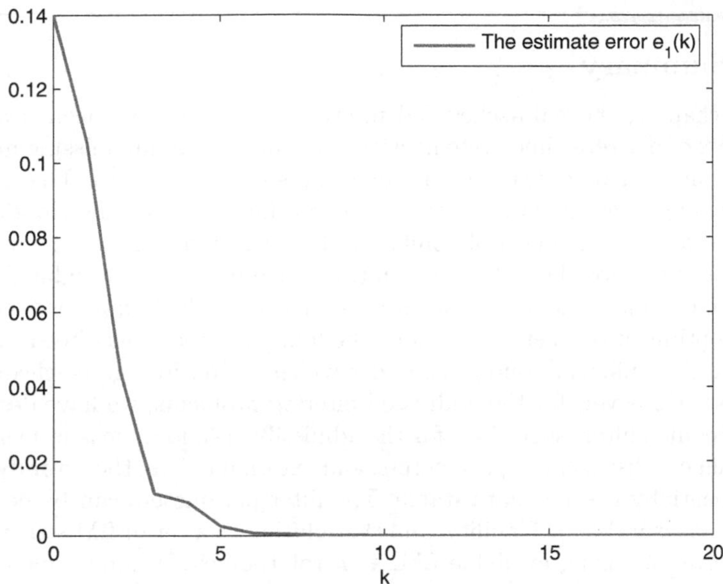

FIGURE 8.1
Estimate error $e_1(k)$.

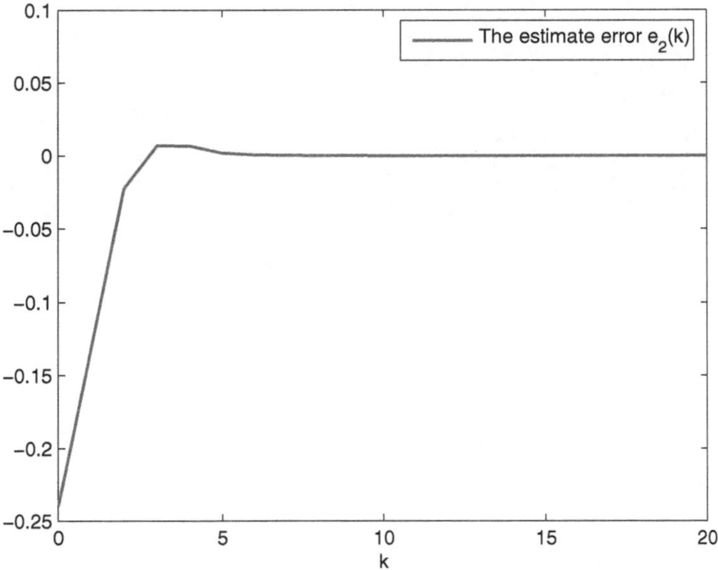

FIGURE 8.2
Estimate error $e_2(k)$.

8.4 Summary

In this chapter, the gain-scheduled filtering problems have been dealt with for a class of discrete-time systems with randomly occurring missing measurements, nonlinear disturbances, and external stochastic noises. The measurement missing is assumed to obey a time-varying Bernoulli distribution, and the time-varying missing probability can be measured in real time with constant upper and low bounds. The multiplicative noises are a scalar Gaussian white noise sequence with known variances. By using the Lyapunov theory and convex optimization method, a gain-scheduling approach has been proposed to derive the sufficient conditions under which the desired gain-scheduled filters exist. Moreover, for the addressed filtering problems, we have designed a gain-scheduled filter, such that, for the admissible random measurement missing, nonlinear disturbance parameters, and external noises, the error dynamics is exponentially mean-square stable. The filter parameters can be derived in terms of measurable probability, and the solution to a set of LMIs that can be easily solved by using available LMI Control Toolbox. A simulation example is exploited to illustrate the effectiveness of the proposed design procedures.

9

Controller Design for 2-D Stochastic Nonlinear Roesser Model

CONTENTS

This chapter is concerned with the static output feedback control problem for two-dimensional (2-D) uncertain stochastic nonlinear systems. The systems under consideration are subjected to time-delays, multiplicative noises, and randomly occurring missing measurements. A random variable sequence following the Bernoulli distribution with time-varying probability is employed to characterize the missing measurements that are assumed to occur in a random way. The gain-scheduled method based on the time-varying probability parameter is proposed to accomplish the design task. By constructing a suitable Lyapunov functional, sufficient conditions to guarantee the systems to be mean-square asymptotically stable are established. The addressed 2-D controller design problem can be reduced to a convex optimization problem by some mathematical techniques. In the last section, a numerical example and the comparative analysis are provided to illustrate the efficiency of our proposed design approach.

9.1 Problem Formulation and Preliminaries

Consider the following 2-D discrete-time nonlinear Roesser model with uncertain parameters and stochastic noises:

$$\begin{aligned}\begin{bmatrix} x^h(k+1,l) \\ x^v(k,l+1) \end{bmatrix} &= (A + H_1F(k,l)E + \eta(k,l)A_s)\begin{bmatrix} x^h(k,l) \\ x^v(k,l) \end{bmatrix} \\ &\quad + Bu(k,l) + Nf\left(\begin{bmatrix} x^h(k,l) \\ x^v(k,l) \end{bmatrix}\right)\end{aligned}$$

$$
\begin{aligned}
& +(A_d + H_{1d}F(k,l)E + \eta_d(k,l)A_{sd}) \\
& \times \begin{bmatrix} x^h(k-d(k,l),l) \\ x^v(k,l-d(k,l)) \end{bmatrix}, \\
y_0(k,l) \quad = \quad & (C + H_2F(k,l)E + \zeta(k,l)C_s) \begin{bmatrix} x^h(k,l) \\ x^v(k,l) \end{bmatrix},
\end{aligned} \tag{9.1}
$$

and the initial boundary conditions are given as

$$
\begin{aligned}
x^{h0}(k,l) &= \begin{cases} h(k,l); & if \ k \in [-d_2, 0], l \in [0, z_1], \\ 0; & if \ k \in [-d_2, 0], l \in (z_1, \infty), \end{cases} \\
x^{v0}(k,l) &= \begin{cases} v(k,l); & if \ k \in [0, z_2], \quad l \in [-d_2, 0], \\ 0; & if \ k \in (z_2, \infty), \ l \in [-d_2, 0], \end{cases}
\end{aligned} \tag{9.2}
$$

where $x^h(k,l) \in \mathbb{R}^{n_1}$ and $x^v(k,l) \in \mathbb{R}^{n_2}$, with $\mathfrak{N} = n_1 + n_2$ are horizontal and vertical state, respectively, $y_0(k,l) \in \mathbb{R}^{\mathfrak{M}}$ is ideal measurement output (without missing measurements) vector, and k, l are nonnegative integers. $A, A_d, A_s, A_{sd}, C, C_s, E, H_1, H_{1d}, H_2$, and N are known real constant matrices with compatible dimensions, and $B \in \mathbb{R}^{\mathfrak{N}\times\mathfrak{N}}$ is of full-column rank. $d(k,l)$ denotes time-delays, which is a sequence of scalar values satisfying $0 \leq d_1 \leq d(k,l) \leq d_2$. $f(\cdot) \in \mathbb{R}^{\mathfrak{N}}$ is nonlinear disturbance. $\eta(k,l) \in \mathbb{R}$, $\eta_d(k,l) \in \mathbb{R}$ and $\zeta(k,l) \in \mathbb{R}$ are multiplicative noises. $z_1 < \infty$ and $z_2 < \infty$ are positive integers, and $h(k,l) \in \mathbb{R}^{n_1}$ and $v(k,l) \in \mathbb{R}^{n_2}$ are given vectors. $F(k,l) \in \mathbb{R}^{\mathfrak{N}\times\mathfrak{N}}$ is an unknown real-valued time-varying matrix, representing the norm-bounded stochastic uncertainty, i.e., $F(k,l)F^T(k,l) \leq I, \forall\, k,l$.

Remark 9.1 *Multiplicative noise is very common in practical project, which is actually a kind of stochastic uncertainty and has been extensively studied in 1-D systems; see, e.g., [212]. In this model, the stochastic uncertainties not only exist in the process, but also in the measurement, which will certainly describe the 2-D stochastic systems more accurately. It's worth mentioning that, since both of the deterministic parametric uncertainty and the stochastic parametric uncertainty are considered, the design process of desired controller will become more complicated. And especially when the two kinds of parametric uncertainties exist in the time-delay 2-D systems, the design task may be more challenging.*

The actual output with missing measurements is described by

$$
y(k,l) = \xi(k,l)y_0(k,l), \tag{9.3}
$$

where $\xi(k,l) \in \mathbb{R}$ is a random white sequence characterizing the phenomenon of probabilistic missing measurements, which obeys the following time-varying Bernoulli distribution,

$$
\begin{aligned}
&\text{Prob}\{\xi(k,l) = 0\} = \mathbb{E}\{\xi(k,l)\} = p(k,l) \\
&\text{Prob}\{\xi(k,l) = 1\} = 1 - \mathbb{E}\{\xi(k,l)\} = 1 - p(k,l),
\end{aligned} \tag{9.4}
$$

where $p(k,l)$ is a time-varying positive scalar sequence that belongs to $[p_1\ p_2] \subseteq [0\ 1]$ with constants p_1 and p_2 being the lower and upper bounds of $p(k,l)$, respectively. Generally speaking, it is said to be admissible in the case of the missing measurements obeying the probability distribution law (9.4).

Remark 9.2 *The intermittent phenomenon of missing measurements has drawn considerable research interests, and the stochastic property of which can be described by a Bernoulli distribution model. Numerous associated results in 1-D systems have been published; see, e.g., [30, 131, 174], and in 2-D systems, there are some initial research works; see, e.g., [92]. Unfortunately, the Bernoulli model they have adopted is a constant, but in reality, statistical property of the missing measurements may change all the time, and a time-varying Bernoulli model will reflect the real situation more closely. According to this reason, we utilize a stochastic variable sequence obeying the time-varying Bernoulli distributions in (9.4) to describe the randomly occurring missing measurements.*

The white Gaussian noise sequences $\eta(k,l)$, $\eta_d(k,l)$, and $\zeta(k,l)$ have the following statistical properties

$$\begin{aligned} &\mathbb{E}\{\eta(k,l)\} = 0, \quad \mathbb{E}\{\eta_d(k,l)\} = 0, \quad \mathbb{E}\{\zeta(k,l)\} = 0, \\ &\mathbb{E}\left\{ \begin{bmatrix} \eta(k,l) \\ \eta_d(k,l) \\ \zeta(k,l) \end{bmatrix} \begin{bmatrix} \eta(i,j) \\ \eta_d(k,l) \\ \zeta(i,j) \end{bmatrix}^T \right\} \\ &= \begin{bmatrix} \sigma_\eta^2 \delta_{(k,l),(i,j)} & 0 & 0 \\ 0 & \sigma_{\eta_d}^2 \delta_{(k,l),(i,j)} & 0 \\ 0 & 0 & \sigma_\zeta^2 \delta_{(k,l),(i,j)} \end{bmatrix}, \end{aligned} \tag{9.5}$$

where $\sigma_\eta, \sigma_{\eta_d}, \sigma_\zeta$ are the given real constants.

Definition 11 *[65] A nonlinearity $\Gamma(\cdot)$ is said to satisfy the sector-bounded condition if*

$$[\Gamma(x(k,l)) - \Xi_1 x(k,l)]^T [\Gamma(x(k,l)) - \Xi_2 x(k,l)] \le 0, \tag{9.6}$$

where Ξ_1 and Ξ_2 are constant real matrices of appropriate dimensions with $\Xi = \Xi_2 - \Xi_1 > 0$. In this case, we say that $\Gamma(\cdot)$ belongs to the sector $[\Xi_1, \Xi_2]$.

Assumption 9.1 *All the noise signals $\eta(k,l)$, $\eta_d(k,l)$, and $\zeta(k,l)$ are uncorrelated to each other. The random variables $\xi(k,l)$ are independent of all the noise signals.*

Assumption 9.2 *[217] The boundary condition is independent of $\eta(k,l)$, $\eta_d(k,l)$, $\zeta(k,l)$, and $\xi(i,j)$ with the following properties*

$$\lim_{N\to\infty} \mathbb{E}\left\{ \sum_{l=-d_2}^{0} \sum_{k=0}^{N} \left(\left\| x^h(l,k) \right\|^2 + \left\| x^v(k,l) \right\|^2 \right) \right\} < \infty.$$

Assumption 9.3 *The nonlinear disturbance function $f(\cdot)$ in (9.1) belongs to sector $[F_1, F_2]$ with $f(0) = 0$, $\forall x \in R^N$, and $F = F_2 - F_1$ is a positive-definite matrix of appropriate dimension. At the same time, $f(\cdot)$ can be divided into linear and nonlinear parts as*

$$f\big(x(k,l)\big) = F_1 x(k,l) + f_s\big(x(k,l)\big). \tag{9.7}$$

Remark 9.3 *Since the nonlinearity appears in many practical engineering systems, we introduce the sector nonlinearity to describe this phenomena, which is a more useful way than the usual Lipschitz condition. Due to its convenience and validity, the sector nonlinearity has been widely used in neural networks, switching bio-process, and gene regulatory networks. Moreover, this kind of nonlinearity has extensive application in the field of control engineering, and attracts much attention from researchers. A rich body of research results has been reported in the literature; see, e.g., [33, 96].*

In this chapter, the following probability-dependent gain-scheduled controller is adopted for the 2-D systems (9.1),

$$u(k,l) = K\big(p(k,l)\big)y(k,l), \tag{9.8}$$

where $K\big(p(k,l)\big)$ is the controller gain sequence to be designed, whose structure is constructed as the following form,

$$K\big(p(k,l)\big) = K_a + p(k,l)K_b, \tag{9.9}$$

where K_a and K_b are the constant controller parameters to be designed.

Remark 9.4 *Compared with traditional controller, which possesses only fixed gain, the gains of the proposed 2-D controller exhibit flexibility, which are actually a set of sequence rather than a constant. From the structure of controller* (9.9)*, we know the gains consist of two parts, one is constant, and the other is the time-variant probability of the randomly occurring missing measurements, which can be measured via statistical tests in real time. The parameters K_a and K_b will be easily obtained by our design method. It's worth pointing out that the holistic alteration of the gain is achieved through the statistical probability $p(k,l)$, but not the fixed parameters K_a and K_b. Thanks to the time-varying characteristic of the gains, the 2-D controller for the uncertain stochastic Roesser model can be scheduled online to resist the adverse effect of missing measurements.*

Let $x^o(k,l) = \begin{bmatrix} x^h(k+1,l) \\ x^v(k,l+1) \end{bmatrix}$, $x(k,l) = \begin{bmatrix} x^h(k,l) \\ x^v(k,l) \end{bmatrix}$, and $x_d(k,l) = \begin{bmatrix} x^h\big(k-d(k,l),l\big) \\ x^v\big(k,l-d(k,l)\big) \end{bmatrix}$, then we have the following closed-loop control system with the static output feedback 2-D controller (9.8),

$$\begin{aligned} x^o(k,l) \;=\; & \big(A + H_1F(k,l)E + \eta(k,l)A_s\big)x(k,l) + Nf\big(x(k,l)\big) \\ & +\xi(k,l)BK\big(p(k,l)\big)\big(C + H_2F(k,l)E + \zeta(k,l)C_s\big)x(k,l) \\ & +\big(A_d + H_{1d}F(k,l)E + \eta_d(k,l)A_{sd}\big)x_d(k,l). \end{aligned} \tag{9.10}$$

Lemma 9.1 *[141] Let $\mathcal{U}, \mathcal{V}$, and $\mathcal{W}$ be real matrices of appropriate dimensions, with $\mathcal{V}$ satisfying $\mathcal{V}^T\mathcal{V} \leqslant I$. Then for any positive scalar ε,*

$$\mathcal{U}\mathcal{V}\mathcal{W} + \mathcal{W}^T\mathcal{V}^T\mathcal{U}^T < \varepsilon\mathcal{U}\mathcal{U}^T + \varepsilon^{-1}\mathcal{W}^T\mathcal{W}. \tag{9.11}$$

Lemma 9.2 *[214] For the matrix B of full-column rank, there always exist two orthogonal matrices $U = [U_1^T, U_2^T]^T \in \mathbb{R}^{\mathfrak{N}\times\mathfrak{N}}$, and $V \in \mathbb{R}^{\mathfrak{M}\times\mathfrak{M}}$, such that*

$$UBV = \begin{bmatrix} U_1 \\ U_2 \end{bmatrix} BV = \begin{bmatrix} \Sigma \\ 0 \end{bmatrix}, \tag{9.12}$$

where $\Sigma = \mathrm{diag}\{\sigma_1, \sigma_2, \cdots, \sigma_{\mathrm{m}}\}$ and σ_i are nonzero singular values of B. If the matrix Λ has the following structure

$$\Lambda = U^T \begin{bmatrix} \lambda_1 & 0 \\ 0 & \lambda_2 \end{bmatrix} U, \tag{9.13}$$

where $\lambda_1 \in \mathbb{R}^{\mathfrak{M}\times\mathfrak{M}}$ and $\lambda_2 \in \mathbb{R}^{(\mathfrak{N}-\mathfrak{M})\times(\mathfrak{N}-\mathfrak{M})}$ are positive-definite matrices, then there exists a nonsingular matrix $L \in \mathbb{R}^{\mathfrak{M}\times\mathfrak{M}}$, such that $BL = \Lambda B$, and then we can easily obtain $L^{-1} = V\Sigma^{-1}\lambda_1^{-1}\Sigma V^T$.

Definition 12 *The 2-D closed-loop systems (9.10) are said to be mean-square asymptotically stable, if*

$$\lim_{k+l\to\infty} \mathbb{E}\left\{\left\|x(k,l)\right\|^2\right\} = 0, \quad k, l \in \mathbb{I}^+ \tag{9.14}$$

holds for all the the initial condition (9.2) under the assumption 9.2.

9.2 Main Results

In the following, we will study the robust stability of the 2-D closed-loop control system (9.10). By constructing a suitable Lyapunov functional, sufficient conditions to guarantee the discussed systems asymptotically stable in the mean-square sense can be derived, where the desired controller is constrained. With the help of Schur complement, the design task of controller for the discussed 2-D systems can be transformed into a convex optimization problem. Gains of the proposed 2-D controller can be obtained in the case of solving a set of LMIs.

Theorem 20 *Consider the closed-loop control system (9.10) with the given gain structure of controller. If there exist positive-definite matrices P and Q,* *matrices K_a and K_b, and two positive scalars ε_1 and ε_2, such that the following matrix inequalities hold*

$$\begin{bmatrix} \Psi & * & * & * & * & * & * & * & * & * \\ 0 & \bar{Q} & * & * & * & * & * & * & * & * \\ F & 0 & -2I & * & * & * & * & * & * & * \\ \overline{\Psi}_2 & 0 & 0 & -\Psi_3 & * & * & * & * & * & * \\ A_s & 0 & 0 & 0 & -\Psi_4 & * & * & * & * & * \\ \Psi_5 & 0 & 0 & 0 & 0 & -\Psi_6 & * & * & * & * \\ 0 & A_{sd} & 0 & 0 & 0 & 0 & -\Psi_7 & * & * & * \\ \overline{\Psi}_8 & A_d & N & 0 & 0 & 0 & 0 & -P^{-1} & * & * \\ 0 & 0 & 0 & \Psi_{10} & 0 & 0 & 0 & \Psi_{11} & -\varepsilon_1^{-1}I & * \\ 0 & 0 & 0 & 0 & 0 & 0 & 0 & H_{1d}^T & 0 & -\varepsilon_2^{-1}I \end{bmatrix} < 0, \tag{9.15}$$

where

$$\begin{aligned} &P = \mathrm{diag}\{P^h, P^v\},\ Q = \mathrm{diag}\{Q^h, Q^v\}, \\ &\Psi = -\Psi_1 + \varepsilon_1^{-1}E^TE, \\ &\bar{Q} = -Q + \varepsilon_2^{-1}E^TE, \\ &\Psi_1 = P - (d_2 - d_1 + 1)Q, \\ &\overline{\Psi}_2 = BK\big(p(k,l)\big)C, \\ &\Psi_3 = \Big(p(k,l)\big(1 - p(k,l)\big)\Big)^{-1}P^{-1}, \\ &\Psi_4 = \sigma_\eta^{-2}P^{-1}, \\ &\Psi_5 = BK\big(p(k,l)\big)C_s, \\ &\Psi_6 = p^{-1}(k,l)\sigma_\zeta^{-2}P^{-1}, \\ &\Psi_7 = \sigma_{\eta_d}^{-2}P^{-1}, \\ &\overline{\Psi}_8 = A + NF_1 + p(k,l)BK\big(p(k,l)\big)C, \\ &\Psi_{10} = \Big(BK\big(p(k,l)\big)H_2\Big)^T, \\ &\Psi_{11} = H_1^T + p(k,l)\Psi_{10}, \end{aligned} \tag{9.16}$$

then the closed-loop system (9.10) is mean-square asymptotically stable for all $p(k,l) \in [p_1\, p_2]$.

Proof 23 *Construct the Lyapunov functional as*

$$V(k,l) = V^h(k,l) + V^v(k,l), \tag{9.17}$$

where

$$\begin{aligned} V^h(k,l) &= V_1^h(k,l) + V_2^h(k,l) + V_3^h(k,l), \\ V^v(k,l) &= V_1^v(k,l) + V_2^v(k,l) + V_3^v(k,l), \end{aligned} \tag{9.18}$$

and

$$
\begin{aligned}
V_1^h(k,l) &= x^{h^T}(k,l)P^h x^h(k,l),\\
V_1^v(k,l) &= x^{vT}(k,l)P^v x^v(k,l),\\
V_2^h(k,l) &= \sum_{r=-d(k,l)}^{-1} x^{hT}(k+r,l)Q^h x^h(k+r,l),\\
V_2^v(k,l) &= \sum_{t=-d(k,l)}^{-1} x^{vT}(k,l+t)Q^v x^v(k,l+t),\\
V_3^h(k,l) &= \sum_{m=-d_2+2}^{-d_1+1}\sum_{r=m-1}^{-1} x^{h^T}(k+r,l)Q^h x^h(k+r,l),\\
V_3^v(k,l) &= \sum_{m=-d_2+2}^{-d_1+1}\sum_{t=m-1}^{-1} x^{vT}(k,l+t)Q^v x^v(k,l+t).
\end{aligned}
\tag{9.19}
$$

Noting $0 \le d_1 \le d(k,l) \le d_2$, from (9.17) to (9.19), it is not difficult to find the following inequalities,

$$
\begin{aligned}
&\mathbb{E}\left\{V^o(k,l)\Big|x(k,l)\right\} - V(k,l)\\
\le\ &\mathbb{E}\Big\{x^{oT}(k,l)Px^o(k,l) - x^T(k,l)Px(k,l) - x_d^T(k,l)Qx_d(k,l)\\
&+(d_2-d_1+1)x^T(k,l)Qx(k,l)\Big|x(k,l)\Big\},
\end{aligned}
\tag{9.20}
$$

where $V^o(k,l) = V^h(k+1,l) + V^v(k,l+1)$.

Substituting (9.10) into (9.20), one can easily obtain

$$
\begin{aligned}
&\mathbb{E}\left\{V^o(k,l)\Big|x(k,l)\right\} - V(k,l)\\
\le\ &\mathbb{E}\Big\{\Big[(A+H_1F(k,l)E)x(k,l) + Nf\big(x(k,l)\big)\\
&+p(k,l)BK\big(p(k,l)\big)\big(C+H_2F(k,l)E\big)x(k,l)\\
&+\Big(\xi(k,l)-p(k,l)\Big)BK\big(p(k,l)\big)\big(C+H_2F(k,l)E\big)x(k,l)\\
&+\big(A_d+H_{1d}F(k,l)E\big)x_d(k,l)\\
&+\Big(\eta(k,l)A_s+\zeta(k,l)p(k,l)BK\big(p(k,l)\big)C_s\Big)x(k,l)\\
&+\zeta(k,l)\big(\xi(k,l)-p(k,l)\big)BK\big(p(k,l)\big)C_s x(k,l) + \eta_d(k,l)A_{sd}x_d(k,l)\Big]^T P\\
&\times\Big[(A+H_1F(k,l)E)x(k,l) + Nf\big(x(k,l)\big)\\
&+p(k,l)BK\big(p(k,l)\big)\big(C+H_2F(k,l)E\big)x(k,l)\\
&+\Big(\xi(k,l)-p(k,l)\Big)BK\big(p(k,l)\big)\big(C+H_2F(k,l)E\big)x(k,l)\\
&+\big(A_d+H_{1d}F(k,l)E\big)x_d(k,l)\\
&+\Big(\eta(k,l)A_s+\zeta(k,l)p(k,l)BK\big(p(k,l)\big)C_s\Big)x(k,l)
\end{aligned}
$$

$$
\begin{aligned}
& +\zeta(k,l)\big(\xi(k,l)-p(k,l)\big)BK\big(p(k,l)\big)C_s x(k,l)+\eta_d(k,l)A_{sd}x_d(k,l)\Big] \\
& -x^T(k,l)Px(k,l)-x_d^T(k,l)Qx_d(k,l) \\
& +(d_2-d_1+1)x^T(k,l)Qx(k,l)\Big|x(k,l)\Big\} \\
= \;& \mathbb{E}\Big\{\Big[\Big(A+H_1F(k,l)E+p(k,l)BK\big(p(k,l)\big)\big(C+H_2F(k,l)E\big)\Big)x(k,l) \\
& +\big(A_d+H_{1d}F(k,l)E\big)x_d(k,l)+Nf\big(x(k,l)\big)\Big]^T P \\
& \times\Big[\Big(A+H_1F(k,l)E+p(k,l)BK\big(p(k,l)\big)\big(C+H_2F(k,l)E\big)\Big)x(k,l) \\
& +\big(A_d+H_{1d}F(k,l)E\big)x_d(k,l)+Nf\big(x(k,l)\big)\Big] \\
& -x^T(k,l)Px(k,l)-x_d^T(k,l)Qx_d(k,l) \\
& +(d_2-d_1+1)x^T(k,l)Qx(k,l)+p(k,l)\big(1-p(k,l)\big) \\
& \times\big[BK\big(p(k,l)\big)\big(C+H_2F(k,l)E\big)x(k,l)\big]^T P \\
& \times BK\big(p(k,l)\big)\big(C+H_2F(k,l)E\big)x(k,l) \\
& +\sigma_\eta^2 x^T(k,l)A_s^TPA_sx(k,l)+\sigma_\zeta^2 p(k,l)x^T(k,l)C_s^T \\
& \cdot K^T\big(p(k,l)\big)B^TPBK\big(p(k,l)\big)C_sx(k,l) \\
& +\sigma_{\eta_d}^2 x_d^T(k,l)A_{sd}^TPA_{sd}x_d(k,l)\Big|x(k,l)\Big\}. \qquad (9.21)
\end{aligned}
$$

From (9.6) and (9.7), it follows that

$$
\mathbb{E}\left\{V^o(k,l)\Big|x(k,l)\right\}-V(k,l) \;\leq\; \mathbb{E}\left\{\bar{x}^T(k,l)\Omega\bar{x}(l,k)\right\}, \qquad (9.22)
$$

where $\bar{x}(k,l)=[x^T(k,l)\;\; x_d^T(k,l)\;\; f_s^T\big(x(k,l)\big)]^T$ *and*

$$
\Omega \;=\; \begin{bmatrix} \Omega_1 & * & * \\ \Omega_2 & \Omega_3 & * \\ \Omega_4 & \Omega_5 & \Omega_6 \end{bmatrix}, \qquad (9.23)
$$

with

$$
\begin{aligned}
\Omega_1 \;=\; & \Big(A+H_1F(k,l)E+NF_1+p(k,l)BK\big(p(k,l)\big)\big(C+H_2F(k,l)E\big)\Big)^T P \\
& \times\Big(A+H_1F(k,l)E+NF_1+p(k,l)BK\big(p(k,l)\big)\big(C+H_2F(k,l)E\big)\Big) \\
& -P+(d_2-d_1+1)Q \\
& +p(k,l)\big(1-p(k,l)\big)\big[BK\big(p(k,l)\big)\big(C+H_2F(k,l)E\big)\big]^T \\
& \times PBK\big(p(k,l)\big)\big(C+H_2F(k,l)E\big) \\
& +\sigma_\eta^2A_s^TPA_s+\sigma_\zeta^2p(k,l)C_s^TK^T\big(p(k,l)\big)B^T \\
& \times PBK\big(p(k,l)\big)C_s,
\end{aligned}
$$

$$\begin{aligned}
\Omega_2 &= \left(A_d + H_{1d}F(k,l)E\right)^T P\Big(A + H_1F(k,l)E + NF_1 + p(k,l) \\
&\quad \times BK\big(p(k,l)\big)\big(C + H_2F(k,l)E\big)\Big), \\
\Omega_3 &= \left(A_d + H_{1d}F(k,l)E\right)^T P\big(A_d + H_{1d}F(k,l)E\big) - Q + \sigma_{\eta_d}^2 A_{sd}^T P A_{sd}, \\
\Omega_4 &= N^T P\Big(A + H_1F(k,l)E + NF_1 + p(k,l)BK\big(p(k,l)\big) \\
&\quad \times \big(C + H_2F(k,l)E\big)\Big) + F, \\
\Omega_5 &= N^T P\big(A_d + H_{1d}F(k,l)E\big), \\
\Omega_6 &= -2I. \qquad (9.24)
\end{aligned}$$

By the Schur complement in [13], it is easy to find that $\Omega \leq 0$ can be inferred from the following matrix inequalities,

$$\begin{bmatrix}
-\Psi_1 & * & * & * & * & * & * & * \\
0 & -Q & * & * & * & * & * & * \\
F & 0 & -2I & * & * & * & * & * \\
\Psi_2 & 0 & 0 & -\Psi_3 & * & * & * & * \\
A_s & 0 & 0 & 0 & -\Psi_4 & * & * & * \\
\Psi_5 & 0 & 0 & 0 & 0 & -\Psi_6 & * & * \\
0 & A_{sd} & 0 & 0 & 0 & 0 & -\Psi_7 & * \\
\Psi_8 & A_d + H_{1d}F(k,l)E & N & 0 & 0 & 0 & 0 & -P^{-1}
\end{bmatrix} < 0, \qquad (9.25)$$

where,

$$\begin{aligned}
&\Psi_2 = BK\big(p(k,l)\big)\big(C + H_2F(k,l)E\big), \\
&\Psi_8 = A + H_1F(k,l)E + NF_1 + p(k,l)BK\big(p(k,l)\big) \\
&\quad \times \big(C + H_2F(k,l)E\big). \qquad (9.26)
\end{aligned}$$

In order to use the Lemma 9.1, we rewrite (9.25) as follows

$$\mathscr{X} + \mathscr{U}_1\mathscr{V}\mathscr{W}_1 + \mathscr{W}_1^T\mathscr{V}^T\mathscr{U}_1^T + \mathscr{U}_2\mathscr{V}\mathscr{W}_2 + \mathscr{W}_2^T\mathscr{V}^T\mathscr{U}_2^T < 0, \qquad (9.27)$$

where

$$\mathscr{X} = \begin{bmatrix}
-\Psi_1 & * & * & * & * & * & * & * \\
0 & -Q & * & * & * & * & * & * \\
F & 0 & -2I & * & * & * & * & * \\
\overline{\Psi}_2 & 0 & 0 & -\Psi_3 & * & * & * & * \\
A_s & 0 & 0 & 0 & -\Psi_4 & * & * & * \\
\Psi_5 & 0 & 0 & 0 & 0 & -\Psi_6 & * & * \\
0 & A_{sd} & 0 & 0 & 0 & 0 & -\Psi_7 & * \\
\overline{\Psi}_8 & A_d & N & 0 & 0 & 0 & 0 & -P^{-1}
\end{bmatrix},$$

$$\mathscr{U}_1 = \begin{bmatrix}\mathscr{U}_{11} & \mathscr{U}_{10} & \mathscr{U}_{10} & \mathscr{U}_{10} & \mathscr{U}_{10} & \mathscr{U}_{10} & \mathscr{U}_{10} & \mathscr{U}_{10}\end{bmatrix},$$

$$\mathscr{U}_{10} = \begin{bmatrix}0 & 0 & 0 & 0 & 0 & 0 & 0 & 0\end{bmatrix}^T,$$

$$\begin{aligned}
\mathscr{U}_2 &= \begin{bmatrix} \mathscr{U}_{10} & \mathscr{U}_{21} & \mathscr{U}_{10} & \mathscr{U}_{10} & \mathscr{U}_{10} & \mathscr{U}_{10} & \mathscr{U}_{10} & \mathscr{U}_{10} \end{bmatrix}, \\
\mathscr{U}_{21} &= \begin{bmatrix} 0 & 0 & 0 & 0 & 0 & 0 & 0 & H_{1d}^T \end{bmatrix}^T, \\
\mathscr{U}_{11} &= \begin{bmatrix} 0 & 0 & 0 & \mathscr{U}_{12} & 0 & 0 & 0 & \mathscr{U}_{12} \end{bmatrix}^T, \\
\mathscr{V} &= \mathrm{diag}_8 \left\{ F(k,l) \right\}, \\
\mathscr{W}_1 &= \mathrm{diag} \left\{ E, 0, 0, 0, 0, 0, 0, 0 \right\}, \\
\mathscr{W}_2 &= \mathrm{diag} \left\{ 0, E, 0, 0, 0, 0, 0, 0 \right\}, \\
\mathscr{U}_{12} &= \Big(H_1 + p(k,l) B K \big(p(k,l) \big) H_2 \Big)^T. \qquad (9.28)
\end{aligned}$$

According to Lemma 9.1, the inequality (9.27) holds, if and only if there exist two positive scalars ε_1 *and* ε_2*, such that*

$$\mathscr{X} + \varepsilon_1 \mathscr{U}_1 \mathscr{U}_1^T + \varepsilon_1^{-1} \mathscr{W}_1^T \mathscr{W}_1 + \varepsilon_2 \mathscr{U}_2 \mathscr{U}_2^T + \varepsilon_2^{-1} \mathscr{W}_2^T \mathscr{W}_2 < 0. \qquad (9.29)$$

Making use of the Schur complement in [13] again, (9.29) is equivalent to (9.15), which implies $\Omega < 0$*. Subsequently,*

$$\begin{aligned}
& \mathbb{E} \left\{ V^o(k,l) \middle| x(k,l) \right\} - V(k,l) \leq \mathbb{E} \left\{ \bar{x}^T(k,l) \Omega \bar{x}(l,k) \right\} \\
\leq \; & -\lambda_{\min}(-\Omega) \, \mathbb{E} \left\{ \left\| x(k,l) \right\|^2 \right\}, \qquad (9.30)
\end{aligned}$$

where $\lambda_{\min}(-\Omega)$ *is the minimum eigenvalue of* $(-\Omega)$*.*

For any integer $N > 1$*, summing up the right and left sides of (9.30) from 0 to N, with respect to k and l, results in*

$$\begin{aligned}
& \mathbb{E} \left\{ \sum_{k=0}^{N} \sum_{l=0}^{N} \left\{ V^h(k+1,l) - V^h(k,l) + V^v(k,l+1) - V^v(k,l) \right\} \right\}, \\
= \; & \mathbb{E} \left\{ V^h(N+1,N) + \cdots + V^h(N+1,0) \right\} \\
& + \mathbb{E} \Big\{ V^v(N,N+1) + \cdots + V^v(0,N+1) \Big\} \\
& - \mathbb{E} \left\{ V^h(0,N) + \cdots + V^h(0,0) \right\} - \mathbb{E} \Big\{ V^v(N,0) + \cdots + V^v(0,0) \Big\}, \\
\leq \; & -\lambda_{\min}(-\Omega) \, \mathbb{E} \left\{ \sum_{k=0}^{N} \sum_{l=0}^{N} \left\| x(k,l) \right\|^2 \right\}. \qquad (9.31)
\end{aligned}$$

Since the Lyapunov functional $V(k,l) \geq 0$*, then the following inequality is true from the assumption 9.2:*

$$\begin{aligned}
& \sum_{k=0}^{N} \sum_{l=0}^{N} \mathbb{E} \left\{ \left\| x(k,l) \right\|^2 \right\} \\
\leq \; & \frac{1}{\lambda_{\min}(-\Omega)} \sum_{l=0}^{N} \mathbb{E} \left\{ V^h(0,l) + V^v(l,0) \right\}
\end{aligned}$$

$$
\begin{aligned}
\leq\ & \frac{1}{\lambda_{\min}(-\Omega)} \max\left\{\lambda_{\max}(P),\ \lambda_{\max}(Q)\right\} \\
& \times\mathbb{E}\left\{ \left\|x^h(0,N)\right\|^2 + \sum_{r=d(0,N)}^{-1} \left\|x^h(0+r,N)\right\|^2 \right. \\
& + \sum_{m=-d_2+2}^{-d_1+1} \sum_{r=m-1}^{-1} \left\|x^h(0+r,N)\right\|^2 + \\
& \qquad \left\|x^v(N,0)\right\|^2 + \sum_{t=d(0,N)}^{-1} \left\|x^v(N,0+t)\right\|^2 \\
& + \sum_{m=-d_2+2}^{-d_1+1} \sum_{t=m-1}^{-1} \left\|x^v(N,0+t)\right\|^2 + \\
& \qquad \vdots \\
& \qquad \left\|x^h(0,0)\right\|^2 + \sum_{r=d(0,0)}^{-1} \left\|x^h(0+r,0)\right\|^2 \\
& + \sum_{m=-d_2+2}^{-d_1+1} \sum_{r=m-1}^{-1} \left\|x^h(0+r,0)\right\|^2 + \\
& \qquad \left\|x^v(0,0)\right\|^2 + \sum_{t=d(0,0)}^{-1} \left\|x^v(0,0+t)\right\|^2 \\
& \left. + \sum_{m=-d_2+2}^{-d_1+1} \sum_{t=m-1}^{-1} \left\|x^v(0,0+t)\right\|^2 \right\} \\
<\ & \frac{(d_2-d_1+2)}{\lambda_{\min}(-\Omega)} \max\left\{\lambda_{\max}(P),\ \lambda_{\max}(Q)\right\} \\
& \times\mathbb{E}\left\{ \sum_{l=-d_2}^{0} \sum_{k=0}^{N} \left(\left\|x^h(l,k)\right\|^2 + \left\|x^v(k,l)\right\|^2 \right) \right\} < \infty. \qquad (9.32)
\end{aligned}
$$

It can be inferred from (9.32) that

$$
\lim_{k+l\to\infty} \mathbb{E}\left\{\left\|x(k,l)\right\|^2\right\} = 0. \qquad (9.33)
$$

By the definition 12, it is easy to draw the conclusion that the closed-loop system (9.10) is asymptotically stable in mean-square sense.

In Theorem 20, a sufficient condition to guarantee the stabilization of the considered closed-loop systems is provided. In the following section, we will discuss how to solve the matrix equalities constrained in (9.15) with the aid of existing LMI technique.

Theorem 21 *For the closed-loop system (9.10), with given controller (9.8) and (9.9), if there exist positive-definite matrices ϕ_1, ϕ_2, P, and Q, matrices $\mathfrak{K}_a$ and $\mathfrak{K}_b$, and two positive scalars ε_1 and ε_2, such that*

$$
\begin{bmatrix}
-\ominus_1 & * & * & * & * & * & * & * & * & \\
0 & -Q+\varepsilon_2^{-1}E^TE & * & * & * & * & * & * & * & * \\
F & 0 & -2I & * & * & * & * & * & * & * \\
\ominus_2 & 0 & 0 & -\ominus_3 & * & * & * & * & * & * \\
\Phi A_s & 0 & 0 & 0 & -\ominus_4 & * & * & * & * & * \\
\ominus_5 & 0 & 0 & 0 & 0 & -\ominus_6 & * & * & * & * \\
0 & \Phi A_{sd} & 0 & 0 & 0 & 0 & -\ominus_7 & * & * & * \\
\ominus_8 & \Phi A_d & \Phi N & 0 & 0 & 0 & 0 & -\ominus_9 & * & * \\
0 & 0 & 0 & \ominus_{10} & 0 & 0 & 0 & \ominus_{11} & -\varepsilon_1^{-1}I & * \\
0 & 0 & 0 & 0 & 0 & 0 & 0 & H_{1d}^T\Phi^T & 0 & -\varepsilon_2^{-1}I
\end{bmatrix} < 0, \tag{9.34}
$$

hold, where

$$
\begin{aligned}
&\Phi = U_1^T\phi_1U_1 + U_2^T\phi_2U_2, \\
&\ominus_1 = \Psi_1 - \varepsilon_1^{-1}E^TE, \\
&\ominus_2 = B\mathfrak{K}\big(p(k,l)\big)C, \\
&\ominus_3 = \Big(p(k,l)\big(1-p(k,l)\big)\Big)^{-1}\ominus_9, \\
&\ominus_4 = \sigma_\eta^{-2}\ominus_9, \\
&\ominus_5 = B\mathfrak{K}\big(p(k,l)\big)C_s, \\
&\ominus_6 = \sigma_\zeta^{-2}p^{-1}(k,l)\ominus_9, \\
&\ominus_7 = \sigma_{\eta d}^{-2}\ominus_9, \\
&\ominus_8 = \Phi A + \Phi NF_1 + p(k,l)B\mathfrak{K}\big(p(k,l)\big)C, \\
&\ominus_9 = -P + \Phi + \Phi^T, \\
&\ominus_{10} = \big(B\mathfrak{K}\big(p(k,l)\big)H_2\big)^T, \\
&\ominus_{11} = H_1^T\Phi^T + p(k,l)\ominus_{10}, \\
&\mathfrak{K}\big(p(k,l)\big) = \mathfrak{K}_a + p(k,l)\mathfrak{K}_b,
\end{aligned} \tag{9.35}
$$

with U_1 and U_2 are obtained from (9.12), then the closed-loop systems (9.10) *is mean-square asymptotically stable for all $p(k,l) \in [p_1\, p_2]$. Moreover, the constant gains of the designed controller can be obtained as follows*

$$K_a = V\Sigma^{-1}\phi_1^{-1}\Sigma V^T \mathfrak{K}_a,$$
$$K_b = V\Sigma^{-1}\phi_1^{-1}\Sigma V^T \mathfrak{K}_b, \tag{9.36}$$

where V and Σ can be chosen out according to (9.12).

Proof 24 *Since P^{-1} is positive definite, and we have*

$$\begin{aligned} & P^{-1} + \Phi^{-1}P\Phi^{-T} - \Phi^{-T} - \Phi^{-1} \\ = & [\Phi^{-1} - P^{-1}]P\Phi^{-T} - [\Phi^{-1} - P^{-1}] \\ = & [\Phi^{-T} - P^{-1}]^T P[\Phi^{-T} - P^{-1}] \geq 0, \end{aligned} \tag{9.37}$$

therefore, $-P^{-1} \leq \Phi^{-1}P\Phi^{-T} - \Phi^{-T} - \Phi^{-1}$, it is easy to verify that (9.15) can be inferred from the following inequalities,

$$\begin{bmatrix} -\Psi_1 + \varepsilon_1^{-1}E^TE & * & * & * & * & * & * & * & * & * \\ 0 & -Q + \varepsilon_2^{-1}E^TE & * & * & * & * & * & * & * & * \\ F & 0 & -2I & * & * & * & * & * & * & * \\ \overline{\Psi}_2 & 0 & 0 & -\overline{\ominus}_3 & * & * & * & * & * & * \\ A_s & 0 & 0 & 0 & -\overline{\ominus}_4 & * & * & * & * & * \\ \Psi_5 & 0 & 0 & 0 & 0 & -\overline{\ominus}_6 & * & * & * & * \\ 0 & A_{sd} & 0 & 0 & 0 & 0 & -\overline{\ominus}_7 & * & * & * \\ \overline{\Psi}_8 & A_d & N & 0 & 0 & 0 & 0 & -\overline{\ominus}_9 & * & * \\ 0 & 0 & 0 & \Psi_{10} & 0 & 0 & 0 & \Psi_{11} & -\varepsilon_1^{-1}I & * \\ 0 & 0 & 0 & 0 & 0 & 0 & 0 & H_{1d}^T & 0 & -\varepsilon_2^{-1}I \end{bmatrix} < 0, \tag{9.38}$$

where

$$\begin{aligned} \overline{\ominus}_3 &= \Big(p(k,l)\big(1 - p(k,l)\big)\Big)^{-1}\overline{\ominus}_9, \\ \overline{\ominus}_4 &= \sigma_\eta^{-2}\overline{\ominus}_9, \\ \overline{\ominus}_6 &= p^{-1}(k,l)\sigma_\zeta^{-2}\overline{\ominus}_9, \\ \overline{\ominus}_7 &= \sigma_{\eta_d}^{-2}\overline{\ominus}_9, \\ \overline{\ominus}_9 &= -\Phi^{-1}P\Phi^{-T} + \Phi^{-T} + \Phi^{-1}. \end{aligned} \tag{9.39}$$

By performing the congruence transformation diag$\{I, I, I, \Phi^T, \Phi^T, \Phi^T, \Phi^T, \Phi^T, I, I\}$ *to (9.38), we can have*

$$\begin{bmatrix} -\Psi_1+\varepsilon_1^{-1}E^TE & * & * & * & * & * & * & * & * & * \\ 0 & -Q+\varepsilon_2^{-1}E^TE & * & * & * & * & * & * & * & * \\ F & 0 & -2I & * & * & * & * & * & * & * \\ \Phi\overline{\Psi}_2 & 0 & 0 & -\ominus_3 & * & * & * & * & * & * \\ \Phi A_s & 0 & 0 & 0 & -\ominus_4 & * & * & * & * & * \\ \Phi\Psi_5 & 0 & 0 & 0 & 0 & -\ominus_6 & * & * & * & * \\ 0 & \Phi A_{sd} & 0 & 0 & 0 & 0 & -\ominus_7 & * & * & * \\ \Phi\overline{\Psi}_8 & \Phi A_d & \Phi N & 0 & 0 & 0 & 0 & -\ominus_9 & * & * \\ 0 & 0 & 0 & \Psi_{10}\Phi^T & 0 & 0 & 0 & \Psi_{11}\Phi^T & -\varepsilon_1^{-1}I & * \\ 0 & 0 & 0 & 0 & 0 & 0 & 0 & H_{1d}^T\Phi^T & 0 & -\varepsilon_2^{-1}I \end{bmatrix} < 0. \tag{9.40}$$

Since $\Phi = U_1^T\phi_1U_1 + U_2^T\phi_2U_2$, *it is not difficult to find that* Φ *has the same structure as that of* Λ. *By Lemma 9.2, we can easily draw the conclusion that there exists a nonsingular matrix* J, *such that* $BJ = \Phi B$. *Hence, the linear matrix inequalities (9.34) can be obtained with* $JK\big(p(k,l)\big) = \mathfrak{K}\big(p(k,l)\big)$. *Subsequently, the following relationship can be established,*

$$K\big(p(k,l)\big) = J^{-1}\mathfrak{K}\big(p(k,l)\big) = V\Sigma^{-1}\phi_1^{-1}\Sigma V^T\mathfrak{K}\big(p(k,l)\big), \tag{9.41}$$

then we can figure out parameters of the probability-dependent gain-scheduled controller as (9.36). The proof of this theorem is now completed.

Remark 9.5 *With the help of MATLAB,* U_1, U_2, V, *and* Σ *in Theorem 21 can be obtained by the method of singular values decomposition. Although a sufficient condition based on a set of linear matrix inequalities has been derived out, the number of the matrix inequalities constrained in (9.34) is actually infinite, owing to the constantly changing probability* $p(k,l)$. *In the following theorem, we will manage to convert these matrix inequalities to a solvable form with only a limited number of LMIs.*

Theorem 22 *Consider the closed-loop system (9.10) with given controller gain structure (9.9). If there exist positive-definite matrices* ϕ_1, ϕ_2, P, *and* Q, *matrices $\mathfrak{K}_a$ and $\mathfrak{K}_b$, and two positive scalars ε_1 and ε_2, such that the following linear matrix inequalities*

$$
\prod^{q,r,s} = \begin{bmatrix}
-\Theta_1 & * & * & * & * & * & * & * & * & * \\
0 & -Q+\varepsilon_2^{-1}E^TE & * & * & * & * & * & * & * & * \\
F & 0 & -2I & * & * & * & * & * & * & * \\
\Theta_2^q & 0 & 0 & -\Theta_3^{r,s} & * & * & * & * & * & * \\
\Phi A_s & 0 & 0 & 0 & -\Theta_4 & * & * & * & * & * \\
\Theta_5^q & 0 & 0 & 0 & 0 & -\Theta_6^r & * & * & * & * \\
0 & \Phi A_{sd} & 0 & 0 & 0 & 0 & -\Theta_7 & * & * & * \\
\Theta_8^{q,r} & \Phi A_d & \Phi N & 0 & 0 & 0 & 0 & -\Theta_9 & * & * \\
0 & 0 & 0 & \Theta_{10}^q & 0 & 0 & 0 & \Theta_{11}^{q,r} & -\varepsilon_1^{-1}I & * \\
0 & 0 & 0 & 0 & 0 & 0 & 0 & H_{1d}^T\Phi^T & 0 & -\varepsilon_2^{-1}I
\end{bmatrix} < 0, \tag{9.42}
$$

hold for q, r, $s = 1, 2$, where

$$
\begin{aligned}
\mathfrak{K}^q &= \mathfrak{K}_a + p_q\mathfrak{K}_b, \\
\Theta_2^q &= B\mathfrak{K}^qC, \\
\Theta_3^{r,s} &= \big(p_r(1-p_s)\big)^{-1}\Theta_9, \\
\Theta_5^q &= B\mathfrak{K}^qC_s, \\
\Theta_6^r &= \sigma_\zeta^{-2}p_r^{-1}\Theta_9, \\
\Theta_8^{q,r} &= \Phi A + \Phi NF_1 + p_rB\mathfrak{K}^qC, \\
\Theta_{10}^q &= \big(B\mathfrak{K}^qH_2\big)^T, \\
\Theta_{11}^{q,r} &= \Big(\Phi H_1 + p_rB\mathfrak{K}^qH_2\Big)^T
\end{aligned} \tag{9.43}
$$

and $\mathfrak{K}_a$, $\mathfrak{K}_b$ have been defined in (9.36), then there exists a controller in the form of (9.9), such that the closed-loop system (9.10) is mean-square asymptotically stable.

Proof 25 *First, set*

$$
\varkappa_1(k,l) = \frac{p_2 - p(k,l)}{p_2 - p_1}, \quad \varkappa_2(k,l) = \frac{p(k,l) - p_1}{p_2 - p_1}. \tag{9.44}
$$

Therefore, we have

$$\begin{cases} p(k,l) = \varkappa_1(k,l)p_1 + \varkappa_2(k,l)p_2, \\ \varkappa_1(k,l) + \varkappa_2(k,l) = 1. \end{cases} \tag{9.45}$$

From the above relationship, we can easily derive

$$\mathfrak{K}\big(p(k,l)\big) = \sum_{q=1}^{2} \varkappa_q(k,l)(\mathfrak{K}_a + p_q\mathfrak{K}_b),$$

$$p(k,l)\big(1 - p(k,l)\big) = \sum_{r,s=1}^{2} \varkappa_r(k,l)\varkappa_s(k,l)p_r(1 - p_s). \tag{9.46}$$

It is easy to find from (9.42) that

$$\sum_{q,r,s=1}^{2} \varkappa_q(k,l)\varkappa_r(k,l)\varkappa_s(k,l)\prod^{q,r,s} < 0, \tag{9.47}$$

which is equivalent to (9.16).

Remark 9.6 *Up to this point, the design problem of the gain-scheduled controller has been transformed into a linear convex optimization problem, while a set of finite and solvable linear matrix inequalities has been presented out. With the help of LMI Toolbox of MATLAB, the desired controller parameters can be obtained via solving the above convex optimization problem in Theorem 22.*

Remark 9.7 *It should be noted that the standard LMI systems have a polynomial-time complexity. The number* $\mathscr{N}(\epsilon)$ *of flops needed to compute an* ϵ*-accurate solution is bounded by* $O(\mathcal{M}\mathcal{N}^3 log(\mathcal{V}/\epsilon))$*, where* $\mathcal{M}$ *is the total row size of the LMI system,* $\mathcal{N}$ *is the total number of scalar decision variables, and* $\mathcal{V}$ *is a data-dependent scaling factor. For the system (9.1), dimensions of the variables can be seen from* $x^h(k,l) \in \mathbb{R}^{n_1}$*,* $x^v(k,l) \in \mathbb{R}^{n_2}$ *(with* $\mathfrak{N} = n_1 + n_2$*),* $y_0(k,l) \in \mathbb{R}^{\mathfrak{M}}$*,* $\eta(k,l) \in \mathbb{R}$*,* $\eta_d(k,l) \in \mathbb{R}$*,* $\zeta(k,l) \in \mathbb{R}$*, and* $\xi(k,l) \in \mathbb{R}$*. In Theorem 22, there are 8 standard LMIs for* q*,* r*,* $s = 1, 2$*. For each LMI, one can verify* $\mathcal{M} = 10\mathfrak{N}$ *and* $\mathcal{N} = \frac{7}{2}\mathfrak{M}^2 + \frac{3}{2}\mathfrak{N}^2 + \frac{1}{2}\mathfrak{M} + \frac{3}{2}\mathfrak{N} - \mathfrak{M}\mathfrak{N} + 2$*. Subsequently, the computational complexity of the developed algorithms can be represented as* $O(\mathfrak{N}^3 + \mathfrak{M}^2\mathfrak{N} - \mathfrak{N}^2\mathfrak{M})$*, which indicates that the computational complexity of the algorithm based on the variable dimensions.*

9.3 An Illustrative Example

In this section, we shall give a numerical example to show the validity of the proposed design method of the discussed gain-scheduled controller for a nonlinear Roesser model. Consider the 2-D uncertain stochastic nonlinear systems (9.1), with parameters as

$$E = \begin{bmatrix} 0.460 & 0.014 & 0.120 & 0.017 \\ 0.210 & 0.160 & 0.012 & 0.015 \\ 0.193 & 0.054 & 0.009 & 0 \\ 0.780 & 0.401 & 0.101 & 0.050 \end{bmatrix},$$

$$A_s = \begin{bmatrix} -0.4400 & 0 & 0 & 0.0320 \\ -0.3112 & 0 & 0.2543 & 0 \\ -0.4400 & 0 & 0.1432 & 0 \\ -0.0345 & 0 & 0 & -0.0453 \end{bmatrix},$$

$$H_2 = \begin{bmatrix} 0.03 & 0.034 & -0.052 & 0.030 \\ -0.017 & -0.043 & 0 & 0 \\ 0 & -0.021 & 0.044 & 0.025 \\ 0.085 & -0.056 & 0 & -0.034 \end{bmatrix},$$

$$C_s = \begin{bmatrix} 0.031 & -0.01 & 0 & -0.079 \\ 0 & -0.053 & 0.011 & -0.05 \\ 0 & 0 & 0.040 & 0 \\ 0.28 & 0 & -0.034 & -0.018 \end{bmatrix},$$

$$A = \begin{bmatrix} -0.6010 & 0.4030 & 0.1880 & -0.1030 \\ 0.1090 & -0.1540 & 0.1090 & 0.5450 \\ -0.3010 & 0.4120 & 0.5330 & 0.1010 \\ 0.5010 & -0.3340 & -0.2140 & 0.0020 \end{bmatrix},$$

$$H_1 = \begin{bmatrix} 0.1224 & 0.3860 & 0.2245 & 0 \\ 0.2363 & 0.1563 & 0 & 0.6520 \\ 0.4861 & 0.1130 & -0.1724 & 0.4260 \\ 0.2154 & 0.1213 & 0.0522 & 0.2931 \end{bmatrix},$$

$$B = \begin{bmatrix} 0.0011 & -0.0004 & -0.0030 & 0.0040 \\ 0.0030 & 0 & -0.0080 & 0.0691; \\ 0.0020 & -0.0014 & 0 & -0.0023 \\ 0.0019 & -0.0016 & 0.0040 & -0.0002 \end{bmatrix},$$

$$N = \begin{bmatrix} -0.2020 & -0.0310 & 0 & -0.1100 \\ -0.1270 & 0.0300 & 0.0420 & 0 \\ 0.0418 & 0.0180 & 0 & 0 \\ 0.0417 & -0.0030 & -0.0350 & -0.0460 \end{bmatrix},$$

$$A_{sd} = \begin{bmatrix} -0.3170 & 0.1711 & -0.0530 & -0.0410 \\ 0.4000 & -0.2880 & 0.0710 & -0.0114 \\ -0.1711 & -0.0600 & 0.0453 & 0 \\ -0.3000 & 0.0571 & -0.0780 & 0.0452 \end{bmatrix},$$

$$C = \begin{bmatrix} 2.230 & -2.41 & -3.640 & 1.340 \\ -2.210 & 0.550 & 0 & -1.880 \\ 2.304 & -1.10 & -1.060 & 0 \\ 1.120 & 1.001 & 0.540 & -0.660 \end{bmatrix},$$

$$A_d = \begin{bmatrix} -0.0838 & 0.0944 & -0.1221 & 0.1120 \\ -0.1215 & -0.0910 & 0 & 0 \\ -0.0600 & 0.0460 & 0.0400 & -0.0280 \\ -0.0190 & 0.0080 & 0.0120 & 0.0211 \end{bmatrix},$$

$$H_{1d} = \begin{bmatrix} -0.4210 & 0.0140 & 0.1500 & 0.4230 \\ -0.2000 & 0 & -0.3100 & 0 \\ 0 & -0.3010 & 0.2020 & 0 \\ -0.2410 & 0 & 0 & 0.1530 \end{bmatrix},$$

$$F_1 = \begin{bmatrix} -0.3260 & 0.0560 & 0.1445 & 0 \\ 0.0320 & -0.2545 & 0 & 0.0713 \\ -0.1580 & 0 & -0.2587 & 0 \\ -0.3356 & 0.0751 & 0.1380 & -0.1345 \end{bmatrix},$$

$$F_2 = \begin{bmatrix} 0.2327 & 0.0560 & 0.1445 & 0 \\ 0.0320 & -0.0712 & 0 & 0.0713 \\ -0.1580 & 0 & -0.1587 & 0 \\ -0.3356 & 0.0751 & 0.1380 & -0.0270 \end{bmatrix},$$

and $p_1 = 0.788$, $p_2 = 0.947$, $d_1 = 1$, $d_2 = 6$, $\sigma_\eta = 0.01$, $\sigma_{\eta_d} = 0.02$, $\sigma_{\eta_\varsigma} = 0.03$. The time-varying probability sequences $p(k,l)$ are taken as $p_1+(p_2-p_1)\left|(\sin(\frac{k\pi}{60})\sin(\frac{l\pi}{60})\sin(200\eta)\right|$. Subsequently, the constant controller parameters K_a and K_b can be obtained as follows:

$$K_a = \begin{bmatrix} -4609.7 & -6127.8 & 1758.9 & -2866.5 \\ -6911.4 & -9187.8 & 2538.9 & -4226.6 \\ -538.4 & -736.7 & 75.2 & -288.6 \\ 138.0 & 183.1 & -67.0 & 92.9 \end{bmatrix},$$

$$K_b = \begin{bmatrix} -0.0002 & -0.0021 & -0.0016 & -0.0007 \\ -0.0003 & -0.0031 & -0.0024 & -0.0010 \\ -0.0000 & -0.0003 & -0.0002 & -0.0001 \\ 0.0000 & 0.0001 & 0.0000 & 0.0000 \end{bmatrix}.$$

Figure 9.1 is the time-varying missing probability $p(k,l)$. The state evolutions of the uncontrolled systems are shown in Figure 9.2(a), Figure 9.2(b), Figure 9.2(c) and Figure 9.2(d), namely, the horizontal state $x_1^h(k,l)$ and $x_2^h(k,l)$, and the horizontal ones $x_1^v(k,l)$ and $x_2^v(k,l)$, respectively. The state responses of the controlled systems are shown in Figure 9.3(a), Figure 9.3(b), Figure 9.3(c), and Figure 9.3(d). The simulation results reveal that the designed controller can very well guarantee the closed-loop system to be asymptotically stable in the mean-square sense.

In order to show the advantage of the discussed gain-scheduled approach, which depends on the time-varying probability $p(k,l)$, we will compare the results obtained from Theorem 22 with the results when $p(k,l)$ is time-invaryring. Let the probability of the missing measurements be $p = 0.93$. Subsequently, the state of the controlled system with the constant probability can be seen

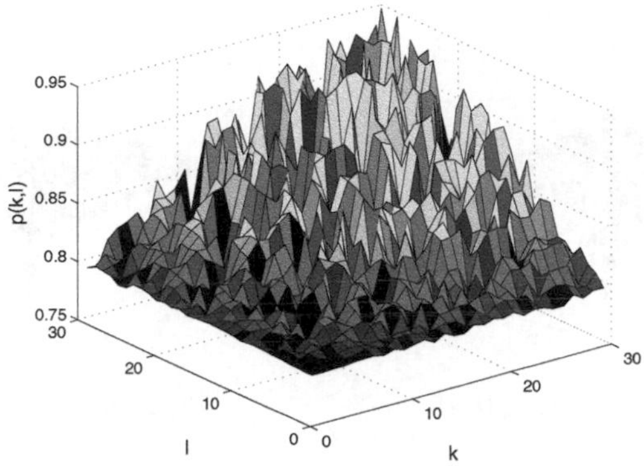

FIGURE 9.1
Time-varying probability $p(k, l)$.

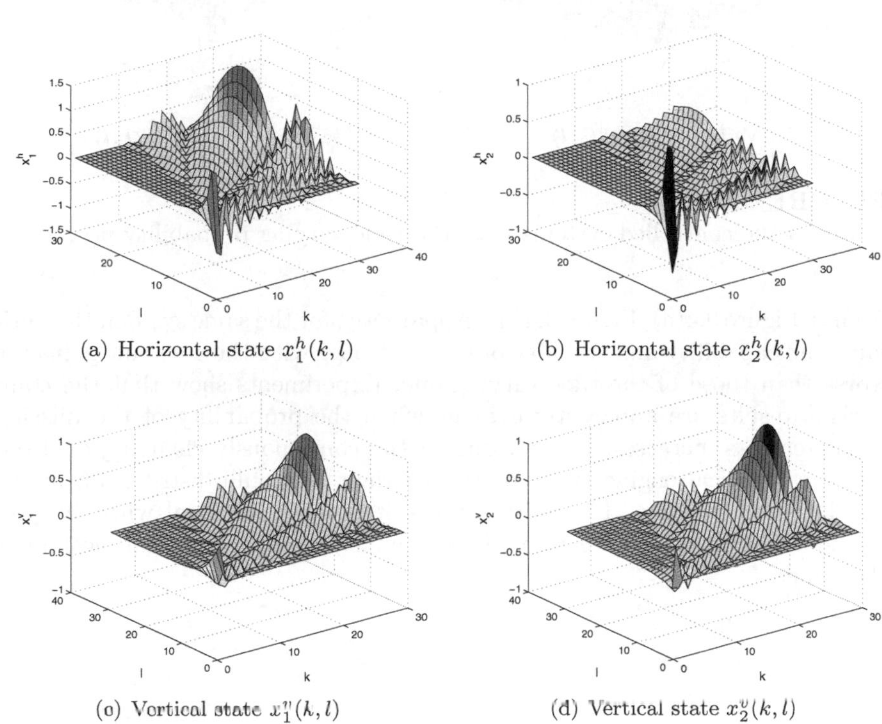

(a) Horizontal state $x_1^h(k, l)$

(b) Horizontal state $x_2^h(k, l)$

(c) Vertical state $x_1^v(k, l)$

(d) Vertical state $x_2^v(k, l)$

FIGURE 9.2
States of the of uncontrolled systems.

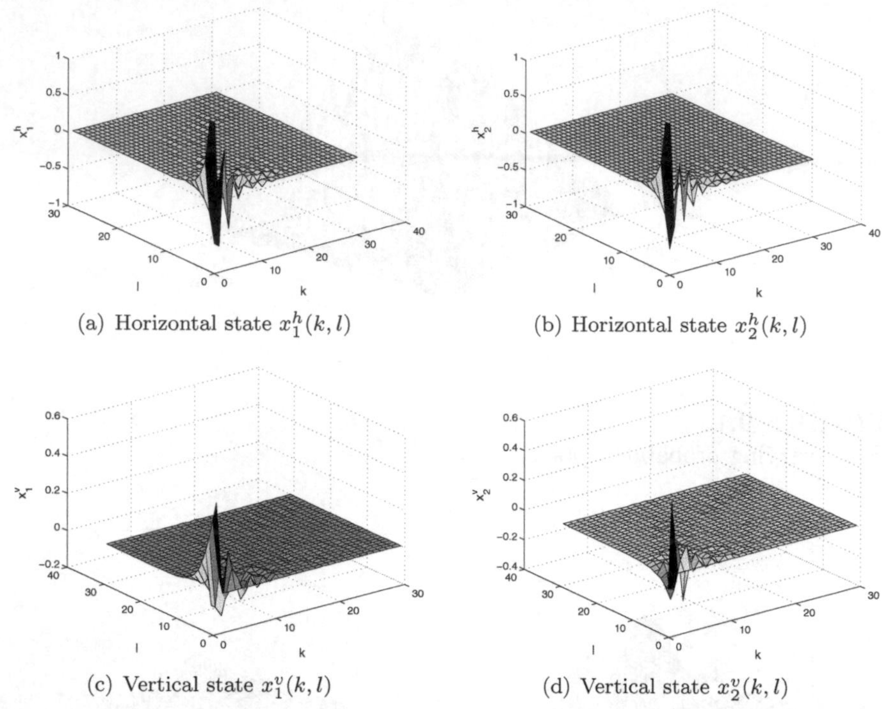

(a) Horizontal state $x_1^h(k,l)$ (b) Horizontal state $x_2^h(k,l)$

(c) Vertical state $x_1^v(k,l)$ (d) Vertical state $x_2^v(k,l)$

FIGURE 9.3
States of the controlled systems with the time-varying probability $p(k,l)$.

from in Figure 9.4(a)–Figure 9.4(d). Apparently, for the same system, the anti-jamming capability and stability of the controller with the constant gains are worse than those of the time-varying one. Experiments show that the comparison results are apparently evident when the probability of the missing measurements increases. In fact, due to the continuously changing environment in practical engineering, the time-variant Bernoulli distribution model can be more effective. The imitation result is fairly identical with the predicted situation, which identifies the rationality of this analytical method in the paper.

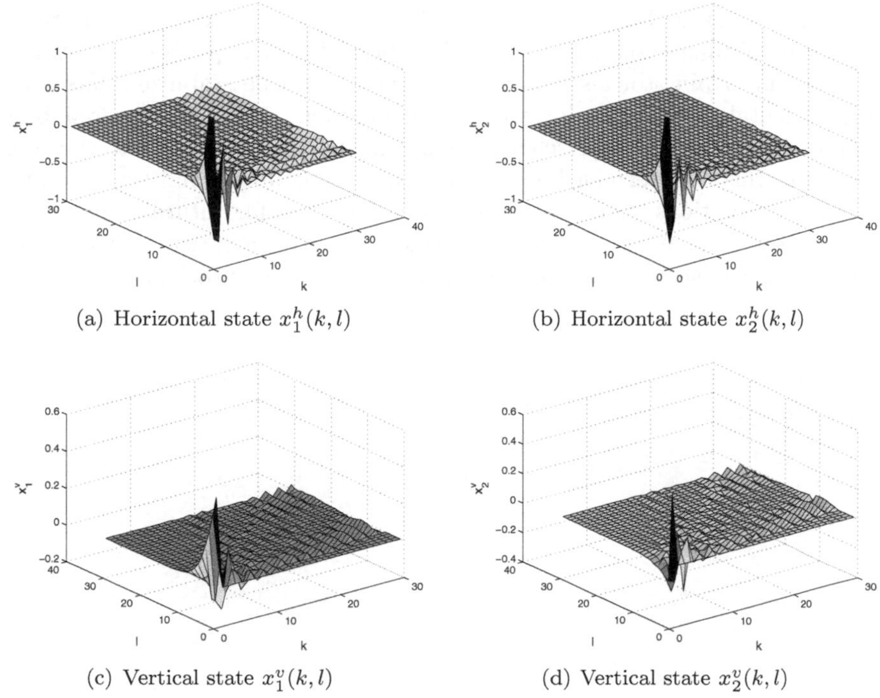

(a) Horizontal state $x_1^h(k,l)$

(b) Horizontal state $x_2^h(k,l)$

(c) Vertical state $x_1^v(k,l)$

(d) Vertical state $x_2^v(k,l)$

FIGURE 9.4
States of the controlled systems with the constant probability p.

9.4 Summary

In this chapter, we have developed a new technique for designing the gain-scheduled static output feedback 2-D controller. The systems governed by a class of 2-D Roesser model involve parameter uncertainty, nonlinearity, time-delays, and stochastic disturbance. Occurrence of the missing measurements has been assumed in a random style and is described by a stochastic variable sequence following the time-varying Bernoulli distribution. In view of the phenomenon of randomly occurring missing measurements, our purpose is to design an adjustable 2-D gain-scheduled controller based on the missing probability to stabilize the discussed 2-D nonlinear systems. In order to derive a sufficient condition for the mean-square asymptotical stability, we have constructed a Lyapunov functional relevant to time-delays. By employing some mathematical techniques, a finite set of linear matrix inequalities have been obtained, and therefore the problem is reduced to a linear convex optimization issue. The parameters of the proposed gain-scheduled controller have been obtained via LMI Control Toolbox. With the proposed controller, the 2-D uncertain stochastic nonlinear systems can be stabilized effectively. Experimental results coincide with our conclusion, and verify the reliability and validity of the design scheme. It should be pointed out that the main results in this chapter can be extended to more complex and realistic measurement situations. Our most promising direction for future research will concern the estimation problem for 2-D systems with a variety of incomplete measurements, such as sensor quantization, sensor saturation, and missing measurement, etc.

10

Filtering for Networked Control Systems with Multiple Delays

CONTENTS

In this chapter, the filtering problem is addressed for a class of discrete-time stochastic nonlinear networked systems with multiple random communication delays and random packet losses. The communication delay and packet loss problems, which are frequently encountered in communication networks with limited digital capacity, are modeled by a stochastic mechanism that combines a certain set of indicator functions dependent on the same stochastic variable. The nonlinear function satisfies the sector-like condition, and the multiplicative stochastic disturbance is in the form of scalar Gaussian white noise with constant variance. We aim to design a linear filter such that, for the admissible random incomplete measurement phenomenon, stochastic disturbances as well as sector nonlinearities, the filtering error dynamics is exponentially mean square stable. By resorting to the linear matrix inequality (LMI) approach and delay-dependent techniques, we first derive sufficient conditions dependent on the occurrence probabilities of both the random communication delay and missing measurement, which ensure the desired filtering performance for the networked system. Then, we characterize the filter gain in terms of the solution to a set of LMIs. A simulation example is exploited for a nonlinear stochastic networked control model in order to demonstrate the effectiveness of the proposed design procedures.

The rest of this chapter is set out as follows. Section 10.1 formulates the filtering problem for a class of discrete-time stochastic nonlinear networked systems with multiple random communication delays and random packet losses. In Section 10.2, filter analysis and synthesis problems are addressed. A simulation example is presented to illustrate the usefulness and flexibility of the filter design method developed in Section 10.3. Our summary are drawn in Section 10.4.

10.1 Problem Formulation

Consider the following discrete-time stochastic nonlinear networked control system with multiple delays in the state:

$$\begin{aligned} (\Sigma): x(k+1) &= Ax(k) + \sum_{i=1}^{q} B_i \phi_i(x(k-d_i)) \\ &\quad +Ex(k)\omega(k), \qquad (10.1) \\ x(k) &= \mu(k), \;\; k=-d_q, -d_q+1, ..., 0, \qquad (10.2) \end{aligned}$$

where $x(k) \in \mathbb{R}^n$ is the state, $d_i \in \mathbb{Z}^+$ $(i = 1, ..., q)$ are known constant time delays that are assumed to satisfy $d_1 < d_2 < ... < d_q$ for simplicity, $\omega(k)$ is a one-dimensional Gaussian white noise sequence satisfying $\mathbb{E}\{\omega(k)\} = 0$, and $\mathbb{E}\{\omega^2(k)\} = \delta^2$. $\phi_i(\cdot)$ is the sector nonlinearity and $\mu(k)$ is the initial state of the system.

The measurement with random communication delays and random data missing is described as

$$\begin{aligned} y(k) &= I_{\{\tau_k=0\}} Cx(k) + \sum_{i=1}^{q} I_{\{\tau_k=d_i\}} D_i \theta_i(x(k-d_i)) \\ &\quad +Fx(k))\omega(k), \qquad (10.3) \end{aligned}$$

where $y(k) \in \mathbb{R}^p$ is the measured output vector. For the indicator functions $I_{\{\tau_k=0\}}$ and $I_{\{\tau_k=d_i\}}$, we assume that $\mathbb{E}\{I_{\{\tau_k=0\}}\} = Pr\{\tau_k = 0\} = p_0$ and $\mathbb{E}\{I_{\{\tau_k=i\}}\} = Pr\{\tau_k = d_i\} = p_i$, where p_i $(i = 0, 1, 2, ..., q)$ are known positive scalars and $\sum_{j=0}^{q} p_j \leq 1$. τ_k is a mutually independent stochastic variable sequence used to describe, at time k, the occurred delays and data missing of the measured output information. $\theta_i(\cdot)$ represents the sector nonlinearity. A, C, E, F, B_i, and D_i $(i = 1, ..., q)$ are constant matrices with appropriate dimensions.

Remark 10.1 *The indicator functions in (10.3) dependent on a stochastic variable sequence τ_k are a random function utilized to describe two kinds of incomplete measurements (random communication delays and random data missing) simultaneously. That is, if $\sum_{j=0}^{q} p_j = 1$, then no data missing occurs in the system; and if $\sum_{j=0}^{q} p_j < 1$, the data missing occurrence probability is $1 - \sum_{j=0}^{q}$, where p_0 and p_i $(i = 1, ..., q)$ denote the occurrence probability of the data missing and the communication delay d_i in the networked system, respectively. These two kinds of incomplete measurements are induced by the limited bandwidth of communication networks, and therefore they must be* *considered at the same time as done in model (10.3). This model can be regarded as an extension of those adopted in [175, 186] and has been used to deal with a robust filtering problem for a class of linear deterministic discrete-time NCSs in [57].*

Assumption 10.1 *The nonlinear functions $\phi_{is}(\cdot)$ and $\theta_{is}(\cdot)$ in stochastic system (10.1)-(10.3) satisfy the following sector conditions for $\forall\ x_s \in \mathbb{R},\ x_s \neq 0,\ i = 1, ..., q,$ and $s = 1, ..., n$:*

$$\begin{aligned} 0 \leq \varphi_{is}(x_s) &:= \frac{\phi_{is}(x_s)}{x_s} \leq k_{is}, \\ 0 \leq \vartheta_{is}(x_s) &:= \frac{\theta_{is}(x_s)}{x_s} \leq l_{is}, \end{aligned} \tag{10.4}$$

which are equivalent to the following matrix inequalities:

$$\phi_i(x)(\phi_i(x) - K_i x) \leq 0, \quad \theta_i(x)(\theta_i(x) - L_i x) \leq 0, \tag{10.5}$$

where

$$\begin{aligned} \phi_i(x) &= [\phi_{i1}(x_1)\ ...\ \phi_{in}(x_n)]^T, \theta_i(x) = [\theta_{i1}(x_1)\ ...\ \theta_{in}(x_n)]^T, \\ K_i &= \text{diag}\{k_{i1},\ k_{i2}, ..., k_{in}\} > 0,\ L_i = \text{diag}\{l_{i1},\ l_{i2}, ..., l_{in}\} > 0. \end{aligned}$$

Remark 10.2 *The sector nonlinearities resulting from the complex environments exist widely in practical systems and often lead to the poor performance of the controlled system. Many researchers have investigated the analysis and synthesis problems for various systems with sector-like nonlinearities; see [15, 54]. The sector-like nonlinear description in (10.4) or (10.5) is quite general and includes the usual Lipschitz condition as a special case, and has been exploited to describe complex networks. Note that both the control analysis and model reduction problems for systems with sector nonlinearities have been intensively studied; see, e.g., [54, 77, 83].*

In this chapter, we are interested in designing a filter of the following structure:

$$(\Sigma_f) : \hat{x}(k+1) = G\hat{x}(k) + Hy(k), \tag{10.6}$$

where $\hat{x}(t) \in \mathbb{R}^n$ is the state estimate, and G and H are filter parameters to be determined.

By augmenting the state variables

$$\xi(k) := \begin{bmatrix} x(k) \\ \hat{x}(k) \end{bmatrix}, \quad \sigma_{di}(k) := \begin{bmatrix} \phi_i(x(k-d_i)) \\ \theta_i(x(k-d_i)) \end{bmatrix}$$

and combining (Σ) and (Σ_f), we obtain the filtering error dynamics as follows:

$$\begin{aligned} \xi(k+1) &= \bar{A}\xi(k) + \sum_{i=1}^{q} \bar{B}_i \sigma_{di}(k) + I^0_{\tau_k} \bar{C} Z\xi(k) \\ &\quad + \sum_{i=1}^{q} I^i_{\tau_k} \bar{D}_i \sigma_{di}(k) + \bar{E} Z\xi(k)\omega(k), \end{aligned} \tag{10.7}$$

where

$$
\begin{aligned}
I_{\tau_k}^i &= I_{\{\tau_k=i\}} - p_i \quad (i=0,1,...,q) \\
\bar{A} &= \begin{bmatrix} A & 0 \\ p_0 HC & G \end{bmatrix}, \ \bar{B}_i = \begin{bmatrix} B_i & 0 \\ 0 & p_i H D_i \end{bmatrix}, \\
\bar{C} &= \begin{bmatrix} 0 \\ HC \end{bmatrix}, \quad \bar{D}_i = \begin{bmatrix} 0 & 0 \\ 0 & HD_i \end{bmatrix}, \\
\bar{E} &= \begin{bmatrix} E \\ HF \end{bmatrix}, \quad Z = \begin{bmatrix} I & 0 \end{bmatrix}.
\end{aligned}
$$

It follows from (10.1), (10.3), (10.4), and (10.7) that

$$
\begin{aligned}
\xi(k+1) &= \bar{A}\xi(k) + \sum_{i=1}^{q} \bar{B}_i \varsigma_{di}(k) Z \xi_{di}(k) + I_{\tau_k}^0 \bar{C} Z \xi(k) \\
&\quad + \sum_{i=1}^{q} I_{\tau_k}^i \bar{D}_i \sigma_{di}(k) + \bar{E} Z \xi(k) \omega(k), \qquad (10.8)
\end{aligned}
$$

where

$$
\xi_{di}(k) := \begin{bmatrix} x(k-d_i) \\ \hat{x}(k-d_i) \end{bmatrix}, \ \varsigma_{di}(k) := \begin{bmatrix} \mathrm{diag}\{\bar{\varphi}_{di1}(k), ..., \bar{\varphi}_{din}(k))\} \\ \mathrm{diag}\{\bar{\vartheta}_{di1}(k), ..., \bar{\vartheta}_{din}(k))\} \end{bmatrix}, \qquad (10.9)
$$

with $\bar{\varphi}_{dij}(k) := \varphi_{ij}(x(k-d_i)$ and $\bar{\vartheta}_{dij}(k) := \vartheta_{ij}(x(k-d_i)$ $(j=1,...,n)$.

Observe the system (10.7) and let $\xi(k;\nu)$ denote the state trajectory from the initial data $\xi(\theta)=\nu(s)$ on $-d_k \le s \le 0$. Obviously, $\xi(k,0) \equiv 0$ is the trivial solution of system (10.7) corresponding to the initial datum $\nu = 0$.

Before formulating the problem to be investigated, we first introduce the following stability concepts for the augmented system (10.7).

Definition 13 For the system (10.7) and every initial conditions ν, the trivial solution is said to be *exponentially mean-square stable* if there exist scalars α $(\alpha > 0)$ and β $(0 < \beta < 1)$, such that

$$
\mathbb{E}|\xi(k,\nu)|^2 \le \alpha\beta^k \sup_{-d_q \le s \le 0} \mathbb{E}|\nu(s)|^2. \qquad (10.10)
$$

The purpose of this chapter is to design an exponential filter of the form (10.6) for the system (10.1)–(10.3), such that, for all admissible random communication time delays, random data missing, and exogenous stochastic disturbances, the filtering error system (10.7) is exponentially mean-square stable. It is expected that the derived stability conditions are dependent on the length of time-delay and the occurrence probability of the random delays, as well as the random packet dropouts.

10.2 Main Results

10.2.1 Filter Analysis

First, let us give the following lemmas that will be used in the proofs of our main results in this chapter.

Lemma 10.1 *[120] Assume that $a \in \mathbb{R}^{n_1}$, $b \in \mathbb{R}^{n_2}$, and $\Xi \in \mathbb{R}^{n_1 \times n_2}$. Then, for any matrices $M_1 \in \mathbb{R}^{n_1 \times n_1}$, $M_2 \in \mathbb{R}^{n_1 \times n_2}$, and $M_3 \in \mathbb{R}^{n_2 \times n_2}$ satisfying*

$$\begin{bmatrix} M_1 & M_2 \\ M_2^T & M_3 \end{bmatrix} \geq 0, \tag{10.11}$$

the following matrix inequality holds:

$$-2a^T \Xi b \leq \begin{bmatrix} a \\ b \end{bmatrix}^T \begin{bmatrix} M_1 & M_2 - \Xi \\ M_2^T - \Xi^T & M_3 \end{bmatrix} \begin{bmatrix} a \\ b \end{bmatrix}. \tag{10.12}$$

In the following theorem, the delay-dependent technique and an LMI-based method are combined together to deal with the stability analysis problem for the filter design of the nonlinear stochastic networked control system (10.1)–(10.3). A sufficient condition is derived to guarantee the solvability of the exponential filtering problem.

Theorem 23 Consider the filtering error system (10.7) with given filter parameters. If there exist positive definite matrices $P > 0$, $Q_j > 0$, $S_j > 0$, $M_j > 0$, and $T_j > 0$ $(j = 1, ..., q)$, such that the following matrix inequalities

$$\begin{bmatrix} \Omega_1 & -M_d & 0 & \Omega_2 & 0 & \Omega_5 \\ * & -Q_d & \bar{K}_d^T & 0 & 0 & 0 \\ * & * & -I_d & \Omega_3 & \Omega_4 & 0 \\ * & * & * & -\Lambda_1 & 0 & 0 \\ * & * & * & * & -P_d & 0 \\ * & * & * & * & * & -\Lambda_2 \end{bmatrix} < 0, \tag{10.13}$$

$$\begin{bmatrix} S_j & M_j \\ * & T_j \end{bmatrix} > 0, \qquad \forall\, j = 1, ..., q, \tag{10.14}$$

hold, where

$$\begin{aligned} \Omega_1 &:= -P + \sum_{j=1}^{q} Z^T Q_j Z + \sum_{j=1}^{q} (Z^T M_j^T + M_j Z) \\ &\quad + \sum_{j=1}^{q} d_j S_j, \ \Omega_2 := [(\bar{A}^T - I) Z^T \Pi \ \ \bar{A}^T P], \\ \Omega_3 &:= [\bar{B}_d^T Z^T \Pi \ \ \bar{B}_d^T P], \ \Omega_4 := \rho_d \bar{D}_d^T P, \\ \Omega_5 &:= [\delta Z^T \bar{E}^T Z^T \Pi \ \ \rho_0 Z^T \bar{C}^T P \ \ \delta Z^T \bar{E}^T P \ \ \bar{A}^T P], \end{aligned}$$

$$\begin{aligned}
P_d &:= \mathrm{diag}_q\{P\},\ \Pi := \sum_{j=1}^{q} d_j T_j,\ I_d := \mathrm{diag}_q\{I\},\\
M_d &:= [M_1, ..., M_q],\ Q_d := \mathrm{diag}\{Q_1, ..., Q_q\},\\
\bar{K}_d &:= \mathrm{diag}\{\bar{K}_1^T, ..., \bar{K}_q^T\},\ \bar{K}_j^T := [K_j\ \ L_j],\\
\bar{B}_d &:= [\bar{B}_1, ..., \bar{B}_q],\ \bar{D}_d := \mathrm{diag}\{\bar{D}_1, ..., \bar{D}_q\},\\
\rho_j &:= \sqrt{p_j},\ \rho_d := \mathrm{diag}\{\rho_1, ..., \rho_q\},\ \rho_0 := \sqrt{p_0},\\
\Lambda_1 &:= \mathrm{diag}\{\Pi,\ P\},\ \Lambda_2 := \mathrm{diag}\{\Pi, P, P, P\},
\end{aligned} \tag{10.15}$$

then the filtering error system (10.7) is exponentially mean-square stable.

Proof 26 *Recalling (10.7), we can write*

$$\xi_{di}(k) = \xi(k) - \sum_{m=k-d_i}^{k-1} (\xi(m+1) - \xi(m)) = \xi(k) - \sum_{m=k-d_i}^{k-1} \zeta(m), \tag{10.16}$$

where

$$\begin{aligned}
\zeta(m) &:= \xi(m+1) - \xi(m)\\
&= (\bar{A} - I)\xi(m) + \sum_{i=1}^{q} \bar{B}_i \sigma_{di}(m) + I_{\tau_m}^0 \bar{C} Z \xi(m)\\
&\quad + \sum_{i=1}^{q} I_{\tau_m}^i \bar{D}_i \sigma_{di}(m) + \bar{E} Z \xi(m) \omega(m).
\end{aligned} \tag{10.17}$$

By substituting (10.16) into (10.8), we can obtain

$$\begin{aligned}
\xi(k+1) &= \left(\bar{A} + \sum_{i=1}^{q} \bar{B}_i \varsigma_{di}(k) Z\right) \xi(k)\\
&\quad - \sum_{j=1}^{q} \sum_{m=k-d_j}^{k-1} \bar{B}_j \varsigma_{dj}(k) Z \zeta(m) + I_{\tau_k}^0 \bar{C} Z \xi(k)\\
&\quad + \sum_{i=1}^{q} I_{\tau_k}^i \bar{D}_i \sigma_{di}(k) + \bar{E} Z \xi(k) \omega(k).
\end{aligned} \tag{10.18}$$

Define the following Lyapunov functional candidate for the system (10.18):

$$\begin{aligned}
V(\tilde{\xi}(k), k) &= V_1(\tilde{\xi}(k), k) + V_2(\tilde{\xi}(k), k) + V_3(\tilde{\xi}(k), k)\\
&= \xi^T(k) P \xi(k) + \sum_{j=1}^{q} \sum_{m=k-d_j}^{k-1} \xi^T(m) Z^T Q_j Z \xi(m)\\
&\quad + \sum_{j=1}^{q} \sum_{r=-d_j}^{-1} \sum_{m=k+r}^{k-1} \zeta^T(m) Z^T T_j Z \zeta(m),
\end{aligned} \tag{10.19}$$

where $\tilde{\xi}(k) := [\xi^T(k)\ \ \xi^T(k-1)\ \ ...\ \ \xi^T(0)]^T$.

Calculating the difference of the Lyapunov functional (10.19) according to (10.7) gives

$$\Delta V(\tilde{\xi}(k),k)=\mathbb{E}\left\{V(\tilde{\xi}(k+1),k+1)|\tilde{\xi}(k)\right\}-V(\tilde{\xi}(k),k). \tag{10.20}$$

First, we obtain

$$\begin{aligned}
&\Delta V_1(\tilde{\xi}(k),k)= \\
&\xi^T(k)(\bar{A}+\sum_{i=1}^{q}\bar{B}_i\varsigma_{di}(k)Z)^TP(\bar{A}+\sum_{i=1}^{q}\bar{B}_i\varsigma_{di}(k)Z)\xi(k) \\
&-\xi^T(k)P\xi(k)+\left[\sum_{j=1}^{q}\sum_{m=k-d_j}^{k-1}\bar{B}_j\varsigma_{dj}(k)Z\zeta(m)\right]^TP \\
&\times\left[\sum_{j=1}^{q}\sum_{m=k-d_j}^{k-1}\bar{B}_j\varsigma_{dj}(k)Z\zeta(m)\right]-2\sum_{j=1}^{q}\sum_{m=k-d_j}^{k-1}\xi^T(k) \\
&\times\left(\bar{A}+\sum_{i=1}^{q}\bar{B}_i\varsigma_{di}(k)Z\right)^TP\bar{B}_j\varsigma_{dj}(k)Z\zeta(m) \\
&+\mathbb{E}\left\{\left[I_{\tau_k}^0\bar{C}Z\xi(k)+\sum_{i=1}^{q}I_{\tau_k}^i\bar{D}_i\sigma_{di}(k)\right]^TP\right. \\
&\left.\times\left[I_{\tau_k}^0\bar{C}Z\xi(k)+\sum_{i=1}^{q}I_{\tau_k}^i\bar{D}_i\sigma_{di}(k)\right]\right\} \\
&+\mathbb{E}\left\{\omega^T(k)\xi^T(k)Z^T\bar{E}^TP\bar{E}Z\xi(k)\omega(k)\right\}.
\end{aligned} \tag{10.21}$$

From Lemma 10.1, we have

$$\begin{aligned}
&-2\xi^T(k)\left(\bar{A}+\sum_{i=1}^{q}\bar{B}_i\varsigma_{di}(k)Z\right)^TP\bar{B}_j\varsigma_{dj}(k)Z\zeta(m) \\
\leq\ &\xi^T(k)S_j\xi(k)+2\zeta^T(m)Z^TM_j^T\xi(k) \\
&-2\zeta^T(m)Z^T\varsigma_{dj}(k)^T\bar{B}_j^TP\left(\bar{A}+\sum_{i=1}^{q}\bar{B}_i\varsigma_{di}(k)Z\right)\xi(k) \\
&+\zeta^T(m)Z^TT_jZ\zeta(m),
\end{aligned} \tag{10.22}$$

with $0\leq S_j^T=S_j\in\mathbb{R}^{2n\times 2n}$, $M_j\in\mathbb{R}^{2n\times n}$, $0\leq T_j^T=T_j\in\mathbb{R}^{n\times n}$ *satisfying (10.14).*

It follows from (10.5) and the notation of $\sigma_{di}(k)$ *that*

$$\sum_{i=1}^{q}\sigma_{di}^T(k)(\sigma_{di}(k)-\bar{K}_jZ\xi_{dj}(k))\leq 0, \tag{10.23}$$

with $\bar{K}_j$ *defined in (10.15).*

Again, we can obtain from the expression (10.3) that, for $0 \leq i \leq q$ *and* $0 \leq j \leq q$

$$\mathbb{E}\{(I^i_{\tau_k})(I^j_{\tau_k})\} = \begin{cases} p_i(1-p_i), & i = j \\ -p_i p_j, & i \neq j \end{cases}. \tag{10.24}$$

In addition, from (10.19), it follows that

$$\begin{aligned} \Delta V_2(\tilde{\xi}(k),k) &= \sum_{j=1}^{q} \xi^T(k) Z^T Q_j Z \xi(k) \\ &\quad - \sum_{j=1}^{q} \xi_{dj}(k)^T Z^T Q_j Z \xi_{dj}(k), \end{aligned} \tag{10.25}$$

$$\begin{aligned} \Delta V_3(\tilde{\xi}(k),k) &= \sum_{j=1}^{q} d_j \zeta^T(k) Z^T T_j Z \zeta(k) \\ &\quad - \sum_{j=1}^{q} \sum_{r=k-d_j}^{k-1} \zeta^T(r) Z^T T_j Z \zeta(r). \end{aligned} \tag{10.26}$$

From (10.17) and (10.21)–(10.26), it can be seen that

$$\begin{aligned} \Delta V(\xi(k),k) \leq\ & \xi^T(k)\left[\bar{A}^T P \bar{A} - P + p_0 Z^T \bar{C}^T P \bar{C} Z\right]\xi(k) \\ & + \left(\sum_{j=1}^{q} \bar{B}_j \sigma_{dj}(k)\right)^T P \left(\sum_{j=1}^{q} \bar{B}_j \sigma_{dj}(k)\right) \\ & + \sum_{i=1}^{q} \sigma^T_{di}(k)(p_i \bar{D}_i^T P \bar{D}_i - 2I)\sigma_{di}(k) \\ & + \delta^2 \xi^T(k) Z^T \bar{E}^T P \bar{E} Z \xi(k) \\ & + \sum_{j=1}^{q} [\xi^T(k) Z^T Q_j Z \xi(k) - \xi^T_{dj}(k) Z^T Q_j Z \xi_{dj}(k)] \\ & + 2\sum_{j=1}^{q} \sigma^T_{dj}(k)[\bar{B}_j^T P \bar{A} \xi(k) + \bar{K}_j Z \xi_{dj}(k)] \\ & + \sum_{j=1}^{q} d_j \xi^T(k) S_j \xi(k) + \sum_{j=1}^{q} d_j \zeta^T(k) Z^T T_j Z \zeta(k) \\ & + 2\sum_{j=1}^{q} \xi^T(k) Z^T M_j^T \xi(k) - 2\sum_{j=1}^{q} \xi^T_{dj}(k) M_j^T \xi(k) \\ & - \left[p_0 \bar{C} Z \xi(k) + \sum_{i=1}^{q} p_i \bar{D}_i \sigma_{di}(k)\right]^T P \\ & \times \left[p_0 \bar{C} Z \xi(k) + \sum_{i=1}^{q} p_i \bar{D}_i \sigma_{di}(k)\right] \\ \leq\ & \bar{\xi}^T(k) \Gamma \bar{\xi}(k), \end{aligned} \tag{10.27}$$

where

$$\begin{aligned}\bar{\xi}(k) &= [\xi^T(k)\ \xi_{d1}^T(k)Z^T \ldots \xi_{dq}^T(k)Z^T\ \sigma_{d1}^T(k) \ldots \sigma_{dq}^T(k)]^T, \\ \Gamma: &= \begin{bmatrix} \Gamma_1 & -M_d & \Gamma_2 \\ * & -Q_d & \bar{K}_d^T \\ * & * & \Gamma_3 \end{bmatrix} \end{aligned} \tag{10.28}$$

with

$$\begin{aligned}\Gamma_1 &:= \Omega_1 + \bar{A}^T P \bar{A} + \rho_0^2 Z^T \bar{C}^T P \bar{C} Z + \delta^2 Z^T \bar{E}^T P \bar{E} Z \\ &\quad + \rho_0^2 Z^T \bar{C}^T Z^T \Pi Z \bar{C} Z + \delta^2 Z^T \bar{E}^T Z^T \Pi Z \bar{E} Z \\ &\quad + (\bar{A}^T - I) Z^T \Pi Z (\bar{A} - I), \end{aligned} \tag{10.29}$$

$$\begin{aligned}\Gamma_2 &:= (\bar{A}^T - I) Z^T \Pi Z \bar{B}_d + \bar{A}^T P \bar{B}_d, \\ \Gamma_3 &:= \bar{B}_d^T (Z^T \Pi Z + P) \bar{B}_d - I_d + \mathrm{diag}\{\rho_1^2 \bar{D}_1^T (P \\ &\quad + Z^T \Pi Z) \bar{D}_1, \ldots, \rho_q^2 \bar{D}_q^T (P + Z^T \Pi Z) \bar{D}_q\}, \end{aligned} \tag{10.30}$$

and Q_d, $\bar{K}_d$, Y_d, $\bar{B}_d$, Π, I_d, ρ_0, and ρ_j $(j = 1, \ldots, q)$ are defined in (10.15).

By Schur complement, we can obtain from (10.13) and (10.14) that $\Gamma < 0$, and, therefore, there always exists a scalar $\alpha > 0$, such that

$$\Gamma < \begin{bmatrix} -\alpha I & 0 & 0 \\ 0 & 0 & 0 \\ 0 & 0 & 0 \end{bmatrix} \tag{10.31}$$

and, subsequently,

$$\mathbb{E}\left\{V(\tilde{\xi}(k+1), k+1) | \tilde{\xi}(k)\right\} - V(\tilde{\xi}(k), k) < -\alpha |\xi(k)|^2. \tag{10.32}$$

Finally, we can confirm from Lemma 1 of [186] that the filtering error system (10.7) is exponentially mean-square stable.

In the next subsection, our attention is focused on the design of filter parameters G and H by using the results in Theorem 23. The explicit expression of the expected filter parameters will be obtained in term of the solution to a set of LMIs.

10.2.2 Filter Synthesis

The following theorem shows that the desired filter parameters can be derived by solving several LMIs.

Theorem 24 Consider the system (10.7). If there exist matrices $X > 0$, $\mathcal{Y} > 0$, $Q_j > 0$, $\bar{S}_{ji} > 0$, $T_j > 0$, $\tilde{G}$, $\tilde{H}$, $\bar{M}_{j1}$, and $\bar{M}_{j2}$ $(j = 1, 2, \ldots, q;\ i = 1, 2, 3)$, such that the following linear matrix inequalities,

$$\begin{bmatrix} -\Xi_1 & \Xi_2 & -\tilde{M}_{d1} & 0 & \Xi_4 & \Xi_7 \\ * & -\Xi_3 & -\tilde{M}_{d2} & 0 & \Xi_5 & \Xi_8 \\ * & * & -Q_d & \bar{K}_d^T & 0 & 0 \\ * & * & * & -I_d & \Xi_6 & 0 \\ * & * & * & * & -\Theta_1 & 0 \\ * & * & * & * & * & -\Theta_2 \end{bmatrix} < 0, \tag{10.33}$$

$$\begin{bmatrix} \bar{S}_{j1} & \bar{S}_{j2} & \bar{M}_{j1} \\ * & \bar{S}_{j3} & \bar{M}_{j2} \\ * & * & T_j \end{bmatrix} > 0, \qquad \forall\, j = 1, ..., q, \tag{10.34}$$

hold, where

$$\begin{aligned}
\Xi_1 &:= \mathcal{Y} - \Sigma_{j=1}^{q}(Q_j + d_j\bar{S}_{j1} + \Sigma_{s=1}^{2}(\bar{M}_{js} + \bar{M}_{js}^T)), \\
\Xi_2 &:= \mathcal{Y} + \Sigma_{j=1}^{q}(Q_j + d_j\bar{S}_{j2} + \bar{M}_{j1} + \bar{M}_{j1}^T + \bar{M}_{j2}), \\
\Xi_3 &:= X - \Sigma_{j=1}^{q}(Q_j + d_j\bar{S}_{j3} + \bar{M}_{j1} + \bar{M}_{j1}^T), \\
\Xi_4 &:= \Big[(A^T - I)\Pi)\ [A^T\mathcal{Y}\ \ A^TX + p_0C^T\tilde{H}^T + \tilde{G}^T] \\
&\qquad 0\ \ \delta E^T\Pi\ \ [0\ \ \rho_0 C^T\tilde{H}^T]\Big], \\
\Xi_5 &:= \Big[(A^T - I)\Pi)\ [A^T\mathcal{Y}\ \ A^TX + p_0C^T\tilde{H}^T] \\
&\qquad 0\ \ \delta E^T\Pi\ \ [0\ \ \rho_0 C^T\tilde{H}^T]\Big], \\
\Xi_6 &:= \Big[\tilde{B}_d^T\Pi\ \ \hat{B}_d\ \ \tilde{D}_d\ \ 0\ \ [0\ \ 0]\Big], \\
\Xi_7 &:= \Big[[\delta E^T\mathcal{Y}\ \ \delta E^TX + \delta F^T\tilde{H}^T] \\
&\qquad [A^T\mathcal{Y}\ \ A^TX + p_0C^T\tilde{H}^T + \tilde{G}^T]\Big], \\
\Xi_8 &:= \Big[[\delta E^T\mathcal{Y}\ \ \delta E^TX + \delta F^T\tilde{H}^T] \\
&\qquad [A^T\mathcal{Y}\ \ A^TX + p_0C^T\tilde{H}^T]\Big], \\
\tilde{M}_{d1} &:= [\bar{M}_{11} + \bar{M}_{12}\ \ ...\ \ \bar{M}_{q1} + \bar{M}_{q2}], \\
\tilde{M}_{d2} &:= [\bar{M}_{11}\ \ ...\ \ \bar{M}_{q1}],\ \ \tilde{B}_d := [B_1, 0, B_2, 0, ..., B_q, 0], \\
\hat{B}_d &:= [\hat{B}_1, ..., \hat{B}_q]^T,\ \hat{B}_j^T := \begin{bmatrix} B_j^T\mathcal{Y} & B_j^TX \\ 0 & p_jD_j^T\tilde{H}^T \end{bmatrix}, \\
\tilde{D}_d &:= \mathrm{diag}\left\{0, D_1^T\tilde{H}^T, 0, D_2^T\tilde{H}^T, ..., 0, D_q^T\tilde{H}^T\right\}, \\
\Theta_1 &:= \mathrm{diag}\left\{\Pi,\ \Delta,\ \mathrm{diag}_q\{\Delta\},\ \Pi,\ \Delta\right\}, \\
\Theta_2 &:= \mathrm{diag}\left\{\Delta,\ \Delta\right\},\ \Delta := \begin{bmatrix} \mathcal{Y} & \mathcal{Y} \\ \mathcal{Y} & X \end{bmatrix},
\end{aligned}$$

then the system (10.7) is exponentially mean-square stable. In this case, the parameters of the desired filter (Σ_f) are given as follows:

$$G := (\mathcal{Y} - X)^{-1}\tilde{G},\ H := (\mathcal{Y} - X)^{-1}\tilde{H}. \tag{10.35}$$

Proof 27 *Let*

$$P = \begin{bmatrix} X & \mathcal{Y} - X \\ \mathcal{Y} - X & X - \mathcal{Y} \end{bmatrix} > 0, \quad \Upsilon = \begin{bmatrix} Y & I \\ Y & 0 \end{bmatrix}, \tag{10.36}$$

where $Y = \mathcal{Y}^{-1} > 0$.

Denote $\Psi = \mathrm{diag}\{Y,\ I\}$. *From (10.34), we set*

$$S_j \ := \ \Upsilon^{-T}\Psi \begin{pmatrix} \bar{S}_{j1} & \bar{S}_{j2} \\ \bar{S}_{j2}^T & \bar{S}_{j3} \end{pmatrix} \Psi\Upsilon^{-1} > 0, \tag{10.37}$$

$$M_j \ := \ \Upsilon^{-T}\Psi \begin{pmatrix} \bar{M}_{j1} \\ \bar{M}_{j2} \end{pmatrix} \Psi\Upsilon^{-1} > 0. \tag{10.38}$$

Premultiplying and postmultiplying the LMIs in (10.33) by

$$\mathrm{diag}\left\{Y, I, I, I, \mathrm{diag}\{I, \Psi, \mathrm{diag}_q\{\Psi\}, I, \Psi\}, \mathrm{diag}\{\Psi, \Psi\}\right\}$$

and (10.34) by $\mathrm{diag}\{Y, I, I\}$, *we have*

$$\begin{bmatrix} -\bar{\Xi}_1 & \bar{\Xi}_2 & -Y\tilde{M}_{d1} & 0 & \bar{\Xi}_4 & \bar{\Xi}_7 \\ * & -\Xi_3 & -\tilde{M}_{d2} & 0 & \bar{\Xi}_5 & \bar{\Xi}_8 \\ * & * & -Q_d & \bar{K}_d^T & 0 & 0 \\ * & * & * & -I_d & \bar{\Xi}_6 & 0 \\ * & * & * & * & -\bar{\Theta}_1 & 0 \\ * & * & * & * & * & -\bar{\Theta}_2 \end{bmatrix} < 0, \tag{10.39}$$

$$\begin{bmatrix} Y\bar{S}_{j1}Y & Y\bar{S}_{j2} & Y\bar{M}_{j1} \\ * & \bar{S}_{j3} & \bar{M}_{j2} \\ * & * & T_j \end{bmatrix} > 0, \ \forall\, j = 1, ..., q, \tag{10.40}$$

where

$$\begin{aligned}
\bar{\Xi}_1 &:= Y - \Sigma_{j=1}^q Y(Q_j + d_j\bar{S}_{j1} + \Sigma_{s=1}^2(\bar{M}_{js} + \bar{M}_{js}^T))Y, \\
\bar{\Xi}_2 &:= I + \Sigma_{j=1}^q Y(Q_j + d_j\bar{S}_{j2} + \bar{M}_{j1} + \bar{M}_{j1}^T + \bar{M}_{j2}), \\
\bar{\Xi}_4 &:= \Big[Y(A^T - I)\Pi)\ \mathcal{P}\ 0\ \ \delta YE^T\Pi\ [0\ \ \rho_0 YC^T\tilde{H}^T]\Big], \\
\bar{\Xi}_5 &:= \Big[Y(A^T - I)\Pi)\ [A^T\ \ A^TX + p_0C^T\tilde{H}^T], \\
&\qquad 0\ \ \delta YE^T\Pi\ [0\ \ \rho_0YC^T\tilde{H}^T]\Big], \\
\bar{\Xi}_6 &:= \Big[\tilde{B}_d^T\Pi\ \check{B}_d\ \tilde{D}_d\ 0\ [0\ \ 0]\Big],\ \bar{\Theta}_2 := \mathrm{diag}\{\bar{\Delta},\ \bar{\Delta}\},
\end{aligned}$$

$$\begin{aligned}
\bar{\Xi}_7 &:= \Big[[\delta Y E^T \;\; \delta Y E^T X + \delta Y F^T \tilde{H}^T] \;\; \mathcal{P}\Big], \\
\bar{\Xi}_8 &:= \Big[[\delta E^T \;\; \delta E^T X + \delta F^T \tilde{H}^T][A^T \;\; A^T X + p_0 C^T \tilde{H}^T]\Big], \\
\mathcal{P} &:= [YA^T \;\; YA^T X + p_0 Y C^T \tilde{H}^T + Y\tilde{G}^T], \\
\bar{\Theta}_1 &:= \mathrm{diag}\left\{\Pi, \bar{\Delta}, \mathrm{diag}_q\{\bar{\Delta}\}, \Pi, \bar{\Delta}\right\}, \; \check{B}_d := [\check{B}_1, ..., \check{B}_q]^T, \\
\bar{\Delta} &:= \begin{bmatrix} Y & I \\ I & X \end{bmatrix}, \; \check{B}_j^T := \begin{bmatrix} B_j^T & B_j^T X \\ 0 & p_j D_j^T \tilde{H}^T \end{bmatrix}.
\end{aligned}$$

From the definitions of P and Υ, the LMIs in (10.39)–(10.40) are equivalent to the following matrix inequalities

$$\begin{bmatrix} \Upsilon^T \Omega_1 \Upsilon & -\Upsilon^T M_d & 0 & \bar{\Omega}_2 & 0 & \bar{\Omega}_5 \\ * & -Q_d & \bar{K}_d^T & 0 & 0 & 0 \\ * & * & -I_d & \bar{\Omega}_3 & \bar{\Omega}_4 & 0 \\ * & * & * & -\bar{\Lambda}_1 & 0 & 0 \\ * & * & * & * & -\bar{P}_d & 0 \\ * & * & * & * & * & -\bar{\Lambda}_2 \end{bmatrix} < 0, \tag{10.41}$$

$$\begin{bmatrix} \Upsilon^T S_j \Upsilon & \Upsilon^T M_j \\ * & T_j \end{bmatrix} > 0, \qquad \forall\, j = 1, ..., q, \tag{10.42}$$

where

$$\begin{aligned}
\bar{\Omega}_2 &:= \Upsilon^T \Omega_2 \mathrm{diag}\{I, \Upsilon\}, \\
\bar{\Omega}_3 &:= \Omega_3 \mathrm{diag}\{I, \Upsilon\}, \; \bar{\Omega}_4 := \Omega_4 \Upsilon, \\
\bar{\Omega}_5 &:= \Upsilon^T [\delta Z^T \bar{E}^T Z^T \Pi \;\; \rho_0 Z^T \bar{C}^T P \;\; \delta Z^T \bar{E}^T P \;\; \bar{A}^T P], \\
\bar{\Lambda}_1 &:= \mathrm{diag}\{\Pi, \;\; \Upsilon^T P \Upsilon\}, \bar{P}_d := \mathrm{diag}_q\{\Upsilon^T P \Upsilon\}, \\
\bar{\Lambda}_2 &:= \mathrm{diag}\{\Pi, \Upsilon^T P \Upsilon, \Upsilon^T P \Upsilon, \Upsilon^T P \Upsilon\}.
\end{aligned}$$

Finally, premultiplying and postmultiplying (10.41) by

$$\mathrm{diag}\{\Upsilon^{-T}, \; I, I, \mathrm{diag}\{I, \Upsilon^{-T}\}, \mathrm{diag}_q\{\Upsilon^{-T}\}, \mathrm{diag}\{I, \Upsilon^{-T}, \Upsilon^{-T}, \Upsilon^{-T}\}\}$$

and its transpose, and (10.42) by $\mathrm{diag}\{\Upsilon^{-T}, I\}$ *and its transpose, we can obtain from Theorem 23 and Schur complement lemma that, with the given filter parameters in (10.35), the system (10.7) is exponentially mean-square stable.*

Remark 10.3 *The desired filter design problem is solved in Theorem 24 for the addressed nonlinear networked control systems with sensor nonlinearities, external stochastic disturbances, multiple random state-delay, and random missing measurements. LMI-based sufficient conditions are obtained for the existence of full-order filters that ensure the exponential mean-square stability of the resulting filtering error system for all admissible time-delays and nonlinearities. The feasibility of the filter design problem can be readily checked by the solvability of two sets of LMIs, which can be determined by using the MATLAB LMI Toolbox in a straightforward way. In the next section, an illustrative example will be provided to show the usefulness of the proposed techniques.*

10.3 An Illustrative Example

In this section, a simulation example is presented to illustrate the usefulness and flexibility of the filter design method developed in this chapter. We are interested in designing the filter for the discrete-time NCS with stochastic disturbances and sensor nonlinearities.

The system data of (10.1)–(10.3) are given as follows:

$$
\begin{aligned}
& q=2,\ d_1=1,\ d_2=3,\ p_0=0.4,\ p_1=0.6,\ p_2=0.9, \\
& A=\begin{bmatrix} 0.1 & 0 \\ -0.02 & 0.2 \end{bmatrix}, \quad B_1=\begin{bmatrix} 0.12 & 0 \\ 0 & 0.1 \end{bmatrix}, \\
& B_2=\begin{bmatrix} 0.1 & 0.0512 \\ 0 & 0.1 \end{bmatrix}, \quad C=\begin{bmatrix} 0.02 & 0 \\ 0 & 0.01 \end{bmatrix}, \\
& D_1=\begin{bmatrix} -0.1 & 0 \\ 0 & 0.1 \end{bmatrix}, \quad D_2=\begin{bmatrix} -0.1 & 0 \\ 0 & -0.1 \end{bmatrix}, \\
& E=\begin{bmatrix} 0.08 & 0.07 \\ 0.01 & 0.14 \end{bmatrix}, \quad F=\begin{bmatrix} 0.13 & 0.09 \\ 0.01 & 0.16 \end{bmatrix}, \\
& K_1=\mathrm{diag}\{0.35, 0.35\}, \quad K_2=\mathrm{diag}\{0.35, 0.35\}, \\
& L_1=\mathrm{diag}\{0.3, 0.3\}, \quad L_2=\mathrm{diag}\{0.3, 0.3\}.
\end{aligned}
$$

Using MATLAB LMI Control Toolbox to solve the LMIs in (10.33) and (10.34), we can calculate that the filter parameters as follows:

$$
G=\begin{bmatrix} 0.1217 & 0.0009 \\ -0.0401 & 0.3059 \end{bmatrix}, \quad H=\begin{bmatrix} 0.4559 & 0.2520 \\ 0.2835 & 0.7126 \end{bmatrix}.
$$

Figure 10.1–Figure 10.2 are the simulation results for the performance of the designed filter, where the sensor nonlinearities are taken as

$$
\begin{aligned}
\phi_1(x) &= \frac{x^2}{2^2+x^2}, \quad \phi_2(x)=\frac{x^2}{2^2+x^2}, \\
\theta_1(x) &= \frac{x^2}{2.3^2+x^2}, \quad \theta_2(x)=\frac{x^2}{2.3^2+x^2},
\end{aligned}
$$

which satisfy (10.5). It is confirmed from the simulation results that all the expected objectives are well achieved.

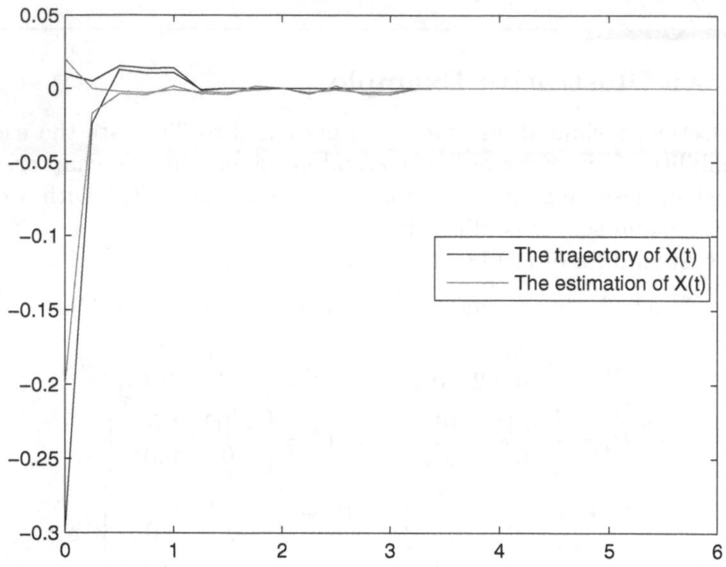

FIGURE 10.1
The state and estimation of $x(k)$.

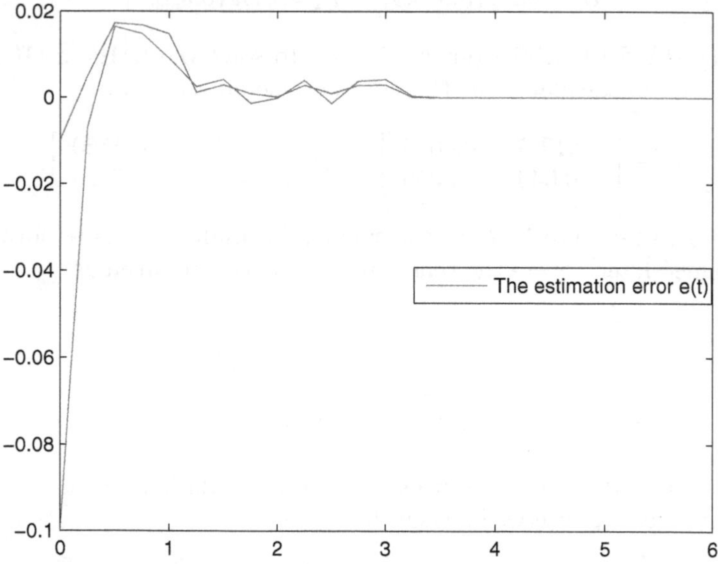

FIGURE 10.2
Estimation error of $e(k)$.

10.4 Summary

In this chapter, the filtering problem has been investigated for a class of discrete-time stochastic nonlinear networked control systems with multiple random communication delays and random packet losses. A linear filter has been designed such that, for the admissible random incomplete measurement phenomenon, stochastic disturbances as well as sector nonlinearities, the filtering error dynamics are exponentially mean square stable. By using the LMI technique and delay-dependent techniques, sufficient conditions have been derived for ensuring the desired filtering performance, which are dependent on the occurrence probability of both the random sensor delay and missing measurement. The filter gain has been characterized in terms of the solution to a set of LMIs, and a simulation example has been exploited for a nonlinear stochastic networked control model in order to demonstrate the effectiveness of the proposed design procedures.

11

State Estimation for Stochastic Delayed Gene Regulatory Networks

CONTENTS

This chapter is concerned with the filtering problem for a class of nonlinear genetic regulatory networks with state-dependent stochastic disturbances as well as state delays. The feedback regulation is described by a sector-like nonlinear function, the stochastic perturbation is a scalar Brownian motion, and the time-delays enter into both the translation process and the feedback regulation process. We aim to estimate the true concentrations of the mRNA and protein by designing a linear filter with guaranteed exponential stability of the filtering augmented systems. By using the linear matrix inequality (LMI) technique, sufficient conditions are first derived for ensuring the exponentially mean-square stable with a prescribed decay rate β for the gene regulatory model, and then the filter gain is characterized in terms of the solution to an LMI, which can be easily solved by using available software packages. A simulation example is employed for a gene expression model.

The rest of this chapter is arranged as follows. Section 11.1 formulates a class of nonlinear genetic regulatory networks with state-dependent stochastic disturbances as well as state delays to be studied. In Section 11.2, filter analysis and synthesis problems are dealt with. A simulation example is presented to illustrate the usefulness and flexibility of the filter design method developed in Section 11.3. Section 11.4 gives our summary.

11.1 Problem Formulation

In this chapter, we consider the following nonlinear time-delay genetic regulatory networks [19, 134], with state-dependent stochastic disturbances:

$$(\Sigma):\begin{cases} dx_m(t) &= [-A_1x_m(t)+Bg(x_p(t-\tau_1))]dt+Ex_m(t)d\omega_1(t)\\ dx_p(t) &= [-A_2x_p(t)+Dx_m(t-\tau_2)]dt+Fx_p(t)d\omega_2(t)\\ y_m(t) &= C_1x_m(t)\\ y_p(t) &= C_2x_p(t)\\ x_m(t) &= \phi_m(t),\ x_p(t)=\phi_p(t), \forall\, t\in[-2\tau,0], \end{cases} \tag{11.1}$$

where $x_m(t)=[x_{m1}(t),\cdots,x_{mn}(t)]^T\in\mathbb{R}^n$, $x_p(t)=[x_{p1}(t),\cdots,x_{pn}(t)]^T\in\mathbb{R}^n$; $x_{mi}(t)$ and $x_{pi}(t)$ $(i=1,\ldots,n)$ denote the concentrations of mRNA and protein of the ith node at time t, respectively; $y_m(t)=[y_{m1}(t)\ y_{m2}(t)\cdots y_{mr}(t)]^T\in\mathbb{R}^r$, $y_p(t)=[y_{p1}(t)\ y_{p2}(t)\cdots y_{pr}(t)]^T\in\mathbb{R}^r$; $y_{mj}(t)$ and $y_{pj}(t)$ $(j=1,\ldots,r)$ represent the expression levels of mRNA and protein of the jth node at time t, respectively; $\phi_m(t)$, $\phi_p(t)$ are the initial functions of $x_m(t)$ and $x_p(t)$, respectively; and $g(x_p(t-\tau_1))=[g_1(x_{p1}(t-\tau_1))\ g_2(x_{p2}(t-\tau_1))\cdots g_n(x_{pn}(t-\tau_1))]^T\in\mathbb{R}^n$, with the function $g_i(\cdot)$ representing the feedback regulation of the protein on the transcription, which is generally a nonlinear function but has a form of monotonicity with each variable [11]. Both $\omega_1(t)$ and $\omega_2(t)$ are scalar Brownian motions, with mean being 0, variance being 1, and are mutually uncorrelated. The constants $\tau_1>0$, $\tau_2>0$ denote, respectively, the translation delay, the feedback regulation delay, and $\tau=\max\{\tau_1,\ \tau_2\}$. The matrix $B=(b_{ij})\in\mathbb{R}^{n\times n}$ is defined as follows:

$$b_{ij}\begin{cases} >0, & \text{if transcription factor } j \text{ is an activator of gene } i;\\ =0, & \text{if there is no link from node } j \text{ to } i;\\ <0, & \text{if transcription factor } j \text{ is a repressor of gene } i. \end{cases} \tag{11.2}$$

The matrices $A_1=\text{diag}\{a_{11}, a_{12},\ldots,a_{1n}\}$, $A_2=\text{diag}\{c_{11}, c_{12},\ldots,c_{1n}\}$, and $D=\text{diag}\{d_1, d_2,\ldots,d_n\}$ are diagonal matrices, with a_{1i}, c_{1i}, d_i $(i=1,\ldots,n)$ being the rate of degradation of mRNA, the rate of degradation of protein, and the translation rate of the ith node, respectively. A_1, A_2, B, C_1, C_2, D, E, and F are all constant matrices with appropriate dimensions.

Assumption 11.1 *The function $g_i(\cdot)$ satisfies the following sector condition:*

$$0\le f_i(x_i):=\frac{g_i(x_i)}{x_i}\le k_i,\ \ \forall\, x_i\in\mathbb{R},\ x_i\neq 0, f_i(0)=0,\ \ g_i(0)=0, \tag{11.3}$$

which is equivalent to

$$g^T(x)(g(x)-Kx)\le 0, \tag{11.4}$$

where $i=1,\ldots,n$, $K=\text{diag}\{k_1, k_2,\ldots,k_n\}>0$.

Remark 11.1 *By Assumption 11.1, the GRN (11.1) can be regarded as a kind of stochastic Lur'e system, in which the fruitful Lur'e system method in control theory [85] could be applied. Notice that the sector-like nonlinear function $g_i(\cdot)$ has been used to model the structure and regulation mechanism of the genetic networks in many references. For example, as a monotonic increasing or decreasing regulatory function, $g_i(\cdot)$ is usually of the Michaelis–Menten or Hill form [85], which can easily be transformed to a nonlinear function satisfying the sector condition.*

The main aim of this chapter is to estimate the concentrations of mRNA and protein through their expression level. The linear filter adopted is of the form

$$(\Sigma_f):\begin{cases} d\hat{x}_m(t) &= \hat{A}\hat{x}_m(t)dt+\hat{B}y_m(t)dt \\ d\hat{x}_p(t) &= \hat{C}\hat{x}_p(t)dt+\hat{D}y_p(t)dt \\ \hat{x}_m(t) &= \psi_m(t),\ \hat{x}_p(t)=\psi_p(t), \forall\, t\in[-2\tau,0], \end{cases} \tag{11.5}$$

where $\hat{x}_m(t)\in\mathbb{R}^n$ and $\hat{x}_p(t)\in\mathbb{R}^n$ are the estimates for $x_m(t)$ and $x_p(t)$, respectively, $\psi_m(t)$, $\psi_p(t)$ are the initial functions of $\hat{x}_m(t)$ and $\hat{x}_p(t)$, respectively, and $\hat{A}$, $\hat{B}$, $\hat{C}$, and $\hat{D}$ are filter parameters to be determined.

By defining

$$\bar{x}_m(t):=\begin{bmatrix} x_m(t) \\ \hat{x}_m(t)\end{bmatrix},\quad \bar{x}_{m\tau_2}:=\begin{bmatrix} x_m(t-\tau_2) \\ \hat{x}_m(t-\tau_2)\end{bmatrix},$$
$$\bar{x}_p(t):=\begin{bmatrix} x_p(t) \\ \hat{x}_p(t)\end{bmatrix},\quad \bar{x}_{p\tau_1}:=\begin{bmatrix} x_p(t-\tau_1) \\ \hat{x}_p(t-\tau_1)\end{bmatrix},$$

and combining (Σ) and (Σ_f), we obtain the augmented filtering dynamics as follows:

$$(\Sigma_e):\begin{cases} d\bar{x}_m(t) &= \bar{A}\bar{x}_m(t)dt+\bar{B}g(Z\bar{x}_{p\tau_1})dt+\bar{E}Z\bar{x}_m(t)d\omega_1(t) \\ d\bar{x}_p(t) &= \bar{C}\bar{x}_p(t)dt+\bar{D}Z\bar{x}_{m\tau_2}dt+\bar{F}Z\bar{x}_p(t)d\omega_2(t), \end{cases} \tag{11.6}$$

where

$$\bar{A} = \begin{bmatrix} -A_1 & 0 \\ \hat{B}C_1 & \hat{A}\end{bmatrix},\ \bar{B}=\begin{bmatrix} B \\ 0\end{bmatrix},\ \bar{E}=\begin{bmatrix} E \\ 0\end{bmatrix},$$
$$\bar{C} = \begin{bmatrix} -A_2 & 0 \\ \hat{D}C_2 & \hat{C}\end{bmatrix},\ \bar{D}=\begin{bmatrix} D \\ 0\end{bmatrix},\ \bar{F}=\begin{bmatrix} F \\ 0\end{bmatrix},\ Z=[I\ \ 0].$$

For presentation convenience, we let

$$\begin{cases} \xi_m(t) &= \bar{A}\bar{x}_m(t)+\bar{B}g(Z\bar{x}_{p\tau_1}) \\ \xi_p(t) &= \bar{C}\bar{x}_p(t)+\bar{D}Z\bar{x}_{m\tau_2}, \end{cases} \tag{11.7}$$

and then (Σ_e) in (11.6) can be rewritten as

$$\begin{cases} d\bar{x}_m(t) &= \xi_m(t)dt + \bar{E}Z\bar{x}_m(t)d\omega_1(t) \\ d\bar{x}_p(t) &= \xi_p(t)dt + \bar{F}Z\bar{x}_p(t)d\omega_2(t). \end{cases} \tag{11.8}$$

Before formulating the problem to be investigated, we first introduce the following stability concept for the augmented system (11.6).

Definition 14 *System (11.6) is said to be exponentially mean-square stable if there exist scalars $\alpha > 0$ and $\beta > 0$, such that*

$$\mathbb{E}|\bar{x}(t,\rho)|^2 \leq \alpha e^{-\beta t} \sup_{-2\tau \leq t \leq 0} \mathbb{E}|\rho(t)|^2, \quad \forall\, \rho(t) \in \mathbb{R}^{4n \times n}, \tag{11.9}$$

or, equivalently,

$$\lim_{t\to\infty} \sup \frac{1}{t} \log(\mathbb{E}|\bar{x}(t,\rho)|^2) \leq -\beta,$$

where $\bar{x}(t) := \left[\bar{x}_m^T(t) \;\; \bar{x}_p^T(t)\right]^T$, and $\rho(t) := \left[\phi_m^T(t) \;\; \psi_m^T(t) \;\; \phi_p^T(t) \;\; \psi_p^T(t)\right]^T$ is the initial function of $\bar{x}(t)$.

The purpose of this chapter is to design a desired filter of the form (11.5) for the system (Σ) in (11.1), such that, for all admissible time-delays, nonlinearities, and stochastic disturbances, the augmented system (11.6) is exponentially mean-square stable.

11.2 Main Results

11.2.1 State Estimator Analysis

First, we give the following lemmas that will be used in the proofs of our main results in this chapter.

Lemma 11.1 *Let $x \in \mathbb{R}^n$, $y \in \mathbb{R}^n$, and matrix $Q > 0$. Then, we have $x^T y + y^T x \leq x^T Q^{-1} x + y^T Q y$.*

In the following theorem, a delay-dependent LMI method is used to deal with the filtering problem for the nonlinear stochastic genetic regulatory model (11.1), and a sufficient condition is derived that ensures the solvability of the filtering problem.

Theorem 25 *Consider system (11.6) with given filter parameters. For a prescribed constant $\beta > 0$, if there exist matrices $P_i > 0$, diagonal matrices $Q_i > 0$, $R_i > 0$, $S_i > 0$, and scalars $\varepsilon_i > 0$ $(i = 1, 2)$, such that the following linear matrix inequality*

$$\begin{bmatrix} \Omega_1 & 0 & 0 & 0 & 0 & \mathbb{A} & 0 & \mathcal{U}_1 & 0 & \hat{E} & 0 \\ * & \Omega_2 & 0 & 0 & 0 & 0 & \mathbb{C} & 0 & \mathcal{U}_2 & 0 & \hat{F} \\ * & * & -Q_1 & 0 & 0 & 0 & \mathbb{D} & 0 & 0 & 0 & 0 \\ * & * & * & -Q_2 & K & 0 & 0 & 0 & 0 & 0 & 0 \\ * & * & * & * & -2I & \mathbb{B} & 0 & 0 & 0 & 0 & 0 \\ * & * & * & * & * & -\mathbb{R}_1 & 0 & 0 & 0 & 0 & 0 \\ * & * & * & * & * & * & -\mathbb{R}_2 & 0 & 0 & 0 & 0 \\ * & * & * & * & * & * & * & -\mathcal{V}_1 & 0 & 0 & 0 \\ * & * & * & * & * & * & * & * & -\mathcal{V}_2 & 0 & 0 \\ * & * & * & * & * & * & * & * & * & -P_1 & 0 \\ * & * & * & * & * & * & * & * & * & * & -P_2 \end{bmatrix} < 0 \tag{11.10}$$

holds, where

$$\begin{aligned} \Omega_1 &:= \bar{A}^T P_1 + P_1 \bar{A} + Z^T S_2 Z + \beta P_1, \\ \Omega_2 &:= \bar{C}^T P_2 + P_2 \bar{C} + Z^T K S_1 K Z + \beta P_2, \\ \hat{E} &:= Z^T \bar{E}^T P_1, \ \hat{F} := Z^T \bar{F}^T P_2, \\ \mathcal{U}_1 &:= [P_1 \bar{B} \ \ \tau_1 P_1 \bar{B} K \ \ P_1 \bar{B} K \ \ \varepsilon_1 Z^T E^T \ \ Z^T Q_1], \\ \mathcal{U}_2 &:= [P_2 \bar{D} \ \ \tau_2 P_2 \bar{D} \ \ P_2 \bar{D} \ \ \varepsilon_2 Z^T F^T \ \ Z^T Q_2], \\ \mathcal{V}_1 &:= [S_1 \ \ \tau_1 R_2 \ \ \varepsilon_2 I \ \ \frac{\varepsilon_1 \beta}{e^{\beta\tau_2} - 1} I \ \ e^{-\beta\tau_2} Q_1], \\ \mathcal{V}_2 &:= [S_2 \ \ \tau_2 R_1 \ \ \varepsilon_1 I \ \ \frac{\varepsilon_2 \beta}{e^{\beta\tau_1} - 1} I \ \ e^{-\beta\tau_1} Q_2], \\ \mathbb{A} &:= \bar{A}^T Z^T R_1, \mathbb{B} := \bar{B}^T Z^T R_1, \\ \mathbb{C} &:= \bar{C}^T Z^T R_2, \ \mathbb{D} := \bar{D}^T Z^T R_2, \\ \mathbb{R}_1 &:= -\frac{\beta}{e^{\beta\tau_2} - 1} R_1, \ \mathbb{R}_2 := \frac{\beta}{e^{\beta\tau_1} - 1} R_2, \end{aligned} \tag{11.11}$$

then the augmented system (11.6) is exponentially mean-square stable.

Proof 28 *Recalling the Newton–Leibniz formula and (11.8), we can write*

$$\begin{cases} \bar{x}_m(t-\tau_2) = \bar{x}_m(t) - \int_{t-\tau_2}^t d\bar{x}_m(t) = \bar{x}_m(t) \\ \qquad - \int_{t-\tau_2}^t \xi_m(s)ds - \int_{t-\tau_2}^t \bar{E} Z \bar{x}_m(s) d\omega_1(s) \\ \bar{x}_p(t-\tau_1) = \bar{x}_p(t) - \int_{t-\tau_1}^t d\bar{x}_p(t) = \bar{x}_p(t) \\ \qquad - \int_{t-\tau_1}^t \xi_p(s)ds - \int_{t-\tau_1}^t \bar{F} Z \bar{x}_p(s) d\omega_2(s). \end{cases} \tag{11.12}$$

It is easy to know from (11.3) and (11.12) that (11.6) is equivalent to the following system:

$$\begin{cases} d\bar{x}_m(t) = \bar{A}\bar{x}_m(t)dt + \bar{B} f_{\tau_1} Z \Big[\bar{x}_p(t) - \int_{t-\tau_1}^t \xi_p(s)ds \\ \qquad - \int_{t-\tau_1}^t \bar{F}\bar{x}_p(s)d\omega_2(s)\Big] dt + \bar{E} Z \bar{x}_m(t) d\omega_1(t) \\ d\bar{x}_p(t) = \bar{C}\bar{x}_p(t)dt + \bar{D} Z \Big[\bar{x}_m(t) - \int_{t-\tau_2}^t \xi_m(s)ds \\ \qquad - \int_{t-\tau_2}^t \bar{E}\bar{x}_m(s)d\omega_1(s)\Big] dt + \bar{F} Z \bar{x}_p(t) d\omega_2(t), \end{cases} \tag{11.13}$$

where

$$f_{\tau_1} = \text{diag}\left\{f_1(\bar{x}_{p1(t-\tau_1)}), f_2(\bar{x}_{p2(t-\tau_1)}), \ldots, f_n(\bar{x}_{pn(t-\tau_1)})\right\}, \tag{11.14}$$

with $\bar{x}_{pi(t-\tau_1)}$ $(i = 1, 2, \ldots, n)$ *denoting the ith element of the vector* $\bar{x}_{p\tau_1}$, *and* $f_i(x)$ *is defined in (11.3). Hence, we only need to confirm that the system (11.13) is exponentially mean-square stable.*

In order to show that system (11.13) is exponentially mean-square stable under condition (11.10), we define the following Lyapunov–Krasovskii functional candidate for system (11.13):

$$\begin{aligned}
V(t) \quad = \quad & e^{\beta t}\bar{x}_m^T(t)P_1\bar{x}_m(t) + e^{\beta t}\bar{x}_p^T(t)P_2\bar{x}_p(t) \\
& + \int_{t-\tau_2}^{t} e^{\beta(s+\tau_2)}\bar{x}_m^T(s)Z^TQ_1Z\bar{x}_m(s)ds \\
& + \int_{t-\tau_1}^{t} e^{\beta(s+\tau_1)}\bar{x}_p^T(s)Z^TQ_2Z\bar{x}_p(s)ds \\
& + \int_{-\tau_2}^{0}\int_{t+s}^{t} e^{\beta(\theta-s)}\xi_m^T(\theta)Z^TR_1Z\xi_m(\theta)d\theta ds \\
& + \int_{-\tau_1}^{0}\int_{t+s}^{t} e^{\beta(\theta-s)}\xi_p^T(\theta)Z^TR_2Z\xi_p(\theta)d\theta ds \\
& + \varepsilon_1\int_{-\tau_2}^{0}\int_{t+s}^{t} e^{\beta(\theta-s)}\bar{x}_m^T(\theta)Z^TE^TEZ\bar{x}_m(\theta)d\theta ds \\
& + \varepsilon_2\int_{-\tau_1}^{0}\int_{t+s}^{t} e^{\beta(\theta-s)}\bar{x}_p^T(\theta)Z^TF^TFZ\bar{x}_p(\theta)d\theta ds.
\end{aligned} \tag{11.15}$$

By Itô's differential formula [81], we have

$$\begin{aligned}
dV(t) \quad = \quad & \mathcal{L}V(t)dt + 2\bar{x}_m^T(t)P_1\bar{E}Z\bar{x}_m(t)d\omega_1(t) \\
& + 2\bar{x}_p^T(t)P_2\bar{F}Z\bar{x}_p(t)d\omega_2(t),
\end{aligned} \tag{11.16}$$

where

$$\begin{aligned}
\mathcal{L}V(t) \quad = \quad & \beta e^{\beta t}\bar{x}_m^T(t)P_1\bar{x}_m(t) + e^{\beta t}\bar{x}_m^T(t)(\bar{A}^TP_1 + P_1\bar{A})\bar{x}_m^T(t) \\
& + e^{\beta t}\bar{x}_m^T(t)Z^T\bar{E}^TP_1\bar{E}Z\bar{x}_m(t) + 2e^{\beta t}\bar{x}_m^T(t)P_1\bar{B}f_{\tau_1}Z\bar{x}_p(t) \\
& - 2e^{\beta t}\bar{x}_m^T(t)P_1\bar{B}f_{\tau_1}Z\left[\int_{t-\tau_1}^{t}\xi_p(s)ds + \int_{t-\tau_1}^{t}\bar{F}Z\bar{x}_p(s)d\omega_2(s)\right] \\
& + \beta e^{\beta t}\bar{x}_p^T(t)P_2\bar{x}_p(t) + e^{\beta t}\bar{x}_p^T(t)(\bar{C}^TP_2 + P_2\bar{C})\bar{x}_p(t) \\
& + e^{\beta t}\bar{x}_p^T(t)Z^T\bar{F}^TP_2\bar{F}Z\bar{x}_p(t) + 2e^{\beta t}\bar{x}_p^T(t)P_2\bar{D}Z\bar{x}_m(t) \\
& - 2e^{\beta t}\bar{x}_p^T(t)P_2\bar{D}Z\left[\int_{t-\tau_2}^{t}\xi_m(s)ds + \int_{t-\tau_2}^{t}\bar{E}Z\bar{x}_m(s)d\omega_1(s)\right] \\
& + e^{\beta(t+\tau_2)}\bar{x}_m^T(t)Z^TQ_1Z\bar{x}_m(t) - e^{\beta t}\bar{x}_m^T(t-\tau_2)Z^TQ_1Z\bar{x}_m(t-\tau_2) \\
& + e^{\beta(t+\tau_1)}\bar{x}_p^T(t)Z^TQ_2Z\bar{x}_p(t) - e^{\beta t}\bar{x}_p^T(t-\tau_1)Z^TQ_2Z\bar{x}_p(t-\tau_1)
\end{aligned}$$

$$
\begin{aligned}
&+e^{\beta t}\frac{e^{\beta\tau_2}-1}{\beta}\xi_m^T(t)Z^TR_1Z\xi_m(t)-e^{\beta t}\int_{t-\tau_2}^{t}\xi_m^T(s)Z^TR_1Z\xi_m(s)ds\\
&+e^{\beta t}\frac{e^{\beta\tau_1}-1}{\beta}\xi_p^T(t)Z^TR_2Z\xi_p(t)-e^{\beta t}\int_{t-\tau_1}^{t}\xi_p^T(s)Z^TR_2Z\xi_p(s)ds\\
&+\varepsilon_1e^{\beta t}\frac{e^{\beta\tau_2}-1}{\beta}\bar{x}_m^T(t)Z^TE^TEZ\bar{x}_m(t)\\
&-\varepsilon_1e^{\beta t}\int_{t-\tau_2}^{t}\bar{x}_m^T(s)Z^TE^TEZ\bar{x}_m(s)ds\\
&+\varepsilon_2e^{\beta t}\frac{e^{\beta\tau_1}-1}{\beta}\bar{x}_p^T(t)Z^TF^TFZ\bar{x}_p(t)\\
&-\varepsilon_2e^{\beta t}\int_{t-\tau_1}^{t}\bar{x}_p^T(s)Z^TF^TFZ\bar{x}_p(s)ds. \qquad (11.17)
\end{aligned}
$$

From (11.3) and the definition of f_{τ_1} in (11.14), it can be easily seen that $f_{\tau_1} \le K$. In terms of the fact that the positive definite matrices Q_i, R_i, and S_i $(i = 1, 2)$ are diagonal, we have from Lemma 11.1 that

$$
\begin{aligned}
& 2\bar{x}_m^T(t)P_1\bar{B}f_{\tau_1}Z\bar{x}_p(t)\\
\le\ & \bar{x}_m^T(t)P_1\bar{B}S_1^{-1}\bar{B}^TP_1\bar{x}_m(t)+\bar{x}_p^T(t)Z^Tf_{\tau_1}^TS_1f_{\tau_1}Z\bar{x}_p(t)\\
\le\ & \bar{x}_m^T(t)P_1\bar{B}S_1^{-1}\bar{B}^TP_1\bar{x}_m(t)+\bar{x}_p^T(t)Z^TKS_1KZ\bar{x}_p(t),\\
& -2\bar{x}_m^T(t)P_1\bar{B}f_{\tau_1}Z\xi_p(s)\\
\le\ & \bar{x}_m^T(t)P_1\bar{B}f_{\tau_1}R_2^{-1}f_{\tau_1}^T\bar{B}^TP_1\bar{x}_m^T(t)+\xi_p^T(s)Z^TR_2Z\xi_p(s) \qquad (11.18)
\end{aligned}
$$

$$
\begin{aligned}
\le\ & \bar{x}_m^T(t)P_1\bar{B}KR_2^{-1}K\bar{B}^TP_1\bar{x}_m^T(t)+\xi_p^T(s)Z^TR_2Z\xi_p(s),\\
& 2\bar{x}_p^T(t)P_2\bar{D}Z\bar{x}_m(t)\\
\le\ & \bar{x}_p^T(t)P_2\bar{D}S_2^{-1}\bar{D}^TP_2\bar{x}_p(t)+\bar{x}_m^T(t)Z^TS_2Z\bar{x}_m(t),\\
& -2\bar{x}_p^T(t)P_2\bar{D}Z\xi_m(s)\\
\le\ & \bar{x}_p^T(t)P_2\bar{D}R_1^{-1}\bar{D}^TP_2\bar{x}_p^T(t)+\xi_m^T(s)Z^TR_1Z\xi_m(s). \qquad (11.19)
\end{aligned}
$$

Similarly, noticing that $Z\bar{E} = E$, $Z\bar{F} = F$, and ε_i $(i = 1, 2)$ are positive scalars, we obtain

$$
\begin{aligned}
& \mathbb{E}\{-2\bar{x}_m^T(t)P_1\bar{B}f_{\tau_1}Z\int_{t-\tau_1}^{t}\bar{F}Z\bar{x}_p(s)d\omega_2(s)\}\\
=\ & \mathbb{E}\{-2\bar{x}_m^T(t)P_1\bar{B}f_{\tau_1}\int_{t-\tau_1}^{t}FZ\bar{x}_p(s)d\omega_2(s)\}\\
\le\ & \varepsilon_2^{-1}\mathbb{E}\{\bar{x}_m^T(t)P_1\bar{B}f_{\tau_1}f_{\tau_1}^T\bar{B}^TP_1\bar{x}_m^T(t)\}\\
& +\varepsilon_2\mathbb{E}|\int_{t-\tau_1}^{t}FZ\bar{x}_p(s)d\omega_2(s)|^2
\end{aligned}
$$

$$\begin{aligned}
&\leq \varepsilon_2^{-1}\mathbb{E}\{\bar{x}_m^T(t)P_1\bar{B}KK\bar{B}^TP_1\bar{x}_m^T(t)\} \\
&\quad +\varepsilon_2\mathbb{E}\int_{t-\tau_1}^{t}\bar{x}_p^T(s)Z^TF^TFZ\bar{x}_p^T(s)ds, \qquad (11.20)
\end{aligned}$$

$$\begin{aligned}
&\mathbb{E}\{-2\bar{x}_p^T(t)P_2\bar{D}Z\int_{t-\tau_2}^{t}\bar{E}Z\bar{x}_m(s)\omega_1(s)ds\} \\
&= \mathbb{E}\{-2\bar{x}_p^T(t)P_2\bar{D}\int_{t-\tau_2}^{t}EZ\bar{x}_m(s)d\omega_1(s)\} \\
&\leq \varepsilon_1^{-1}\mathbb{E}\{\tau_2\bar{x}_p^T(t)P_2\bar{D}\bar{D}^TP_2\bar{x}_p^T(t)\} \\
&\quad +\varepsilon_1\mathbb{E}|\int_{t-\tau_2}^{t}EZ\bar{x}_m(s)d\omega_1(s)|^2 \\
&\leq \varepsilon_1^{-1}\mathbb{E}\{\tau_2\bar{x}_p^T(t)P_2\bar{D}\bar{D}^TP_2\bar{x}_p^T(t)\} \\
&\quad +\varepsilon_1\mathbb{E}\int_{t-\tau_2}^{t}\bar{x}_m^T(s)Z^TE^TEZ\bar{x}_m(s)ds. \qquad (11.21)
\end{aligned}$$

Noting the sector condition (11.4) and substituting (11.18)-(11.21) into (11.17) and taking mathematical expectation on both sides result in

$$\begin{aligned}
&\mathbb{E}\{\mathcal{L}V(t)\} \\
\leq\ & e^{\beta t}\mathbb{E}\{\bar{x}_m^T(t)(\bar{A}^TP_1+P_1\bar{A}+P_1\bar{B}S_1^{-1}\bar{B}^TP_1+e^{\beta\tau_2}Z^TQ_1Z \\
&+\tau_1P_1\bar{B}KR_2^{-1}K\bar{B}^TP_1+\varepsilon_2^{-1}P_1\bar{B}K^2\bar{B}^TP_1+Z^T\bar{E}^TP_1\bar{E}Z \\
&+Z^TS_2Z+\varepsilon_1\frac{e^{\beta\tau_2}-1}{\beta}Z^TE^TEZ+\beta P_1)\bar{x}_m^T(t)+\bar{x}_p^T(t)(\bar{C}^TP_2 \\
&+P_2\bar{C}+P_2\bar{D}S_2^{-1}\bar{D}^TP_2+\tau_2P_2\bar{D}R_1^{-1}\bar{D}^TP_2+\varepsilon_1^{-1}P_2\bar{D}\bar{D}^TP_2 \\
&+e^{\beta\tau_1}Z^TQ_2Z+Z^T\bar{F}^TP_2\bar{F}Z+Z^TKS_1KZ+\beta P_2 \\
&+\varepsilon_2\frac{e^{\beta\tau_1}-1}{\beta}Z^TF^TFZ)\bar{x}_p(t)-\bar{x}_{m\tau_2}^TZ^TQ_1Z\bar{x}_{m\tau_2} \\
&-2g^T(Z\bar{x}_{p\tau_1})(g(Z\bar{x}_{p\tau_1})-KZ\bar{x}_{p\tau_1})-\bar{x}_{p\tau_1}^TZ^TQ_2Z\bar{x}_{p\tau_1} \\
&+\frac{e^{\beta\tau_2}-1}{\beta}[\bar{A}\bar{x}_m(t)+\bar{B}g(Z\bar{x}_{p\tau_1})]^TZ^TR_1Z[\bar{A}\bar{x}_m(t)+\bar{B}g(Z\bar{x}_{p\tau_1})] \\
&+\frac{e^{\beta\tau_1}-1}{\beta}[\bar{C}\bar{x}_p(t)+\bar{D}Z\bar{x}_{m\tau_2}]^TZ^TR_2Z[\bar{C}\bar{x}_p(t)+\bar{D}Z\bar{x}_{m\tau_2})]\}. \\
\leq\ & e^{\beta t}\mathbb{E}\{\eta^T(t)\Gamma\eta(t)\}, \qquad (11.22)
\end{aligned}$$

where

$$\eta(t)=[\bar{x}_m^T(t)\ \ \bar{x}_p^T(t)\ \ Z\bar{x}_{m\tau_2}^T\ \ Z\bar{x}_{p\tau_1}^T\ \ g(Z\bar{x}_{p\tau_1})]^T,$$

$$\Gamma=\begin{bmatrix}\Gamma_1 & 0 & 0 & 0 & \mathbb{N}_1 \\ * & \Gamma_2 & \frac{e^{\beta\tau_1}-1}{\beta}\bar{C}^TZ^TR_2Z\bar{D} & 0 & 0 \\ * & * & -Q_1+\frac{e^{\beta\tau_1}-1}{\beta}\bar{D}^TZ^TR_2Z\bar{D} & 0 & 0 \\ * & * & * & -Q_2 & K \\ * & * & * & * & -\mathbb{N}_2\end{bmatrix}, \qquad (11.23)$$

with

$$
\begin{aligned}
\Gamma_1 &= \bar{A}^T P_1 + P_1\bar{A} + P_1\bar{B}S_1^{-1}\bar{B}^T P_1 + \tau_1 P_1\bar{B}KR_2^{-1}K\bar{B}^T P_1 + e^{\beta\tau_2}Z^T Q_1 Z \\
&\quad + Z^T\bar{E}^T P_1\bar{E}Z + \frac{e^{\beta\tau_2}-1}{\beta}\bar{A}^T Z^T R_1 Z\bar{A} + \varepsilon_2^{-1}P_1\bar{B}K^2\bar{B}^T P_1 \\
&\quad + Z^T S_2 Z + \varepsilon_1\frac{e^{\beta\tau_2}-1}{\beta}Z^T E^T E Z + \beta P_1, \\
\Gamma_2 &= \bar{C}^T P_2 + P_2\bar{C} + P_2\bar{D}S_2^{-1}\bar{D}^T P_2 + \tau_2 P_2\bar{D}R_1^{-1}\bar{D}^T P_2 + \varepsilon_1 P_1^{-1}\bar{D}\bar{D}^T P_2 \\
&\quad + Z^T\bar{F}^T P_2\bar{F}Z + \frac{e^{\beta\tau_1}-1}{\beta}\bar{C}^T Z^T R_2 Z\bar{C} + e^{\beta\tau_1}Z^T Q_2 Z \\
&\quad + Z^T K S_1 K Z + \varepsilon_2\frac{e^{\beta\tau_1}-1}{\beta}Z^T F^T F Z + \beta P_2, \\
\mathbb{N}_1 &= \frac{e^{\beta\tau_2}-1}{\beta}\bar{A}^T Z^T R_1 Z\bar{B},\ \mathbb{N}_2 = 2I - \frac{e^{\beta\tau_2}-1}{\beta}\bar{B}^T Z^T R_1 Z\bar{B}.
\end{aligned}
$$

By Schur complement, we can obtain from (11.10) that $\Gamma < 0$ *and, therefore,* $\mathbb{E}\{\mathcal{L}V(t)\} < 0$.

The mean square exponential stability of system (11.13) can be proved as follows. Define

$$
\begin{aligned}
\lambda_p = \min\{\lambda_{\min}(P_1),\ \lambda_{\min}(P_2)\},\ \|P\| = \max\{\|P_1\|,\ \|P_2\|\}, \\
\|Q\| = \max\{\|Q_1\|,\ \|Q_2\|\},\ \|R\| = \max\{\|R_1\|,\ \|R_2\|\},\ \varepsilon = \max\{\varepsilon_1,\ \varepsilon_2\}.
\end{aligned}
$$

From the definition of $\xi_m(t)$, $\xi_p(t)$, $\bar{x}(t)$, $\rho(t)$, *and (11.15), there exists a positive scalar* δ, *such that*

$$
\mathbb{E}V(t) \geq e^{\beta t}\lambda_p\mathbb{E}|\bar{x}(t)|^2 \tag{11.24}
$$

and

$$
\begin{aligned}
\mathbb{E}V(0) &\leq \{\|P\| + \tau e^{\beta\tau}\|Q\| + \tau^2 e^{\beta\tau}\left[2\|R\|(\|\bar{A}\|^2 + \|K\|^2\|\bar{B}\|^2 \right. \\
&\quad + \|\bar{C}\|^2 + \|\bar{D}\|^2) \\
&\quad \left. + \varepsilon\|E\|^2 + \varepsilon\|F\|^2\right]\} \sup_{-2\tau\leq t\leq 0}|\rho(t)|^2 \\
&\leq \delta \sup_{-2\tau\leq t\leq 0}\mathbb{E}|\rho(t)|^2.
\end{aligned} \tag{11.25}
$$

By Itô's formula [81], we obtain that

$$
\mathbb{E}V(t) = \mathbb{E}V(0) + \mathbb{E}\int_0^t \mathcal{L}V(s)ds \leq \delta \sup_{-2\tau\leq t\leq 0}\mathbb{E}|\rho(t)|^2. \tag{11.26}
$$

It follows from (11.24), (11.25), and (11.26) that

$$
e^{\beta t}\lambda_p\mathbb{E}|\bar{x}(t,\rho)|^2 \leq \delta \sup_{-2\tau\leq t\leq 0}\mathbb{E}|\rho(t)|^2,
$$

or, equivalently,

$$\lim_{t\to\infty} \sup \frac{1}{t} \log(\mathbb{E}|\bar{x}(t,\rho)|^2) \leq -\beta,$$

which indicates that the trivial solution of (11.13) is exponentially mean-square stable, and the exponential decay rate is β*. The proof is now completed.*

Remark 11.2 *In Theorem 25, the decay rate* β *is characterized and is very flexible for adjudgement according to different practical background. Such decay rate characterization approach has been exploited in [148] for a class of stochastic neural networks with time-delays.*

11.2.2 State Estimator Parameter Design

The following theorem shows that the desired filter parameters can be determined by solving an LMI.

Theorem 26 *Consider system (11.6). For a prescribed constant* $\beta > 0$*, if there exist matrices* $X_i > 0$*,* $\mathcal{Y}_i > 0$*, and diagonal matrices* $Q_i > 0$*,* $R_i > 0$*,* $S_i > 0$*, and scalars* $\varepsilon_i > 0$ $(i = 1, 2)$*, such that the following linear matrix inequality,*

$$\left[\begin{array}{ccccccccccccc}
\Pi_{11} & \Pi_{12} & 0 & 0 & 0 & 0 & 0 & -A_1^T R_1 & 0 & \Phi_1 & 0 & \Phi_5 & 0 \\
* & \Pi_{22} & 0 & 0 & 0 & 0 & 0 & -A_1^T R_1 & 0 & \Phi_2 & 0 & \Phi_6 & 0 \\
* & * & \Pi_{33} & \Pi_{34} & 0 & 0 & 0 & 0 & -A_2^T R_2 0 & \Phi_3 & 0 & \Phi_7 & \\
* & * & * & \Pi_{44} & 0 & 0 & 0 & 0 & -A_2^T R_2 & 0 & \Phi_4 & 0 & \Phi_8 \\
* & * & * & * & -Q_1 & 0 & 0 & 0 & D^T R_2 & 0 & 0 & 0 & 0 \\
* & * & * & * & * & -Q_2 & K & 0 & 0 & 0 & 0 & 0 & 0 \\
* & * & * & * & * & * & -2I & B^T R_1 & 0 & 0 & 0 & 0 & 0 \\
* & * & * & * & * & * & * & -\frac{\beta}{e^{\beta\tau_2}-1} R_1 & 0 & 0 & 0 & 0 & 0 \\
* & * & * & * & * & * & * & * & -\frac{\beta}{e^{\beta\tau_1}-1} R_2 & 0 & 0 & 0 & 0 \\
* & * & * & * & * & * & * & * & * & -\Lambda_1 & 0 & 0 & 0 \\
* & * & * & * & * & * & * & * & * & * & -\Lambda_2 & 0 & 0 \\
* & * & * & * & * & * & * & * & * & * & * & -\Lambda_3 & 0 \\
* & * & * & * & * & * & * & * & * & * & * & * & -\Lambda_4
\end{array}\right] < 0, \qquad (11.27)$$

holds, where

$$
\begin{aligned}
\Pi_{11} &:= -A_1^T\mathcal{Y}_1 - \mathcal{Y}_1 A_1 + \beta\mathcal{Y}_1, \\
\Pi_{12} &:= -A_1^T X_1 - \mathcal{Y}_1 A_1 + C_1^T\tilde{B}^T + \tilde{A}^T + \beta\mathcal{Y}_1, \\
\Pi_{33} &:= -A_2^T\mathcal{Y}_2 - \mathcal{Y}_2 A_2 + \beta\mathcal{Y}_2, \\
\Lambda_3 &:= \begin{bmatrix} \mathcal{Y}_1 & \mathcal{Y}_1 \\ \mathcal{Y}_1 & X_1 \end{bmatrix}, \ \Lambda_4 := \begin{bmatrix} \mathcal{Y}_2 & \mathcal{Y}_2 \\ \mathcal{Y}_2 & X_2 \end{bmatrix}, \\
\Pi_{22} &:= -A_1^T X_1 - X_1 A_1 + \tilde{B}C_1 + C_1^T\tilde{B}^T + \beta X_1, \\
\Pi_{34} &:= -A_2^T X_2 - \mathcal{Y}_2 A_2 + C_2^T\tilde{D}^T + \tilde{C}^T + \beta\mathcal{Y}_2, \\
\Pi_{44} &:= -A_2^T X_2 - X_2 A_2 + \tilde{D}C_2 + C_2^T\tilde{D}^T + \beta X_2, \\
\Phi_1 &:= [\mathcal{Y}_1 B \ \ \tau_1\mathcal{Y}_1 BK \ \ \mathcal{Y}_1 BK \ \ \varepsilon_1 E^T \ \ Q_1, \ S_2], \\
\Phi_2 &:= [X_1 B \ \ \tau_1 X_1 BK \ \ X_1 BK \ \ \varepsilon_1 E^T \ \ Q_1, \ S_2], \\
\Phi_3 &:= [\mathcal{Y}_2 D \ \ \tau_2\mathcal{Y}_2 D \ \ \mathcal{Y}_2 D \ \ \varepsilon_2 F^T \ \ Q_2, \ KS_1], \\
\Phi_4 &:= [X_2 D \ \ \tau_2 X_2 D \ \ X_2 D \ \ \varepsilon_2 F^T \ \ Q_2 \ \ KS_1] \\
\Phi_5 &:= [E^T\mathcal{Y}_1 \ \ E^T X_1], \ \Phi_6 := [E^T\mathcal{Y}_1 \ \ E^T X_1], \\
\Phi_7 &:= [F^T\mathcal{Y}_2 \ \ F^T X_2], \ \Phi_8 := [F^T\mathcal{Y}_2 \ \ F^T X_2], \\
\Lambda_1 &:= \text{diag}\{S_1, \tau_1 R_2, \varepsilon_2 I, \frac{\varepsilon_1\beta}{e^{\beta\tau_2}-1}I, e^{-\beta\tau_2}Q_1, S_2\}, \\
\Lambda_2 &:= \text{diag}\{S_2, \tau_2 R_1, \varepsilon_1 I, \frac{\varepsilon_2\beta}{e^{\beta\tau_1}-1}I, e^{-\beta\tau_1}Q_2, S_1\}, \qquad (11.28)
\end{aligned}
$$

then system (11.6) is exponentially mean-square stable.

In this case, the parameters of the desired filter (Σ_f) are given as follows:

$$
\begin{aligned}
\hat{A} &:= (\mathcal{Y}_1 - X_1)^{-1}\tilde{A}, \ \hat{B} := (\mathcal{Y}_1 - X_1)^{-1}\tilde{B}, \\
\hat{C} &:= (\mathcal{Y}_2 - X_2)^{-1}\tilde{C}, \ \hat{D} := (\mathcal{Y}_2 - X_2)^{-1}\tilde{D}. \qquad (11.29)
\end{aligned}
$$

Proof 29 *Define*

$$
P_i = \begin{bmatrix} X_i & \mathcal{Y}_i - X_i \\ \mathcal{Y}_i - X_i & X_i - \mathcal{Y}_i \end{bmatrix} > 0, \quad \Upsilon_i = \begin{bmatrix} Y_i & I \\ Y_i & 0 \end{bmatrix}, \quad i = 1, 2, \quad (11.30)
$$

where $Y_i = \mathcal{Y}_i^{-1} > 0$.

Premultiplying and postmultiplying the LMI in (11.27) by

$$
\text{diag}\{Y_1, I, Y_2, I, I, I, I, I, I, \text{diag}_6\{I\}, \text{diag}_6\{I\}, \text{diag}\{Y_1, I\}, \text{diag}\{Y_2, I\}\},
$$

we have

$$
\left[\begin{array}{cccccccc}
\bar{\Pi}_{11} & \bar{\Pi}_{12} & 0 & 0 & 0 & 0 & 0 & -Y_1A_1^TR_1 \\
* & \Pi_{22} & 0 & 0 & 0 & 0 & 0 & -A_1^TR_1 \\
* & * & \bar{\Pi}_{33} & \bar{\Pi}_{34} & 0 & 0 & 0 & 0 \\
* & * & * & \Pi_{44} & 0 & 0 & 0 & 0 \\
* & * & * & * & -Q_1 & 0 & 0 & 0 \\
* & * & * & * & * & -Q_2 & K & 0 \\
* & * & * & * & * & * & -2I & B^TR_1 \\
* & * & * & * & * & * & * & -\frac{\beta}{e^{\beta\tau_2}-1}R_1 \\
* & * & * & * & * & * & * & * \\
* & * & * & * & * & * & * & * \\
* & * & * & * & * & * & * & * \\
* & * & * & * & * & * & * & * \\
* & * & * & * & * & * & * & *
\end{array}\right.
$$

$$
\left.\begin{array}{cccccc}
0 & \bar{\Phi}_1 & 0 & \bar{\Phi}_5 & 0 & \\
0 & \Phi_2 & 0 & \bar{\Phi}_6 & 0 & \\
-Y_2A_2^TR_2 & 0 & \bar{\Phi}_3 & 0 & \bar{\Phi}_7 & \\
-A_2^TR_2 & 0 & \Phi_4 & 0 & \bar{\Phi}_8 & \\
D^TR_2 & 0 & 0 & 0 & 0 & \\
0 & 0 & 0 & 0 & 0 & \\
0 & 0 & 0 & 0 & 0 & \\
0 & 0 & 0 & 0 & 0 & \\
 & -\frac{\beta}{e^{\beta\tau_1}-1}R_2 & 0 & 0 & 0 & 0 \\
* & -\Lambda_1 & 0 & 0 & 0 & \\
* & * & -\Lambda_2 & 0 & 0 & \\
* & * & * & -\bar{\Lambda}_3 & 0 & \\
* & * & * & * & -\bar{\Lambda}_4 &
\end{array}\right] < 0, \qquad (11.31)
$$

where

$$
\begin{aligned}
\bar{\Pi}_{11} &:= -Y_1A_1^T - A_1Y_1 + \beta Y_1, \bar{\Phi}_5 := [Y_1E^T \;\; Y_1E^TX_1], \\
\bar{\Pi}_{12} &:= -Y_1A_1^TX_1 - A_1 + Y_1C_1^T\tilde{B}^T + Y_1\tilde{A}^T + \beta I, \\
\bar{\Pi}_{33} &:= -Y_2A_2^T - A_2Y_2 + \beta Y_2, \; \bar{\Phi}_6 := [E^T \;\; E^TX_1], \\
\bar{\Pi}_{34} &:= -Y_2A_2^TX_2 - A_2 + Y_2C_2^T\tilde{D}^T + Y_2\tilde{C}^T + \beta I, \\
\bar{\Phi}_1 &:= [B \;\; \tau_1BK \;\; BK \;\; \varepsilon_1Y_1E^T \;\; Y_1Q_1, \; Y_1S_2], \\
\bar{\Phi}_3 &:= [D \;\; \tau_2D \;\; D \;\; \varepsilon_2Y_2F^T \;\; Y_2Q_2, \; Y_2KS_1], \\
\bar{\Phi}_7 &:= [Y_2F^T \;\; Y_2F^TX_2], \; \bar{\Phi}_8 := [F^T \;\; F^TX_2], \\
\bar{\Lambda}_3 &:= \begin{bmatrix} Y_1 & I \\ I & X_1 \end{bmatrix}, \; \bar{\Lambda}_4 := \begin{bmatrix} Y_2 & I \\ I & X_2 \end{bmatrix}.
\end{aligned}
$$

It can be seen from the definitions of P_1, P_2, Υ_1, and Υ_2 that the LMI in

(11.31) is equivalent to

$$\begin{bmatrix} \Upsilon_1^T\Omega_1\Upsilon_1 & 0 & 0 & 0 & 0 & \Psi_1 & 0 & \Upsilon_1^T\mathcal{U}_1 & 0 & \Psi_5 & 0 \\ * & \Upsilon_2^T\Omega_2\Upsilon_2 & 0 & 0 & 0 & 0 & \Psi_3 & 0 & \Upsilon_2^T\mathcal{U}_2 & 0 & \Psi_6 \\ * & * & -Q_1 & 0 & 0 & 0 & \Psi_4 & 0 & 0 & 0 & 0 \\ * & * & * & -Q_2 & K & 0 & 0 & 0 & 0 & 0 & 0 \\ * & * & * & * & -2I & \Psi_2 & 0 & 0 & 0 & 0 & 0 \\ * & * & * & * & * & -\frac{\beta}{e^{\beta\tau_2}-1}R_1 & 0 & 0 & 0 & 0 & 0 \\ * & * & * & * & * & * & -\frac{\beta}{e^{\beta\tau_2}-1}R_2 & 0 & 0 & 0 & 0 \\ * & * & * & * & * & * & * & -\mathcal{V}_1 & 0 & 0 & 0 \\ * & * & * & * & * & * & * & * & -\mathcal{V}_2 & 0 & 0 \\ * & * & * & * & * & * & * & * & * & -\Delta_1 & 0 \\ * & * & * & * & * & * & * & * & * & * & -\Delta_2 \end{bmatrix} < 0, \quad (11.32)$$

where

$$\begin{aligned} \Psi_1 &:= \Upsilon_1^T\bar{A}^TZ^TR_1,\ \Psi_2 := \bar{B}^TZ^TR_1, \\ \Psi_3 &:= \Upsilon_2^T\bar{C}^TZ^TR_2,\ \Psi_4 := \bar{D}^TZ^TR_2 \\ \Psi_5 &:= \Upsilon_1^TZ^T\bar{E}^TP_1\Upsilon_1,\ \Psi_6 := \Upsilon_2^TZ^T\bar{F}^TP_2\Upsilon_2, \\ \Delta_1 &:= \Upsilon_1^TP_1\Upsilon_1,\ \Delta_2 := \Upsilon_2^TP_2\Upsilon_2. \end{aligned} \quad (11.33)$$

Finally, premultiplying and postmultiplying (11.32) by

$$\text{diag}\{\Upsilon_1^{-T}, \Upsilon_2^{-T}, I, I, I, I, I, \text{diag}_5\{I\}, \text{diag}_5\{I\}, \Upsilon_1^{-T}, \Upsilon_2^{-T}\}$$

and its transpose, we can obtain from Theorem 25 and Schur complement that system (11.6) is exponentially mean square stable with the given filter parameters in (11.29).

11.3 An Illustrative Example

In this section, a simulation example is presented to illustrate the usefulness and flexibility of the filter design method developed in this chapter.

The dynamics of repressilator has been theoretically predicted and experimentally investigated in Escherichia coli [36]. The repressilator is a cyclic negative-feedback loop comprising three repressor genes (`lacl`, `tetR`, and `cl`) and their promoters. The kinetics of the system are described as follows:

$$\begin{cases} \dot{x}_{m_i} &= -x_{m_i} + \frac{a}{1+x_{p_j}^n} \\ \dot{x}_{p_i} &= -b(x_{p_i} - x_{m_i}), \end{cases}$$

where $i =$`lacl`, `tetR`, and `cl`; $j =$ `cl`, `lacl`, `tetR`. x_{m_i}, and x_{p_i} are the concentrations of the three mRNA and repressor protein, $b > 0$ denotes the ratio of the protein decay rate to mRNA decay rate, and a is the feedback regulation coefficient. Taking into account the time-delay and stochastic disturbance, we consider the following compact matrix form nonlinear genetic regulatory network model (11.1):

$$\begin{cases} dx_m(t) &= [-A_1x_m(t) + Bg(x_p(t-\tau_1))]dt + Ex_m(t)d\omega_1(t) \\ dx_p(t) &= [-A_2x_p(t) + Dx_m(t-\tau_2)]dt + Fx_p(t)d\omega_2(t) \\ y_m(t) &= C_1x_m(t) \\ y_p(t) &= C_2x_p(t), \end{cases}$$

with transcriptional time-delays $\tau_1 = 0.3$, $\tau_2 = 0.7$, exponential decay rate $\beta = 1.0$, as well as the following data:

$$A_1 = \text{diag}\{3,\ 3,\ 3\},\ C_1 = \text{diag}\{0.72,\ 0.1,\ 1.3\},$$

$$B = \begin{bmatrix} 0 & 0 & -1.2 \\ -0.02 & 0 & 0 \\ 0 & -0.2 & 0 \end{bmatrix},$$

$$A_2 = \text{diag}\{3.5,\ 3.5,\ 3.5\},\ C_2 = \text{diag}\{1.2,\ 3.1,\ 0.03\},$$

$$D = \text{diag}\{1.21,\ 2.12,\ 0.22\},\ E = \text{diag}\{0.53,\ 0.41,\ 0.42\},$$

$$F = \text{diag}\{0.51,\ 0.41,\ 0.42\},\ K = \text{diag}\{0.6,\ 0.6,\ 0.6\},$$

where the sector-like nonlinearities are taken as

$$g_i(x_{p_i}) = x_{p_i}^2/(1 + x_{p_i}^2), \quad i = 1,\ 2, \ldots, n.$$

It is easy to check that the maximal value of the derivative of $g_i(x_{p_i})$ is less than $k_i = 0.6$.

Using the LMI toolbox, we solve (11.27) and obtain

$$X_1 = \text{diag}\{89.2144, 899.4217, 436.4357\},$$

$$X_2 = \text{diag}\{3.5167, 28.2677, 492.5005\},$$

$$Y_1 = \text{diag}\{29.1878, 419.5382, 186.9026\},$$

$$Y_2 = \text{diag}\{1.1156, 8.5315, 231.5299\},$$

$$Q_1 = \text{diag}\{1.1973, 109.4880, 47.2546\},$$

$$Q_2 = \text{diag}\{0.2548, 2.3778, 97.5095\},$$

$$R_1 = \text{diag}\{0.7410, 55.6729, 12.1232\},$$

$$R_2 = \text{diag}\{0.0717, 1.5388, 62.9470\}.$$

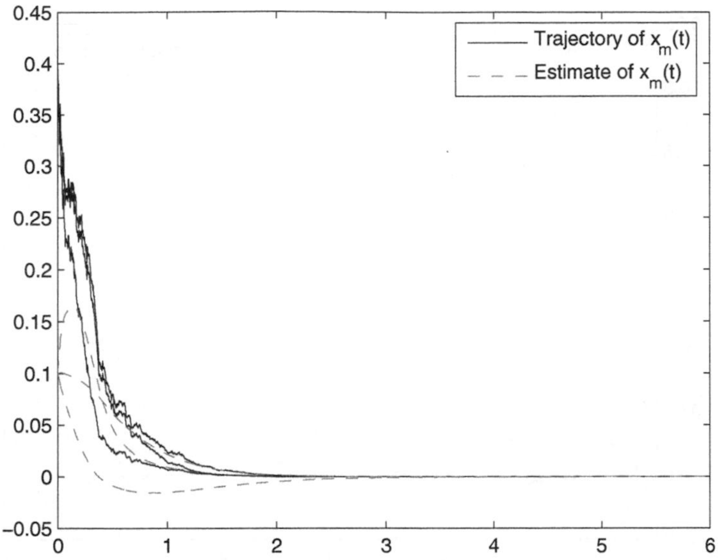

FIGURE 11.1
The state and estimate of $x_m(t)$.

$$\begin{aligned} S_1 &= \mathrm{diag}\{0.6005, 16.8998, 593.8720\}, \\ S_2 &= \mathrm{diag}\{6.8134, 256.8588, 103.3940\}, \\ \varepsilon_1 &= 22.2904,\ \varepsilon_2 = 13.0905. \end{aligned}$$

According to Theorem 26, the filter parameters can be calculated as follows:

$$\begin{aligned} \hat{A} &= \mathrm{diag}\{-8.0781, -1.8193, -2.9005\}, \\ \hat{B} &= \mathrm{diag}\{8.8710, -7.9237, 0.6790\}, \\ \hat{C} &= \mathrm{diag}\{-159.4923, -22.3503, -2.7487\}, \\ \hat{D} &= \mathrm{diag}\{131.5371, 6.8160, 6.7016\}. \end{aligned}$$

Figures 11.1-11.4 give the simulation results for the performance of the designed filter, Figure 11.5 gives the simulation for the actual decay rate of augmented systems state $\bar{x}(t)$ via $-\sup \frac{1}{t} \log(\mathbb{E}|\bar{x}(t,\rho)|^2)$ and the decay rate estimation β, where the initial condition $\rho(t) = [0.4\ 0.4\ 0.4\ 0.1\ 0.1\ 0.1\ 0.2\ 0.2\ 0.2\ 0.1\ 0.1\ 0.1]^T$ $(-2\tau \le t \le 0)$.

It is confirmed from the simulation results that the expected exponentially mean-square stable performance is well achieved.

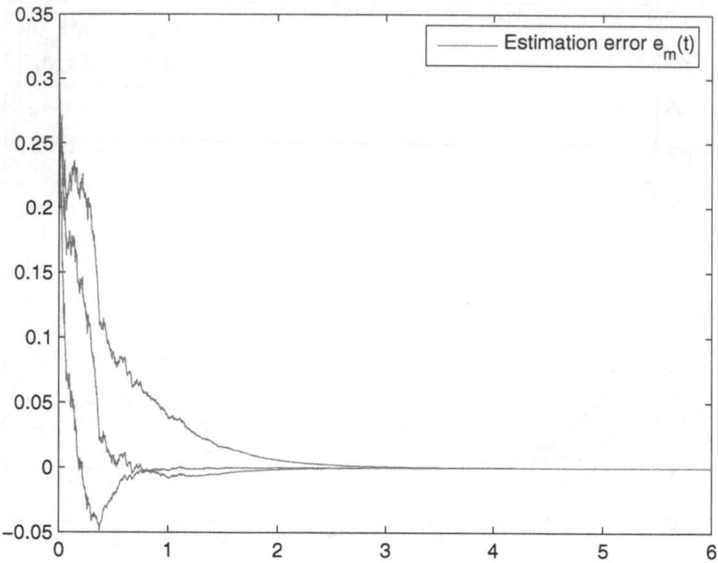

FIGURE 11.2
Estimation error $e_m(t)$.

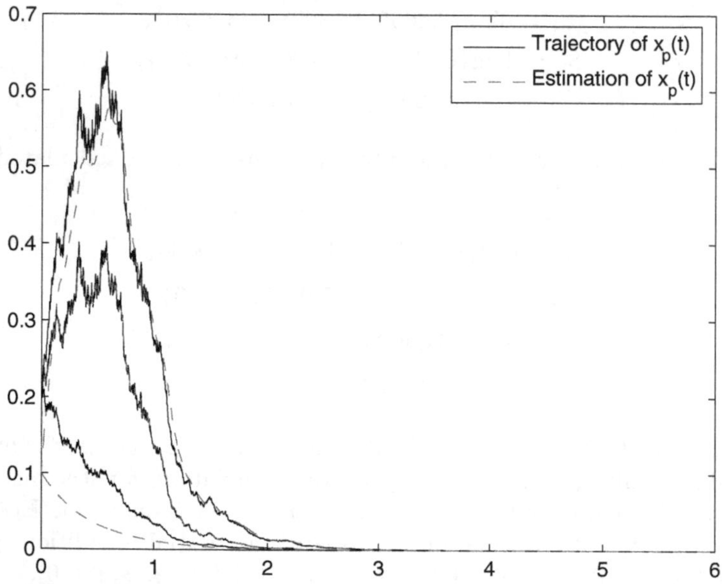

FIGURE 11.3
The state and estimate of $x_p(t)$.

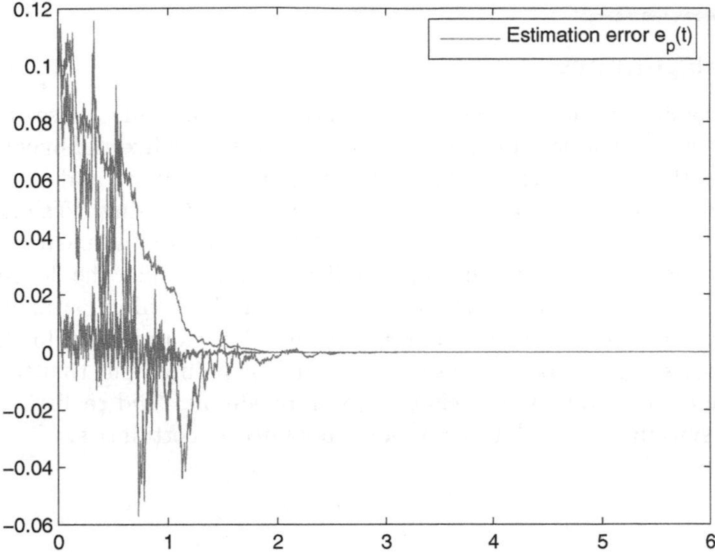

FIGURE 11.4
Estimation error $e_p(t)$.

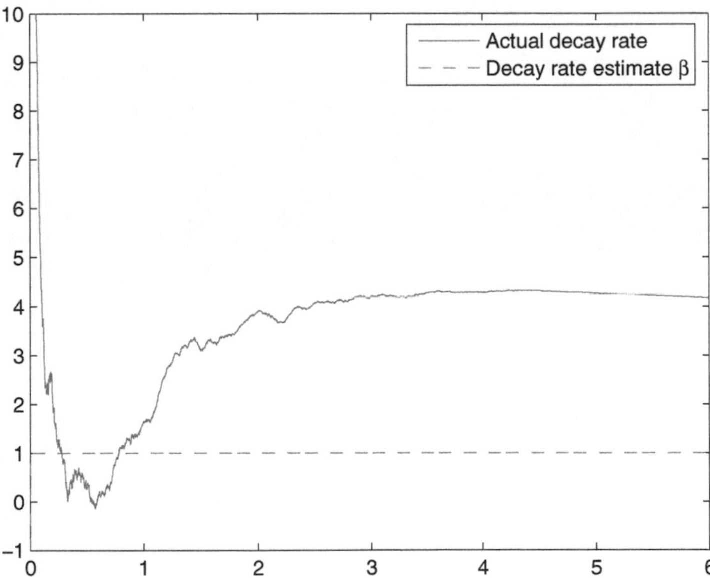

FIGURE 11.5
Actual decay rate of $\bar{x}(t)$ and estimated decay rate β.

11.4 Summary

In this chapter, we have investigated the filtering problem on a class of stochastic time-delay genetic regulatory networks with sector-like nonlinearity used to model the genetic regulatory model. The time-delays τ_1 and τ_2 represent the translation delay and feedback regulation delay. By using Itô's differential formula and Lyapunov stability theory, we have proposed a linear matrix inequality method to derive sufficient conditions under which the desired filters exist. We have also characterized the expression of the filter parameters and the decay rate $\beta > 0$, and employed a simulation example to illustrate the effectiveness of the proposed results. It should be pointed out that we can extend the main results in this chapter to more complex and realistic systems, such as systems with polytopic or norm-bounded uncertainties.

12

State Estimation for Complex Networks with ROCD

CONTENTS

In this chapter, the H_∞ state estimation problem is investigated for a class of discrete-time complex networks with randomly occurring phenomena. The proposed randomly occurring phenomena include both probabilistic missing measurements and randomly occurring coupling delays, which are described by two random variable sequences satisfying individual probability distributions, respectively. Rather than the common Lipschitz-type function, a more general sector-like nonlinear function is employed to characterize the nonlinearities in the networks. The purpose of the addressed H_∞ state estimation problem is to design a state estimator, such that, for all admissible nonlinear disturbances, missing measurements as well as coupling delays, the dynamics of the augmented systems is guaranteed to be exponentially mean-square stable and attenuated to a given H_∞ performance level. By constructing a novel Lyapunov–Krasovskii functional and utilizing convex optimization method as well as Kronecker product, we derive the sufficient conditions under which the desired state estimator exists. An illustrative example is exploited to show the effectiveness of the proposed state estimation scheme.

In this chapter, we consider the state estimation problem for a class of complex networks with randomly occurring phenomena involving missing measurements and randomly occurring coupling delays. The main contributions are as follows: *1) instead of the simple Bernoulli distribution, we introduce a novel model by using a certain discrete distribution sequence on the interval [0 1] to describe the randomly occurring missing measurements; 2) a new model is exploited to model the randomly occurring coupling delays in complex networks; and 3) a new state estimation scheme is designed to deal with the state estimation problem for complex networks with missing measurements and randomly occurring coupling delays.*

The rest of the chapter is organized as follows: In Section 12.2, a discrete complex network model with both randomly occurring coupling delays and missing measurements is proposed, and the considered H_∞ state estimation problem for this model is formulated. In Section 12.3, by employing the Lyapunov stability theory and convex program method, some sufficient conditions are derived and the design scheme of the estimator gains is derived. In Section 12.4, a numerical example is presented to demonstrate the effectiveness of the results achieved. Finally, a conclusion is drawn in Section 12.5.

12.1 Problem Formulation

Consider the following complex networks with N coupled nodes:

$$\begin{cases} x_i(k+1) = Ax_i(k) + Bf(x_i(k)) + Df(x_i(k-d(k))) + \sum\limits_{j=1}^{N} w_{ij}\Gamma_1 x_j(k) \\ \qquad\qquad +\varphi_i(k) \sum\limits_{j=1}^{N} \bar{w}_{ij}\Gamma_2 x_j(k-\tau) + E_1 v_1(k) , \\ z_i(k) = Mx_i(k), \\ x_i(s) = \phi_i(s), \ \forall s \in [-d_M, 0], i = 1, 2, ..., N, \end{cases} \tag{12.1}$$

where $x_i(k) \in \mathbb{R}^n$ is the state vector of the ith node, $z_i(k) \in \mathbb{R}^r$ is the regulated output of the ith node, and $v_1(k)$ is the disturbance input belonging to $l_2([0,+\infty); \mathbb{R}^q)$. $\Gamma_1 = \text{diag}(\gamma_1, \gamma_2, \cdots, \gamma_n)$ and $\Gamma_2 = \text{diag}(\bar{\gamma}_1, \bar{\gamma}_2, \cdots, \bar{\gamma}_n)$ are the inner coupling matrices between two connected nodes for all $1 \le i, j \le N$. $W = (w_{ij})_{N\times N}$ and $\bar{W} = (\bar{w}_{ij})_{N\times N}$ are the outer coupled matrices representing the coupling structure of the complex networks. If there is a connection between node i and node j ($i \neq j$), $w_{ij} > 0$, and $\bar{w}_{ij} > 0$. Otherwise, $w_{ij} = \bar{w}_{ij} = 0$. In this chapter, we assume W and $\bar{W}$ to be symmetric matrices and satisfy the condition $\sum_{j=1,i\neq j}^{N} w_{ij} = -w_{ii}$ and $\sum_{j=1,i\neq j}^{N} \bar{w}_{ij} = -\bar{w}_{ii}$, ($i = 1, 2, ..., N$), respectively. A, B, D, E_1, and M are constant matrices with appropriate dimensions, and $\phi_i(s)$ is a given initial condition sequence. $d(k) \in \mathbb{Z}^+$ is the discrete time-varying delays satisfying $d_m \le \tau, d(k) \le d_M$, where d_m, d_M and τ are known positive integers.

The stochastic variable sequence $\varphi_i(k)$ satisfies the following Bernoulli distribution to describe the randomly occurring coupling delays:

$$\text{Prob}\{\varphi_i(k) = 1\} = \bar{\varphi}_i, \ \text{Prob}\{\varphi_i(k) = 0\} = 1 - \bar{\varphi}_i, \tag{12.2}$$

where $\varphi_i(k) = 0$ expresses that there is no coupling delay, and $\varphi_i(k) = 1$ means that coupling delays take place in the complex networks. $\bar{\varphi}_i$ and $\bar{\sigma}_i^2 = \bar{\varphi}_i(1-\bar{\varphi}_i)$ denote the mathematical expectation and variance of $\varphi_i(k)$.

Remark 12.1 *Coupling delays have extensively been considered in dynamical complex networks and usually assumed to take place continuously; see, e.g., [39, 103]. However, in practical systems, coupling delays may randomly occur because of some external influences, such as large environment noises, the random change of transmission capacity among nodes. In this chapter, we describe such a randomly occurring phenomenon by a stochastic variable sequence* $\varphi_i(k)$, $(i = 1, \cdots, N)$ *satisfying Bernoulli distribution. In fact, the Bernoulli distribution sequence has been widely applied to model some randomly occurring phenomena, such as randomly occurring nonlinearity, randomly occurring network-induced delay, randomly occurring sensor saturation, randomly occurring quantization, etc.; see, e.g., [29, 143].*

The nonlinear vector-valued functions $f(\cdot)$ $(f(0) = 0)$ are assumed to satisfy the following sector-bounded conditions:

$$[f(x) - L_1 x]^T [f(x) - L_2 x] \leq 0, \ \forall x \in \mathbb{R}^n, \tag{12.3}$$

where $L_2 - L_1 > 0$, with L_1 and L_2 being constant real matrices of appropriate dimensions. $f(\cdot)$ satisfying the above condition is said to belong to the sector $[L_1, \ L_2]$.

The measurement output with data missing is considered as:

$$y_i(k) = \xi_i(k) C x_i(k) + E_2 v_2(k), \tag{12.4}$$

where $y_i(k) \in \mathbb{R}^m$ is the measurement output, and $\xi_i(k)$ represents the status of the data missing. Here, it is assumed that $\xi_i(k)$ has the probabilistic density function $h_i(s)$ on interval [0 1] with mathematical expectation $\bar{\xi}_i$, and variance σ_i^2. $v_2(k)$ is the disturbance input belonging to $l_2([0,+\infty); \mathbb{R}^p)$.

Remark 12.2 *(12.4) describes the measurement output of ith node, in which the random variable sequence* $\xi_i(k)$ *depicts the data missing of the ith node* $(i = 1, 2, ..., N)$. *The measurement missing phenomenon has been extensively considered, and several models have been set up; see, e.g., [91, 186]. The most popular and effective model is the Bernoulli distribution model in which* 0 *represents the whole measurement missing and* 1 *denotes that the measurement information is fully accessible. Unfortunately, the real-world system is more complicated, and the measurement information may be neither entirely missing nor completely available, i.e., only partial information could be obtained. Therefore, the Bernoulli distribution is too simple to exactly deal with such complex cases. In (12.4), the* $\xi_i(k)$ *can be a random variable sequence satisfying any discrete probability distribution on interval [0 1], hence, it is more general than the Bernoulli distribution model.*

Design the state estimator for the ith node with the following form:

$$\begin{cases} \hat{x}_i(k+1) = A\hat{x}_i(k) + Bf(\hat{x}_i(k)) + Df(\hat{x}_i(k - d(k))) \\ \qquad\qquad\quad + K_i[y_i(k) - \bar{\xi}_i C\hat{x}_i(k)], \\ \hat{z}_i(k) = M\hat{x}_i(k) \ , \\ \hat{x}_i(s) = 0 \ , s \in [-d_M, 0], i = 1, 2, ..., N, \end{cases} \tag{12.5}$$

where $\hat{x}_i(k) \in \mathbb{R}^n$ and K_i are, respectively, the filter state and the filter parameter of the ith node to be determined.

For convenience, we use the following notations:

$$\begin{aligned}
&x_k = [x_1^T(k), \cdots, x_N^T(k)]^T, \; \hat{x}_k = [\hat{x}_1^T(k), \cdots, \hat{x}_N^T(k)]^T, \\
&z_k = [z_1^T(k), \cdots, z_N^T(k)]^T, \;\; \hat{z}_k = [\hat{z}_1^T(k), \cdots, \hat{z}_N^T(k)]^T, \\
&v_{1k} = [\underbrace{v_1^T(k), \cdots, v_1^T(k)}_{N}]^T, v_{2k} = [\underbrace{v_2^T(k), \cdots, v_2^T(k)}_{N}]^T, \\
&\tilde{x}_k = [x_k^T, \hat{x}_k^T]^T, \;\; \tilde{z}_k = z_k - \hat{z}_k \;, \; \tilde{v}_k = [v_{1k}^T, v_{2k}^T]^T, \\
&f(x_k) = [f^T(x_1(k)), f^T(x_2(k)), \cdots, f^T(x_N(k))]^T, \\
&\tilde{f}(x_{k-d(k)}) = [f^T(x_{k-d(k)}), f^T(\hat{x}_{k-d(k)})]^T, \\
&\tilde{f}(x_k) = [f^T(x_k), f^T(\hat{x}_k)]^T, \\
&\widetilde{C} = I \otimes C, \widetilde{E}_1 = I \otimes E_1, \; \widetilde{E}_2 = I \otimes E_2, \\
&\tilde{M} = I \otimes M \;, \; K = \operatorname{diag}\{K_1, K_2, \cdots, K_N\} \;, \\
&N_i = \operatorname{diag}\{\underbrace{0, \cdots, 0}_{i-1}, I, \underbrace{0, \cdots, 0}_{N-i}\}.
\end{aligned} \tag{12.6}$$

By the Kcronecker product, the dynamics of the augmented systems is obtained as follows:

$$\begin{cases} \tilde{x}_{k+1} & = \mathcal{A}\tilde{x}_k + \mathcal{B}\tilde{f}(x_k) + \mathcal{A}_\tau\tilde{x}_{k-\tau} + \bar{\mathcal{A}}\tilde{x}_k + \bar{\mathcal{A}}_\tau\tilde{x}_{k-\tau} + \mathcal{D}\tilde{f}(x_{k-d(k)}) + \mathcal{E}\tilde{v}_k, \\ \tilde{z}_k & = \tilde{\mathcal{M}}\tilde{x}_k, \end{cases} \tag{12.7}$$

where

$$\begin{aligned}
\mathcal{A} &= \begin{bmatrix} I \otimes A + W \otimes \Gamma_1 & 0 \\ K\sum_{i=1}^{N} N_i\bar{\xi}_i\widetilde{C} & I \otimes A - K\sum_{i=1}^{N} N_i\bar{\xi}_i\widetilde{C} \end{bmatrix}, \\
\mathcal{A}_\tau &= \begin{bmatrix} \sum_{i=1}^{N} N_i\bar{\varphi}_i\bar{W} \otimes \Gamma_2 & 0 \\ 0 & 0 \end{bmatrix}, \; \mathcal{B} = \begin{bmatrix} I \otimes B & 0 \\ 0 & I \otimes B \end{bmatrix}, \\
\bar{\mathcal{A}} &= \begin{bmatrix} 0 & 0 \\ K\sum_{i=1}^{N} N_i(\xi_i(k) - \bar{\xi}_i)\widetilde{C} & 0 \end{bmatrix}, \; \mathcal{E} = \begin{bmatrix} \widetilde{E}_1 & 0 \\ 0 & K\widetilde{E}_2 \end{bmatrix}, \\
\bar{\mathcal{A}}_\tau &= \begin{bmatrix} \sum_{i=1}^{N} N_i(\varphi_i(k) - \bar{\varphi}_i)\bar{W} \otimes \Gamma_2 & 0 \\ 0 & 0 \end{bmatrix}, \; \tilde{\mathcal{M}} = \begin{bmatrix} \tilde{M} & 0 \\ 0 & -\tilde{M} \end{bmatrix} \\
\mathcal{D} &= \begin{bmatrix} I \otimes D & 0 \\ 0 & I \otimes D \end{bmatrix}.
\end{aligned} \tag{12.8}$$

Definition 15 *[186] The dynamics of the augmented systems (12.7) is said to be exponentially mean-square stable if, with* $\tilde{v}_k = 0$*, there exist constants* $\alpha > 0$ *and* $\tau \in (0,1)$*, such that* $\mathbb{E}\{\|\tilde{x}_k\|^2\} \leq \alpha\tau^k \sup_{-d_M \leq i \leq 0} \mathbb{E}\{\|\tilde{x}_i\|^2\}$, $k \in \mathbb{Z}^+$.

In the following section, a set of state estimators proposed in (12.5) is designed by means of Lyapunov stability theorem and the convex programming method, for all allowable randomly occurring coupling delays, missing measurements, and external disturbance noises, the following requirements are to be met simultaneously:

1) the augmented system (12.7) is exponentially mean-square stable;

2) under the zero initial condition, for a given disturbance attenuation level $\gamma > 0$ and all nonzero $\tilde{v}_k$, the output error $\tilde{z}(k)$ satisfies H_∞ performance constraint:

$$\sum_{k=0}^{\infty} \mathbb{E}\left\{\tilde{z}_k^T \tilde{z}_k\right\} \leq \gamma^2 \sum_{k=0}^{\infty} \mathbb{E}\left\{\tilde{v}_k^T \tilde{v}_k\right\}. \tag{12.9}$$

12.2 Main Results

12.2.1 Stability Analysis

In the following theorem, the Lyapunov stability theorem and LMI-based method are used to guarantee the augmented system (12.7) to be exponentially mean-square stable and satisfy the H_∞ performance constraint simultaneously, some sufficient conditions are derived.

Theorem 27 *Suppose that the state estimator gains* K_i $(i = 1, 2, ..., N)$ *are given. If there exist matrices* $P_1 > 0$, $P_2 > 0$, $Q_1 > 0$, $Q_2 > 0$, $R_1 > 0$, *and* $R_2 > 0$*, such that the matrix inequality*

$$\begin{bmatrix} \Omega_1 & 0 & \tilde{L}_2 & 0 & 0 & \mathscr{C}^T & 0 & \mathcal{A}^T P \\ * & -R-\tilde{L}_{1d} & 0 & \tilde{L}_{2d} & 0 & 0 & 0 & 0 \\ * & * & -\tilde{I} & 0 & 0 & 0 & 0 & \mathcal{B}^T P \\ * & * & * & -\tilde{I} & 0 & 0 & 0 & \mathcal{D}^T P \\ * & * & * & * & -Q & 0 & \bar{\mathscr{C}}^T & \mathcal{A}_\tau^T P \\ * & * & * & * & * & -\mathcal{P} & 0 & 0 \\ * & * & * & * & * & * & -\mathcal{P} & 0 \\ * & * & * & * & * & * & * & -P \end{bmatrix} < 0 \tag{12.10}$$

holds, then the augmented system (12.7) is exponentially mean-square stable,

where

$$\Omega_1 = -P + Q + (d_M - d_m + 1)R - \tilde{L}_1,\ P = \mathrm{diag}\{I \otimes P_1, I \otimes P_2\},$$
$$Q = \mathrm{diag}\{I \otimes Q_1, I \otimes Q_2\}, R = \mathrm{diag}\{I \otimes R_1, I \otimes R_2\},$$
$$\mathscr{C} := [\sigma_1 C_1^T P, \cdots, \sigma_N C_N^T P]^T,\ \bar{\mathscr{C}} := [\bar{\sigma}_1 \bar{C}_1^T P, \cdots, \bar{\sigma}_N \bar{C}_N^T P]^T,$$
$$\mathcal{P} := \mathrm{diag}\{\underbrace{P, P, \cdots, P}_{N}\},\ C_i = \begin{bmatrix} 0 & 0 \\ N_i K \tilde{C} & 0 \end{bmatrix},$$
$$\bar{C}_i = \begin{bmatrix} N_i \bar{W} \otimes \Gamma_2 & 0 \\ 0 & 0 \end{bmatrix},\ \bar{\mathcal{A}}_i = \begin{bmatrix} 0 & 0 \\ K N_i(\xi_i(k) - \bar{\xi}_i)\widetilde{C} & 0 \end{bmatrix},$$
$$\bar{\mathcal{A}}_{\tau i} = \begin{bmatrix} N_i(\varphi_i(k) - \bar{\varphi}_i)\bar{W} \otimes \Gamma_2 & 0 \\ 0 & 0 \end{bmatrix},\ \tilde{I} = \begin{bmatrix} I & 0 \\ 0 & I \end{bmatrix}. \tag{12.11}$$

Proof 30 *First, we consider the stability of the augmented system (12.7) without disturbance input $\tilde{v}_k$. Define the Lyapunov–Krasovsikii functional candidate:*

$$V(k) = V_1(k) + V_2(k) + V_3(k) + V_4(k)\,, \tag{12.12}$$

where

$$\begin{aligned} V_1(k) &= \tilde{x}_k^T P \tilde{x}_k,\ V_2(k) = \sum_{i=k-\tau}^{k-1} \tilde{x}_i^T Q x_i, \\ V_3(k) &= \sum_{j=k-d_M+1}^{k-d_m} \sum_{i=j}^{k-1} \tilde{x}_i^T R \tilde{x}_i,\ V_4(k) = \sum_{i=k-d(k)}^{k-1} \tilde{x}_i^T R \tilde{x}_i. \end{aligned} \tag{12.13}$$

Calculating the difference of $V_1(k)$ along the trajectory of (12.7) gives

$$\begin{aligned} \mathbb{E}\{\Delta V_1(k)\} =& \mathbb{E}\{V_1(k+1) - V_1(k)\} = \mathbb{E}\{\tilde{x}_{k+1}^T P \tilde{x}_{k+1} - \tilde{x}_k^T P \tilde{x}_k\} \\ =& \mathbb{E}\Big\{[\mathcal{A}\tilde{x}_k + \mathcal{B}\tilde{f}(x_k) + \mathcal{A}_\tau \tilde{x}_{k-\tau} + \bar{\mathcal{A}}\tilde{x}_k + \bar{\mathcal{A}}_\tau \tilde{x}_{k-\tau} \\ &+ \mathcal{D}\tilde{f}(x_{k-d(k)})]^T P [\mathcal{A}\tilde{x}_k + \mathcal{B}\tilde{f}(x_k) + \mathcal{A}_\tau \tilde{x}_{k-\tau} \\ &+ \bar{\mathcal{A}}\tilde{x}_k + \bar{\mathcal{A}}_\tau \tilde{x}_{k-\tau} + \mathcal{D}\tilde{f}(x_{k-d(k)})] - \tilde{x}_k^T P \tilde{x}_k\Big\} \\ =& \mathbb{E}\Big\{[\mathcal{A}\tilde{x}_k + \mathcal{B}\tilde{f}(x_k) + \mathcal{A}_\tau \tilde{x}_{k-\tau} + \mathcal{D}\tilde{f}(x_{k-d(k)})]^T P \\ &\times [\mathcal{A}\tilde{x}_k + \mathcal{B}\tilde{f}(x_k) + \mathcal{A}_\tau \tilde{x}_{k-\tau} + \mathcal{D}\tilde{f}(x_{k-d(k)})]^T \\ &- \tilde{x}_k^T \bar{\mathcal{A}}_\tau^T P \bar{\mathcal{A}}_\tau \tilde{x}_k - \tilde{x}_{k-\tau}^T \bar{\mathcal{A}}_\tau^T P \bar{\mathcal{A}}_\tau \tilde{x}_{k-\tau} - \tilde{x}_k^T P \tilde{x}_k\}. \end{aligned} \tag{12.14}$$

Similarly, we can easily achieve

$$\begin{aligned} \mathbb{E}\{\Delta V_2(k)\} =& \mathbb{E}\{V_2(k+1) - V_2(k)\} = \mathbb{E}\left\{\tilde{x}_k^T Q \tilde{x}_k - \tilde{x}_{k-\tau}^T Q \tilde{x}_{k-\tau}\right\}, \quad (12.15) \\ \mathbb{E}\{\Delta V_3(k)\} =& \mathbb{E}\{V_3(k+1) - V_3(k)\} \\ =& \mathbb{E}\left\{(d_M - d_m)\tilde{x}_k^T R \tilde{x}_k - \sum_{i=k-d_M+1}^{k-d_m} \tilde{x}_i^T R \tilde{x}_i\right\}, \quad (12.16) \end{aligned}$$

$$
\begin{aligned}
\mathbb{E}\{\Delta V_4(k)\} =& \mathbb{E}\{V_4(k+1) - V_4(k)\} \\
=& \mathbb{E}\left\{ \tilde{x}_k^T R \tilde{x}_k - \tilde{x}_{k-d(k)}^T R \tilde{x}_{k-d(k)} + \sum_{i=k-d(k+1)+1}^{k-d_m} \tilde{x}_i^T R \tilde{x}_i \right. \\
& \left. + \sum_{i=k-d_m+1}^{k-1} \tilde{x}_i^T R \tilde{x}_i - \sum_{i=k-d(k)+1}^{k-1} \tilde{x}_i^T R \tilde{x}_i \right\} \\
\leq& \mathbb{E}\left\{ \tilde{x}_k^T R \tilde{x}_k - \tilde{x}_{k-d(k)}^T R \tilde{x}_{k-d(k)} + \sum_{i=k-d_M+1}^{k-d_m} \tilde{x}_i^T R \tilde{x}_i \right\}.
\end{aligned} \tag{12.17}
$$

According to (12.2), (12.4), and (12.8), $\mathbb{E}\{\bar{\mathcal{A}}\} = 0$ *and* $\mathbb{E}\{\bar{\mathcal{A}}_\tau\} = 0$ *are easily achieved. Then we can have the two facts of*

$$
\begin{aligned}
\mathbb{E}\left\{\bar{\mathcal{A}}_i^T P \bar{\mathcal{A}}_j\right\} &= \begin{cases} 0, \ i \neq j, \\ \sigma_i^2 C_i^T P C_i, \ i = j, \end{cases} \\
\mathbb{E}\left\{\bar{\mathcal{A}}_{\tau i}^T P \bar{\mathcal{A}}_{\tau j}\right\} &= \begin{cases} 0, \ i \neq j, \\ \bar{\sigma_i}^2 \bar{C_i}^T P \bar{C}_i, \ i = j, \end{cases}
\end{aligned} \tag{12.18}
$$

then it is easy to conclude that

$$
\mathbb{E}\left\{\bar{\mathcal{A}}^T P \bar{\mathcal{A}}\right\} = \sum_{i=1}^{N} \sigma_i^2 C_i^T P C_i, \ \mathbb{E}\left\{\bar{\mathcal{A}}_\tau^T P \bar{\mathcal{A}}_\tau\right\} = \sum_{i=1}^{N} \bar{\sigma_i}^2 \bar{C_i}^T P \bar{C}_i, \tag{12.19}
$$

where $\bar{\mathcal{A}}_i$ *and* $\bar{\mathcal{A}}_{\tau i}$ *have been defined in (12.11).*

For notation simplicity, we denote

$$
\begin{aligned}
\zeta_k &= [\tilde{x}_k^T \ \ \tilde{x}_{k-d(k)}^T \ \ \tilde{f}^T(x_k) \ \ \tilde{f}^T(x_{k-d(k)}) \ \ \tilde{x}_{k-\tau}^T]^T, \\
\eta &= [\mathcal{A} \ \ 0 \ \ \mathcal{B} \ \ \mathcal{D} \ \ \mathcal{A}_\tau].
\end{aligned} \tag{12.20}
$$

Then we can conclude that

$$
\begin{aligned}
\mathbb{E}\{\Delta V(k)\} =& \mathbb{E}\{\Delta V_1(k) + \Delta V_2(k) + \Delta V_3(k) + \Delta V_4(k)\} \\
\leq& \mathbb{E}\left\{ \zeta_k^T \eta^T P \eta \zeta_k + \tilde{x}_k^T [-P + \sum_{i=1}^{N} \sigma_i^2 C_i^T P C_i \right. \\
& + Q + (d_M - d_m + 1)R]\tilde{x}_k - \tilde{x}_{k-d(k)}^T R \tilde{x}_{k-d(k)} \\
& \left. \tilde{x}_{k-\tau}^T (-Q + \sum_{i=1}^{N} \bar{\sigma_i}^2 \bar{C_i}^T P \bar{C}_i) \tilde{x}_{k-\tau} \right\},
\end{aligned} \tag{12.21}
$$

and from (12.3), we can obtain that

$$[\tilde{x}_k^T \ \tilde{f}^T(x_k)] \begin{bmatrix} \tilde{L}_1 & -\tilde{L}_2 \\ -\tilde{L}_2^T & \tilde{I} \end{bmatrix} \begin{bmatrix} \tilde{x}_k \\ \tilde{f}(x_k) \end{bmatrix} \leq 0, \tag{12.22}$$

$$\begin{bmatrix} \tilde{x}_{k-d(k)} \\ \tilde{f}(\tilde{x}_{k-d(k)}) \end{bmatrix}^T \begin{bmatrix} \tilde{L}_{1d} & -\tilde{L}_{2d} \\ -\tilde{L}_{2d}^T & \tilde{I} \end{bmatrix} \begin{bmatrix} \tilde{x}_{k-d(k)} \\ \tilde{f}(\tilde{x}_{k-d(k)}) \end{bmatrix} \leq 0, \tag{12.23}$$

where $\tilde{I}$ has been defined in (12.11), and

$$\begin{aligned}
\tilde{L}_1 &= \begin{bmatrix} I \otimes \frac{L_1^T L_2 + L_2^T L_1}{2} & 0 \\ 0 & \frac{L_1^T L_2 + L_2^T L_1}{2} \end{bmatrix}, \\
\tilde{L}_2 &= \begin{bmatrix} I \otimes \frac{L_1^T + L_2^T}{2} & 0 \\ 0 & I \otimes \frac{L_1^T + L_2^T}{2} \end{bmatrix}, \\
\tilde{L}_{1d} &= \begin{bmatrix} I \otimes \frac{L_{1d}^T L_{2d} + L_{2d}^T L_{1d}}{2} & 0 \\ 0 & I \otimes \frac{L_{1d}^T L_{2d} + L_{2d}^T L_{1d}}{2} \end{bmatrix}, \\
\tilde{L}_{2d} &= \begin{bmatrix} I \otimes \frac{L_{1d}^T + L_{2d}^T}{2} & 0 \\ 0 & I \otimes \frac{L_{1d}^T + L_{2d}^T}{2} \end{bmatrix}.
\end{aligned} \tag{12.24}$$

Letting $\Theta = \psi + \eta^T P \eta$, where

$$\psi = \begin{bmatrix} \Omega_1 + \sum\limits_{i=1}^{N} \sigma_i^2 C_i^T P C_i & 0 & \tilde{L}_2 & 0 & 0 \\ * & -R - \tilde{L}_{1d} & 0 & \tilde{L}_{2d} & 0 \\ * & * & -\tilde{I} & 0 & 0 \\ * & * & * & -\tilde{I} & 0 \\ * & * & * & * & -Q + \sum\limits_{i=1}^{N} \bar{\sigma}_i{}^2 \bar{C}_i{}^T P \bar{C}_i \end{bmatrix}, \tag{12.25}$$

and then substituting (12.22)–(12.23) into (12.21), it is not difficult to see that

$$\mathbb{E}\{\Delta V(k)\} \leq \mathbb{E}\{\zeta_k^T \Theta \zeta_k\}. \tag{12.26}$$

By (12.10) and Schur complement, we have $\Theta < 0$ and, subsequently, $\mathbb{E}\{\Delta V(k)\} \leq -\lambda_{\min}(-\Theta)\mathbb{E}\{\|\tilde{x}_k\|^2\}$. Finally, we can confirm from Definition 15 and [186] that the augmented system (12.7) is exponentially mean-square stable.

Remark 12.3 *It is well known that delay-independent condition tends to be conservative, especially in the case of the delay being very small. Meanwhile, the LMI method also would bring the extra conservative. However, in this chapter, the delay-dependent term $V_3(k)$ is introduced in the proposed Lyapunov functional, which contains the information of lower and upper bounds* *of time-varying delay, therefore, it has potential to yield less conservative results. As mentioned above, it is not difficult to see that less conservatism will result in choosing the novel Lyapunov functional (13).*

12.2.2 H_∞ Performance Analysis

In Theorem 27, a sufficient condition is established to guarantee the augmented filtering dynamics (12.7) without external disturbance noises exponentially mean-square stable. In the following theorem, we will discuss the H_∞ performance problem for (12.7) to derive the sufficient criteria, such that the resulting dynamics satisfying a given H_∞ performance index.

Theorem 28 *Suppose that the state estimator gains K_i $(i = 1, 2, ..., N)$ are given. Under the zero initial condition, the augmented dynamics (12.7) satisfies the H_∞ performance constraint (12.9) for all nonzero $\tilde{v}_k$ and the prescribed constant $\gamma > 0$, if there exist matrices $P_1 > 0$, $P_2 > 0$, $Q_1 > 0$, $Q_2 > 0$, $R_1 > 0$, and $R_2 > 0$, such that*

$$\begin{bmatrix} \bar{\Omega}_1 & 0 & \tilde{L}_2 & 0 & 0 & 0 & \mathscr{C}^T & 0 & \mathcal{A}^T P \\ * & -R-\tilde{L}_{1d} & 0 & \tilde{L}_{2d} & 0 & 0 & 0 & 0 & 0 \\ * & * & -\tilde{I} & 0 & 0 & 0 & 0 & 0 & \mathcal{B}^T P \\ * & * & * & -\tilde{I} & 0 & 0 & 0 & 0 & \mathcal{D}^T P \\ * & * & * & * & -Q & 0 & 0 & \bar{\mathscr{C}}^T & {\mathcal{A}_\tau}^T P \\ * & * & * & * & * & -\gamma^2 I & 0 & 0 & \mathcal{E}^T P \\ * & * & * & * & * & * & -\mathcal{P} & 0 & 0 \\ * & * & * & * & * & * & * & -\mathcal{P} & 0 \\ * & * & * & * & * & * & * & * & -P \end{bmatrix} < 0 \quad (12.27)$$

holds, where

$$\mathcal{M} = [\tilde{M} \quad -\tilde{M}], \quad \bar{\Omega}_1 = \Omega_1 + \mathcal{M}^T \mathcal{M}. \quad (12.28)$$

Proof 31 *Choosing the Lyapunov function $V(k)$ defined in the proof of Theorem 1 and the augmented systems (12.7), we can conclude as follows:*

$$\begin{aligned} \mathbb{E}\{\Delta V(k)\} &= \mathbb{E}\{\Delta V_1(k) + \Delta V_2(k) + \Delta V_3(k) + \Delta V_4(k)\} \\ &\leq \mathbb{E}\left\{\zeta_k^T \Theta \zeta_k + 2\tilde{v}_k^T \mathcal{E}^T P \mathcal{A} \tilde{x}_k + 2\tilde{v}_k^T \mathcal{E}^T P \mathcal{B} \tilde{f}(x_k) + 2\tilde{v}_k^T \mathcal{E}^T P \mathcal{A}_\tau \tilde{x}_{k-\tau} \right. \\ &\quad \left. + 2\tilde{v}_k^T \mathcal{E}^T P \mathcal{D} \tilde{f}(x_{k-d(k)}) + \tilde{v}_k^T \mathcal{E}^T P \mathcal{E} \tilde{v}_k^T \right\}, \end{aligned} \quad (12.29)$$

where ζ_k and Θ are defined in the Theorem 1 previously. Denoting $\tilde{\zeta}_k = [\zeta_k^T \quad \tilde{v}_k^T]^T$ and $\tilde{\mathcal{E}} = [\mathcal{E}^T P \mathcal{A} \quad 0 \quad \mathcal{E}^T P \mathcal{B} \; \mathcal{E}^T P \mathcal{D} \quad \mathcal{E}^T P \mathcal{A}_\tau]$, inequality (12.29) is equivalent to

$$\mathbb{E}\{\Delta V(k)\} \leq \mathbb{E}\left\{ \tilde{\zeta}_k^T \begin{bmatrix} \Theta & \tilde{\mathcal{E}}^T \\ \tilde{\mathcal{E}} & \mathcal{E}^T P \mathcal{E} \end{bmatrix} \tilde{\zeta}_k \right\}. \quad (12.30)$$

Define $\tilde{\Theta} = \Theta + \bar{Z}^T\bar{Z}$, $\bar{Z} = [\mathcal{M} \quad 0 \quad 0 \quad 0 \quad 0]$, *and let us now analyze the* H_∞ *performance. Under the assumption of the zero initial condition and from (12.30), we have*

$$\begin{aligned}
\mathcal{J} &= \sum_{k=0}^{s} \mathbb{E}\left\{\tilde{z}_k^T\tilde{z}_k - \gamma^2\tilde{v}_k^T\tilde{v}_k + \Delta V(k)\right\} - V(s+1) \\
&\leq \sum_{k=0}^{s} \mathbb{E}\left\{\tilde{z}_k^T\tilde{z}_k - \gamma^2\tilde{v}_k^T\tilde{v}_k + \Delta V(k)\right\} \\
&\leq \sum_{k=0}^{s} \mathbb{E}\left\{\zeta_k^T \begin{bmatrix} \tilde{\Theta} & \tilde{\mathcal{E}}^T \\ \tilde{\mathcal{E}} & \mathcal{E}^T P \mathcal{E} - \gamma^2 I \end{bmatrix} \zeta_k\right\}.
\end{aligned} \tag{12.31}$$

Letting $s \to \infty$, *it follows from Schur complement and (12.27), then we can obtain directly from the above inequality that*

$$\sum_{k=0}^{\infty} \mathbb{E}\left\{\tilde{z}_k^T\tilde{z}_k\right\} \leq \sum_{k=0}^{\infty} \mathbb{E}\left\{\gamma^2\tilde{v}_k^T\tilde{v}_k\right\}, \tag{12.32}$$

from which the H_∞ *performance constraint can be guaranteed, and proof of this theorem is therefore completed.*

Remark 12.4 *For Theorem 27 and Theorem 28, some sufficient conditions are given in terms of LMIs, which can be easily checked by MATLAB Toolbox via semidefinite programming method. The provided sufficient conditions are of certain conservatism, however, by constructing a novel Lyapunov functional, the existed conservatism has been reduced relatively, therefore, the conditions derived in Theorems 27 and 28 have less conservatism.*

12.2.3 State Estimator Synthesis

Based on the established analysis results in Theorem 28, we are now to deal with the design problem of the H_∞ state estimation for the complex networks (12.1).

Theorem 29 *Consider the augmented dynamics (12.7), with the prescribed disturbance attenuation level* $\gamma > 0$, *and under the zero initial condition. The augmented dynamics (12.7) satisfies the* H_∞ *performance for all nonzero* $\tilde{v}_k$, *if there exist following matrices* $P_1 > 0$, $P_2 > 0$, $Q_1 > 0$, $Q_2 > 0$, $R_1 > 0$, *and* $R_2 > 0$, *such that*

$$\begin{bmatrix}
\bar{\Omega}_1 & 0 & \tilde{L}_2 & 0 & 0 & 0 & \mathbb{C}^T & 0 & \mathbb{A}^T \\
* & -R-\tilde{L}_{1d} & 0 & \tilde{L}_{2d} & 0 & 0 & 0 & 0 & 0 \\
* & * & -\tilde{I} & 0 & 0 & 0 & 0 & 0 & \mathcal{B}^T P \\
* & * & * & -\tilde{I} & 0 & 0 & 0 & 0 & \mathcal{D}^T P \\
* & * & * & * & -Q & 0 & 0 & \bar{\mathscr{C}}^T & \mathcal{A}_\tau^T P \\
* & * & * & * & * & -\gamma^2 I & 0 & 0 & \mathbb{E}^T \\
* & * & * & * & * & * & -\mathcal{P} & 0 & 0 \\
* & * & * & * & * & * & * & -\mathcal{P} & 0 \\
* & * & * & * & * & * & * & * & -P
\end{bmatrix} < 0 \tag{12.33}$$

holds, where

$$\mathbb{C} = \left[\begin{bmatrix} 0 & 0 \\ \sigma_1 N_1(I \otimes X_1 C) & 0 \end{bmatrix}^T, \ldots, \begin{bmatrix} 0 & 0 \\ \sigma_N N_N(I \otimes X_N C) & 0 \end{bmatrix}^T\right]^T,$$

$$\mathbb{A} = \begin{bmatrix} I \otimes P_1 A + W \otimes P_1 \Gamma & 0 \\ \sum_{i=1}^{N} \xi_i N_i(I \otimes X_i C) & I \otimes P_2 A - \sum_{i=1}^{N} \xi_i N_i(I \otimes X_i C) \end{bmatrix},$$

$$\mathbb{E} = \begin{bmatrix} \sum_{i=1}^{N} N_i(I \otimes P_1 E_1) & 0 \\ 0 & \sum_{i=1}^{N} N_i(I \otimes X_i E_2) \end{bmatrix}. \tag{12.34}$$

Furthermore, the desired estimator parameters can be derived by

$$K_i = P_2^{-1} X_i. \tag{12.35}$$

Proof 32 *By partitioning* $P = \mathrm{diag}\{I \otimes P_1, I \otimes P_2\}$, $Q = \mathrm{diag}\{I \otimes Q_1, I \otimes Q_2\}$, *and taking notice of* $K_i = P_2^{-1} X_i$, *according to Schur complement, it follows that (12.33) is equivalent to (12.27). Then from Theorem 28, the proof is easily completed.*

In theorem 29, an effective criterion is set up to guarantee that the desired estimator gains can be designed properly, and estimator gains are characterized in terms of the solution to a normal LMI, which can easily be solved by the software toolbox.

12.3 An Illustrative Example

In this section, a numerical example is presented to illustrate the effectiveness and usefulness of our state estimator design scheme for the discussed discrete-time complex network with measurement missing and randomly occurring coupling delays.

The parameters of system are given as follows:

$$A = \begin{bmatrix} 0.45 & 0 \\ 0 & 0.21 \end{bmatrix}, \; B = \begin{bmatrix} 0.13 & 0.21 \\ 0.28 & 0.33 \end{bmatrix},$$

$$C = \begin{bmatrix} 0.1 & 0.2 \\ 0.15 & 0.23 \end{bmatrix}, \; D = \begin{bmatrix} 0.03 & 0.19 \\ 0.21 & 0.33 \end{bmatrix},$$

$$\tau = 1, \; d_m = 2, \; d_M = 3, \; \gamma = 0.8.$$

Assume that there are three nodes in the complex networks. The coupling configuration matrices are considered as

$$W = \bar{W} = \begin{bmatrix} -0.8 & 0.8 & 0 \\ 0.8 & -1.5 & 0.7 \\ 0 & 0.7 & -0.7 \end{bmatrix},$$

and the inner coupling matrices are given as $\Gamma_1 = \Gamma_2 = \text{diag}\{0.15, 0.15\}$. The disturbance and the output matrices are taken as

$$E_1 = \begin{bmatrix} 0.4 \\ 0.3 \end{bmatrix}, \ E_2 = \begin{bmatrix} 0.2 \\ 0.4 \end{bmatrix}, \ M = \begin{bmatrix} 0.7 \\ 0.2 \end{bmatrix}.$$

The nonlinear vector-valued functions $f(x_i(k))$ are assumed as

$$f(x_i(k)) = \begin{bmatrix} -0.6x_{i1}(k) + 0.3x_{i2}(k) + \tanh(0.3x_{i1}(k)) \\ 0.6x_{i2}(k) - \tanh(0.2x_{i2}(k)) \end{bmatrix}.$$

Then, it is easily verified that

$$L_1 = \begin{bmatrix} -0.6 & 0.3 \\ 0 & 0.4 \end{bmatrix}, L_2 = \begin{bmatrix} -0.3 & 0.3 \\ 0 & 0.6 \end{bmatrix}.$$

In addition, we select the probability density functions of $\xi_i(k)$ and $\varphi_i(k)$ $(i = 1, 2, 3)$ with following forms:

$$h_1(s_1) = \begin{cases} 0.8, & s_1 = 0 \\ 0.1, & s_1 = 0.5 \\ 0.1, & s_1 = 1 \end{cases}, \quad \bar{h}_1(s_1) = \begin{cases} 0.2, & s_1 = 0 \\ 0.8, & s_1 = 1 \end{cases},$$

$$h_2(s_2) = \begin{cases} 0.7, & s_2 = 0 \\ 0.2, & s_2 = 0.5 \\ 0.1, & s_2 = 1 \end{cases}, \quad \bar{h}_2(s_2) = \begin{cases} 0.7, & s_2 = 0 \\ 0.3, & s_2 = 1 \end{cases},$$

$$h_3(s_3) = \begin{cases} 0.1, & s_3 = 0 \\ 0.1, & s_3 = 0.5 \\ 0.8, & s_3 = 1 \end{cases}, \quad \bar{h}_3(s_3) = \begin{cases} 0.6, & s_3 = 0 \\ 0.4, & s_3 = 1 \end{cases},$$

from which the expectations and variances can be easily calculated as $\bar{\varphi}_1 = 0.15$, $\sigma_1^2 = 0.1025$; $\bar{\varphi}_2 = 0.2$, $\sigma_2^2 = 0.11$; $\bar{\varphi}_3 = 0.85$, $\sigma_3^2 = 0.1025$; $\bar{\xi}_1 = 0.8$, $\bar{\sigma}_1^2 = 0.4$; $\bar{\xi}_2 = 0.7$, $\bar{\sigma}_2^2 = 0.4583$; and $\bar{\xi}_3 = 0.6$, $\bar{\sigma}_3^2 = 0.4899$. By using MATLAB LMI Control Toolbox, from (12.33), we can obtain estimator parameters as follows:

$$K_1 = \begin{bmatrix} 2.4253 & -1.2144 \\ 2.6504 & -1.0632 \end{bmatrix}$$

$$K_2 = \begin{bmatrix} 2.0018 & -1.0034 \\ 2.1148 & -1.0632 \end{bmatrix}$$

$$K_3 = \begin{bmatrix} 0.5982 & -0.3070 \\ 0.6200 & -0.3327 \end{bmatrix}.$$

In the simulation, we assume the exogenous disturbance inputs as $v_1(k) = 5\exp(-0.5k)\sin(k)$, $v_2(k) = \frac{6\sin(0.8k)}{k+1}$. The discrete time-varying delays $d(k)$ satisfy $d(k) = 2 + (1 + (-1)^k)/2$. Figure 12.1 and Figure 12.2 describe the performance of the designed estimators, and Figure 12.3 depicts the estimation

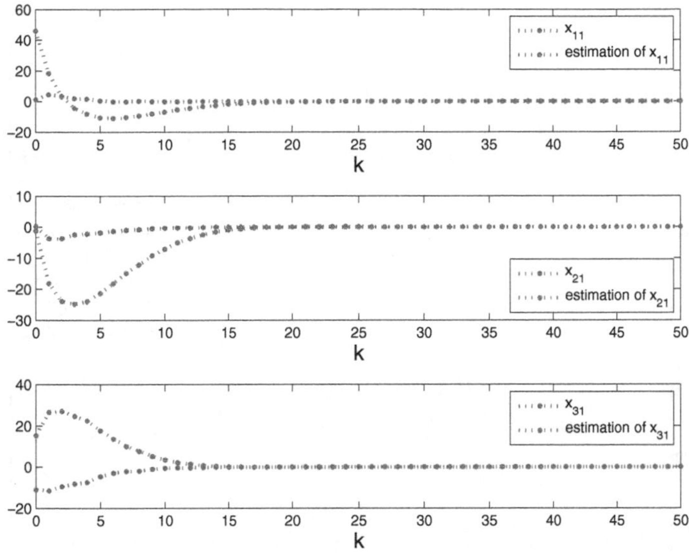

FIGURE 12.1
The state trajectories of $x(k)$ and $\hat{x}(k)$.

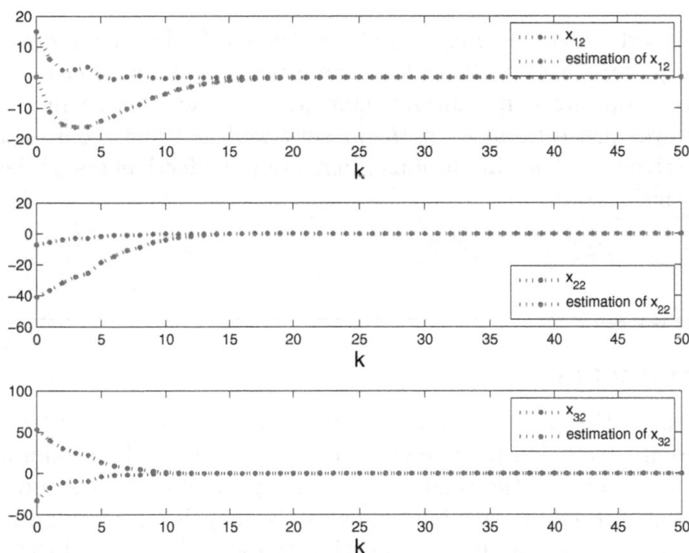

FIGURE 12.2
The state trajectories of $x(k)$ and $\hat{x}(k)$.

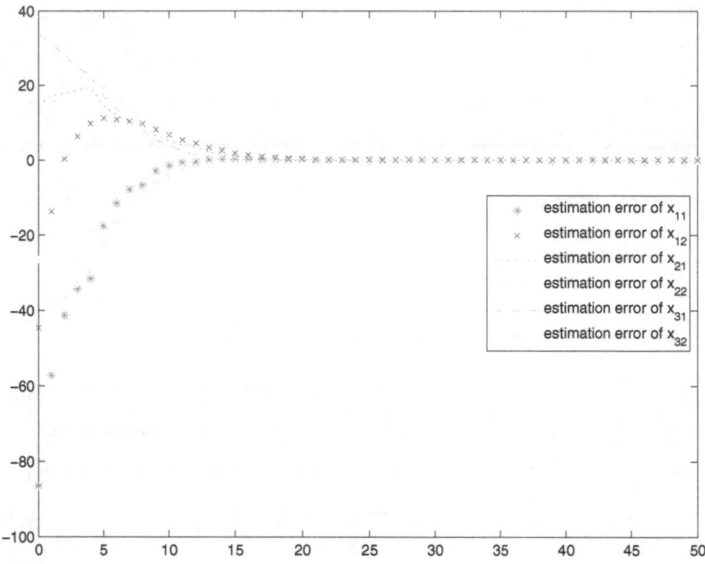

FIGURE 12.3
The estimation error of $x(k)$ and $\hat{x}(k)$.

errors of the estimators. In Figure 12.3, we can confirm that the designed H_∞ estimators perform very well. In the simulation results, obviously, the whole estimation errors are exponentially converged to zero fast, which indicates that the developed estimators perform very well and our theoretical results are of effectiveness. The simulations illustrate the effectiveness of the derived results in this chapter.

12.4 Summary

In this chapter, the state estimation problem for a class of complex networks with randomly occurring phenomena has been studied. The randomly occurring phenomena involve the randomly occurring coupling delays and measurement missing. A random variable sequence obeying the Bernoulli distribution is used to describe the randomly occurring coupling delays, and the measurement missing for every node is depicted by another random variable sequence satisfying a certain discrete probabilistic distribution on interval [0 1]. Then some state estimators are designed to make the augmented system to be stable in mean-square sense, and the estimation error satisfies the specified H_∞ performance requirement. Finally, an illustration example has been exploited to show the effectiveness of the proposed design procedures. Our further research topic would be to investigate the corresponding control problem of complex network with considering the practical factors, such as sensor saturations and the time-varying probabilities of the occurrence of missing measurements and coupling delays. Consequently, it is an extension of our results and seems to be much more interesting and challenging.

13

H_∞ Synchronization for Complex Networks with RVN

CONTENTS

In this chapter, the H_∞ synchronization control problem is investigated for a class of dynamical networks with randomly varying nonlinearities. The time varying nonlinearities of each node are modeled to be randomly switched between two different nonlinear functions by utilizing a Bernoulli distributed variable sequence specified by a randomly varying conditional probability distribution. A probability-dependent gain scheduling method is adopted to handle the time varying characteristic of the switching probability. Attention is focused on the design of a sequence of gain-scheduled controllers, such that the controlled networks are exponentially mean-square stable, and the H_∞ synchronization performance is achieved in the simultaneous presence of randomly varying nonlinearities and external energy bounded disturbances. Except for constant gains, the desired controllers are also composed of time varying parameters, i.e., the time varying switching probability and therefore less conservatism will be resulted compared with traditional controllers. In virtue of semidefinite programming method, controllers parameters are derived in terms of the solutions to a series of linear matrix inequalities (LMIs) that can be easily solved by the MATLAB Toolbox. Finally, a simulation example is exploited to illustrate the effectiveness of the proposed control strategy.

The main contribution of this chapter is highlighted as follows: *1) a new dynamical networks model covers the randomly varying nonlinearities whose occurrence probability is described by a series of varying Bernoulli distributions, yet taking value on a certain interval, which is closer to the practical engineering; 2) the potential conservatism will be reduced, resulting from time varying probability distributions via introducing the parameter dependent Lyapunov function and slack variable; 3) an array of dynamical controller gains for complex networks has been developed, which is scheduled with the changeable probability distributions.*

The remainder of this chapter is organized as follows: In Section 13.2, a stochastic dynamical networks model is formulated in presence of randomly varying nonlinearities and exogenous bounded disturbances. In Section 13.3, a sufficient condition is provided to guarantee the exponentially stable of dynamical networks and, furthermore, the controller gain of each node is derived in terms of the solutions to a sequence of LMIs. An illustrated numerical simulation is given to show the effectiveness and applicability of our proposed algorithm in Section 13.4, and a conclusion is summarized in Section 13.5.

13.1 Problem Formulation

Consider the dynamical networks with N coupled nodes as the following form:

$$\begin{cases} x_i(k+1) = Ax_i(k) + \alpha(k)f(x_i(k),k) + (1-\alpha(k))g(x_i(k),k) \\ \qquad\qquad + \sum\limits_{j=1}^{N} w_{ij}\Gamma x_j(k) + u_i(k) + B_i v(k), \\ z_i(k) = Mx_i(k), \end{cases} \tag{13.1}$$

where $x_i(k) \in \mathbb{R}^n$, $u_i(k) \in \mathbb{R}^n$, and $z_i(k) \in \mathbb{R}^m$ are the state vector, control input, and controlled output of the ith node, respectively. $v(k) \in \mathbb{R}^p$ is the disturbance input belonging to $l_2[0,+\infty)$. $f(\cdot)$ and $g(\cdot)$ are nonlinear vector functions. $\Gamma = \text{diag}(\gamma_1, \gamma_2, \cdots, \gamma_n)$ is the inner coupling matrix between two connected nodes for all $1 \le i, j \le N$. $W = (w_{ij})_{N\times N}$ is the coupled configuration matrix representing the coupling structure of the dynamical networks. If there is a connection between node i and node j $(i \ne j)$, $w_{ij} > 0$, otherwise, $w_{ij} = 0$. In this chapter, as usual, we assume W to be symmetric matrix and satisfy the condition $\sum_{j=1,i\ne j}^{N} w_{ij} = -w_{ii}$, $(i = 1, 2, ..., N)$. A, B_i, and M are constant matrices with appropriate dimensions.

The vector-value functions $f(\cdot)$ and $g(\cdot)$: $\mathbb{R}^n \to \mathbb{R}^n$ represent two different nonlinear disturbances, which are assumed to be continuous and satisfy the following conditions:

$$\|f(x(k))\|^2 \le \delta\|G_1 x(k)\|^2, \tag{13.2}$$

$$\|g(x(k))\|^2 \le \beta\|G_2 x(k)\|^2, \tag{13.3}$$

where δ, β are known positive scalars, and G_1, G_2 are known constant real matrices of appropriate dimensions.

In this chapter, we are interested in steering the dynamical network systems to a desired state $s(k)$, which is described as the solution to the following specified reference model

$$\begin{cases} s(k+1) = As(k), \\ z(k) = Ms(k), \end{cases} \tag{13.4}$$

where $z(k)$ is the output of the target state. Denoting $e_i(k) = x_i(k) - s(k)$, $\tilde{z}_i(k) = z_i(k) - z(k)$, respectively, the following systems that govern the synchronization error dynamics can be obtained

$$\begin{cases} e_i(k+1) = Ae_i(k) + \alpha(k)f(e_i(k)+s(k),k) + (1-\alpha(k))g(e_i(k)+s(k),k) \\ \qquad\qquad + \sum\limits_{j=1}^{N} w_{ij}\Gamma e_j(k) + u_i(k) + B_i v(k), \\ \tilde{z}_i(k) = Me_i(k), \end{cases} \tag{13.5}$$

for all $i = 1, 2, ...N$. Stochastic variable $\alpha(k)$ in (13.1) is a Bernoulli-distributed sequence that accounts for the randomly varying nonlinearity disturbances satisfying the following probability distribution laws

$$\begin{aligned} &\text{Prob}\{\alpha(k)=1\} = \mathbb{E}\{\alpha(k)\} = p(k), \\ &\text{Prob}\{\alpha(k)=0\} = 1-\mathbb{E}\{\alpha(k)\} = 1-p(k), \end{aligned} \tag{13.6}$$

where $p(k)$ is a time-varying positive scalar sequence taking values on interval $[p_1, p_2] \subseteq [0,1]$, with p_1 and p_2 being lower and upper bounds of $p(k)$, separately.

Remark 13.1 *The Bernoulli distributed model is one of the popular models to characterize the randomly occurring phenomenon, including randomly varying nonlinearities, which has been addressed in many literature during the past years; see, e.g., [32, 94]. As most of them mentioned, the occurrence probability p of switching between the two nonlinearities in each sampling period is described by a fixed constant, however, under such assumption, it may possibly bring much conservatism, since in the most real world time varying systems, the probability p is a changeable parameter and can be estimated/measured by statistical test method in real time. Thus, in this chapter, a time varying probability $p(k)$ is introduced to depict the RVNs for each node instead of a time-invariant one with known upper bound and lower bound, which is more suitable for the parameter-varying nature of practical engineering.*

In order to realize H_∞ synchronization, in this technical note, the gain-scheduled controllers are employed as follows:

$$u_i(k) = K_i(p(k))e_i(k), \tag{13.7}$$

where the controller gains $K_i(p(k))$ can be decomposed with the structures of

$$K_i(p(k)) = K_{i0} + p(k)K_{ip}. \tag{13.8}$$

Remark 13.2 *Compared with the most existing designed controllers with traditional structure containing constant or static gains only, gain-scheduled controller has the advantage of adjusting controller gains along with the time varying system parameters in real time, which mainly owes to its unique structure divided into two parts. One includes the fixed gain parameters K_{i0}, and the other possesses time-varying parameter $p(k)$. Since the dynamic parameter $p(k)$ enters into the controller design, more information is utilized to achieve a better control performance and, consequently, less conservatism will be brought about. Note that the gain scheduling method has been extensively used to deal with control/filtering problem of uncertain time-varying systems, and great research attention has been focused on it in the recent years. However, to the best of authors' knowledge, there are few efforts on the synchronization problem of dynamical networks with gain scheduling method due mainly to its high complexity of mathematical derivation, which encourages our current work.*

Afterwards, the closed-loop systems of synchronization error dynamics (13.5) are derived with the following forms:

$$\begin{cases} e_i(k+1) = Ae_i(k) + \alpha(k)f(e_i(k)+s(k),k) + (1-\alpha(k))g(e_i(k)+s(k),k) \\ \qquad\qquad + \sum\limits_{j=1}^{N} w_{ij}\Gamma e_j(k) + K_i(p(k))e_i(k) + B_iv(k), \\ \tilde{z}_i(k) = Me_i(k). \end{cases} \tag{13.9}$$

For convenience, the following notions are introduced

$$\begin{aligned} & e(k) = [e_1^T(k),\ e_2^T(k), \cdots, e_N^T(k)]^T,\ \tilde{z}(k) = [\tilde{z}_1^T(k),\ \tilde{z}_2^T(k), \cdots, \tilde{z}_N^T(k)]^T, \\ & \bar{B} = \mathrm{diag}\{B_1,\ B_2, \cdots B_N\},\ \eta(k) = [\bar{s}^T(k), e^T(k)]^T,\ \tilde{M} = I \otimes M, \\ & \mathcal{F}(\eta(k),k) = [f^T(e_1(k)+s(k),k), \cdots, f^T(e_N(k)+s(k),k))]^T, \\ & \mathcal{G}(\eta(k),k) = [g^T(e_1(k)+s(k),k), \cdots, g^T(e_N(k)+s(k),k))]^T, \\ & K(p(k)) = \mathrm{diag}\{K_1(p(k)),\ K_2(p(k)), \cdots, K_N(p(k))\},\ N_i = [0, E_i^T]^T, \\ & E_i = \mathrm{diag}\{\underbrace{0, \cdots, 0}_{i-1}, I, \underbrace{0, \cdots, 0}_{N-i}\},\ \tilde{v}(k) = [\underbrace{v^T(k),\ v^T(k), \cdots, v^T(k)}_{N}]^T, \\ & \bar{A} = \mathrm{diag}\{\underbrace{A,\ A, \cdots, A}_{N}\},\ \bar{s}(k) = [\underbrace{s^T(k),\ s^T(k), \cdots, s^T(k)}_{N}]^T. \end{aligned} \tag{13.10}$$

By utilizing the Kronecker product, the error dynamics of (13.9) can be rewritten as the following compact form:

$$\begin{cases} e(k+1) = \bar{A}e(k) + p(k)\mathcal{F}(\eta(k),k) + (1-p(k))\mathcal{G}(\eta(k),k) \\ \qquad\qquad + (\alpha(k)-p(k))\mathcal{F}(\eta(k),k) - (\alpha(k)-p(k))\mathcal{G}(\eta(k),k) \\ \qquad\qquad + (W \otimes \Gamma)e(k) + K(p(k))e(k) + \bar{B}\tilde{v}(k), \\ \tilde{z}(k) = \tilde{M}e(k). \end{cases} \tag{13.11}$$

Then, with the combination of (13.4) and (13.11), we can obtain the following augmented systems:

$$\begin{cases} \eta(k+1) = \mathcal{A}(k)\eta(k) + \mathcal{P}(k)\tilde{\mathcal{F}}(\eta(k),k) + (R-\mathcal{P}(k))\tilde{\mathcal{G}}(\eta(k),k) + \mathcal{B}\tilde{v}(k) \\ \qquad\qquad +(\alpha(k)-p(k))R\tilde{\mathcal{F}}(\eta(k),k) - (\alpha(k)-p(k))R\tilde{\mathcal{G}}(\eta(k),k) \\ \tilde{z}(k) = \mathcal{M}\eta(k), \end{cases} \tag{13.12}$$

where

$$\mathcal{A}(k) = \begin{bmatrix} \bar{A} & 0 \\ 0 & \bar{A} + (W \otimes \Gamma) + K(p(k)) \end{bmatrix}, \tag{13.13}$$

$$\mathcal{B} = \begin{bmatrix} 0 \\ \bar{B} \end{bmatrix}, \ \tilde{\mathcal{F}}(\eta(k),k) = \begin{bmatrix} \mathcal{F}(\eta(k),k) \\ 0 \end{bmatrix},$$

$$\tilde{\mathcal{G}}(\eta(k),k) = \begin{bmatrix} \mathcal{G}(\eta(k),k) \\ 0 \end{bmatrix}, \ R = \begin{bmatrix} 0 & 0 \\ I & 0 \end{bmatrix}, \tag{13.14}$$

$$\mathcal{P}(k) = \begin{bmatrix} 0 & 0 \\ p(k)I & 0 \end{bmatrix}, \ \mathcal{M} = [0 \ \ \tilde{M}]. \tag{13.15}$$

Under the assumption of (13.2) and (13.3), the nonlinear conditions can be rewritten as the following compact forms:

$$\tilde{\mathcal{F}}^T(\eta(k),k)\tilde{\mathcal{F}}(\eta(k),k) \le \eta^T(k)\bar{G}_1^T\bar{G}_1\eta(k), \tag{13.16}$$

$$\tilde{\mathcal{G}}^T(\eta(k),k)\tilde{\mathcal{G}}(\eta(k),k) \le \eta^T(k)\bar{G}_2^T\bar{G}_2\eta(k), \tag{13.17}$$

where $\bar{G}_1 = [\hat{G}_1 \ \ \hat{G}_1]$, $\bar{G}_2 = [\hat{G}_2 \ \ \hat{G}_2]$, $\hat{G}_1 = [\underbrace{\sqrt{\delta}G_1, \ \sqrt{\delta}G_1, \cdots, \sqrt{\delta}G_1}_{N}]$, $\hat{G}_2 = [\underbrace{\sqrt{\beta}G_2, \ \sqrt{\beta}G_2, \cdots, \sqrt{\beta}G_2}_{N}]$.

Definition 16 *The dynamical networks (13.12) are said to reach the H_∞ synchronization to the desired state $s(k)$ in the mean-square sense if the following inequality holds:*

$$\mathbb{E}\left\{\sum_{k=1}^{\infty} \| \tilde{z}(k) \|^2\right\} \le \gamma^2 \sum_{k=1}^{\infty} \| \tilde{v}(k) \|^2 . \tag{13.18}$$

In this chapter, we are interested in designing a series of gain-scheduled controllers for a stochastic dynamical networks, such that, for the randomly varying nonlinearities and admissible external bounded disturbances, the dynamical networks are exponentially stable, and H_∞ synchronization performance is guaranteed with a prescribed attenuation level γ.

13.2 Main Results

13.2.1 Stability Analysis

In the following theorem, Lyapunov stability theory and convex programming method are used to deal with the stability analysis problem of the probability-dependent gain-scheduled synchronization controller design for a class of discrete-time stochastic dynamical networks (13.1) with RVNs. A sufficient condition is formulated to guarantee the solvability of the desired stability analysis problem.

Theorem 30 *The augmented closed-loop error system (13.12) with $\tilde{v}(k) = 0$ is exponentially mean-square stable for all $p(k) \in [p_1, p_2]$, if there exist matrix sequences $Q(p(k)) > 0$, slack matrix S, and two positive scalars ε_1 and ε_2, such that the following matrix inequalities*

$$\begin{bmatrix} \Omega_{11} & 0 & 0 & (S^T\mathcal{A}(k))^T & 0 \\ * & -\varepsilon_1 I & 0 & (S^T\mathcal{P}(k))^T & (\theta(k)S^TR)^T \\ * & * & -\varepsilon_2 I & (S^T\bar{\mathcal{P}}(k))^T & -(\theta(k)S^TR)^T \\ * & * & * & -\Gamma_{k+1} & 0 \\ * & * & * & * & -\theta(k)\Gamma_{k+1} \end{bmatrix} < 0 \tag{13.19}$$

are satisfied, where

$$\begin{aligned} \Omega_{11} &= -Q(p(k)) + \varepsilon_1\hat{G}_1^T\hat{G}_1 + \varepsilon_2\hat{G}_2^T\hat{G}_2,\ \bar{\mathcal{P}}(k) = R - \mathcal{P}(k), \\ \Gamma_{k+1} &= -Q(p(k+1)) + S^T + S,\ \theta(k) = p(k)(1-p(k)). \end{aligned} \tag{13.20}$$

Proof 33 *For notation simplicity, we denote*

$$\begin{aligned} \xi(k) &= [\eta^T(k) \quad \tilde{\mathcal{F}}^T(\eta(k),k) \quad \tilde{\mathcal{G}}^T(\eta(k),k)]^T, \\ \mathscr{A}(k) &= [\mathcal{A}(k) \quad \mathcal{P}(k) \quad R-\mathcal{P}(k)]. \end{aligned} \tag{13.21}$$

Define the probability-dependent Lyapunov functional as

$$V(p(k)) = \eta^T(k)Q(p(k))\eta(k).$$

Note that $\mathbb{E}\{\alpha(k) - p(k)\} = 0$, then it can be calculated that with $\tilde{v}(k) = 0$

$$\begin{aligned} \mathbb{E}\{\Delta V(p(k))\} &= \mathbb{E}\left\{\eta^T(k+1)Q(p(k+1))\eta(k+1) - \eta^T(k)Q(p(k))\eta(k)\right\} \\ &= \mathbb{E}\Big\{\Big[\mathcal{A}(k)\eta(k) + \mathcal{P}(k)\tilde{\mathcal{F}}(\eta(k),k) + (R-\mathcal{P}(k))\tilde{\mathcal{G}}(\eta(k),k)\Big]^T \\ &\qquad Q(p(k+1))\Big[\mathcal{A}(k)\eta(k) + \mathcal{P}(k)\tilde{\mathcal{F}}(\eta(k),k) \\ &\qquad + (R-\mathcal{P}(k))\tilde{\mathcal{G}}(\eta(k),k)\Big] - \eta^T(k)Q(p(k))\eta(k) \end{aligned}$$

$$
\begin{aligned}
&+ \theta(k)\tilde{\mathcal{F}}^T(\eta(k),k)R^TQ(p(k+1))R\tilde{\mathcal{F}}(\eta(k),k) \\
&+ \theta(k)\tilde{\mathcal{G}}^T(\eta(k),k)R^TQ(p(k+1))R\tilde{\mathcal{G}}(\eta(k),k) \\
&- 2(k)\tilde{\mathcal{F}}^T(\eta(k),k)R^TQ(p(k+1))R\tilde{\mathcal{G}}(\eta(k),k) \\
=& \mathbb{E}\{\xi^T(k)(\mathscr{A}(k)^TQ(p(k+1))\mathscr{A}(k) + \bar{\Xi}(k))\xi(k)\}, \qquad (13.22)
\end{aligned}
$$

where

$$
\bar{\Xi}(k) = \begin{bmatrix} -Q(p(k)) & 0 & 0 \\ * & \bar{\Xi}_{22} & -\bar{\Xi}_{23} \\ * & * & \bar{\Xi}_{33} \end{bmatrix}, \qquad (13.23)
$$

with $\bar{\Xi}_{22} = \bar{\Xi}_{23} = \bar{\Xi}_{33} = \theta(k)R^TQ(p(k+1))R$.

From (13.16) and (13.17), we have

$$
[\eta^T(k) \;\; \tilde{\mathcal{F}}^T(\eta(k),k)] \begin{bmatrix} -\hat{G}_1^T\hat{G}_1 & 0 \\ 0 & I \end{bmatrix} \begin{bmatrix} \eta(k) \\ \tilde{\mathcal{F}}(\eta(k),k) \end{bmatrix} \leq 0, \qquad (13.24)
$$

and

$$
[\eta^T(k) \;\; \tilde{\mathcal{G}}^T(\eta(k),k)] \begin{bmatrix} -\hat{G}_2^T\hat{G}_2 & 0 \\ 0 & I \end{bmatrix} \begin{bmatrix} \eta(k) \\ \tilde{\mathcal{G}}(\eta(k),k) \end{bmatrix} \leq 0, \qquad (13.25)
$$

respectively.

Subsequently, taking (13.22)–(13.25) into consideration, we can get

$$
\begin{aligned}
\mathbb{E}\{\Delta V(p(k))\} \leq& \mathbb{E}\Bigg\{\xi^T(k)\Big(\mathscr{A}(k)^TQ(p(k+1)\mathscr{A}(k) + \bar{\Xi}(k)\Big)\xi(k) \\
&- \varepsilon_1 \begin{bmatrix} \eta(k) \\ \tilde{\mathcal{F}}(\eta(k),k) \end{bmatrix}^T \begin{bmatrix} -\hat{G}_1^T\hat{G}_1 & 0 \\ 0 & I \end{bmatrix} \begin{bmatrix} \eta(k) \\ \tilde{\mathcal{F}}(\eta(k),k) \end{bmatrix} \\
&- \varepsilon_2 \begin{bmatrix} \eta(k) \\ \tilde{\mathcal{G}}(\eta(k),k) \end{bmatrix}^T \begin{bmatrix} -\hat{G}_2^T\hat{G}_2 & 0 \\ 0 & I \end{bmatrix} \begin{bmatrix} \eta(k) \\ \tilde{\mathcal{G}}(\eta(k),k) \end{bmatrix}\Bigg\} \\
=& \mathbb{E}\Big\{\xi^T(k)\Xi(k)\xi(k)\Big\}, \qquad (13.26)
\end{aligned}
$$

where

$$
\begin{aligned}
\Xi(k) &= \mathscr{A}(k)^TQ(p(k+1)\mathscr{A}(k) + \tilde{\Xi}(k), \\
\tilde{\Xi}(k) &= \begin{bmatrix} \Omega_{11} & 0 & 0 \\ * & \bar{\Xi}_{22} - \varepsilon_1 I & -\bar{\Xi}_{23} \\ * & * & \bar{\Xi}_{33} - \varepsilon_2 I \end{bmatrix}. \qquad (13.27)
\end{aligned}
$$

Next, we will derive that $\Xi(k) < 0$ *from (13.19). First, making congruence transformation* $\mathrm{diag}\{I \; I \; I \; S^{-1} \;\; \theta^{-1}(k)S^{-1}\}$ *to (13.19), it can be obtained that*

$$
\begin{bmatrix}
\Omega_{11} & 0 & 0 & \mathcal{A}^T(k) & 0 \\
* & -\varepsilon_1 I & 0 & \mathcal{P}^T(k) & R^T \\
* & * & -\varepsilon_2 I & \bar{\mathcal{P}}^T(k) & -R^T \\
* & * & * & -\Pi_{k+1} & 0 \\
* & * & * & * & -\theta^{-1}(k)\Pi_{k+1}
\end{bmatrix} < 0 \qquad (13.28)
$$

hold, with $\Pi_{k+1} = -S^{-T}Q(p(k+1))S^{-1} + S^{-1} + S^{-T}$. *By means of inequality* $Q^{-1}(p(k+1)) \geq -S^{-T}Q(p(k+1))S^{-1} + S^{-1} + S^{-T}$, *the following LMIs*

$$\begin{bmatrix} \Omega_{11} & 0 & 0 & \mathcal{A}^T(k) & 0 \\ * & -\varepsilon_1 I & 0 & \mathcal{P}^T(k) & R^T \\ * & * & -\varepsilon_2 I & \bar{\mathcal{P}}^T(k) & -R^T \\ * & * & * & -\Lambda_{k+1} & 0 \\ * & * & * & * & -\theta^{-1}(k)\Lambda_{k+1} \end{bmatrix} < 0 \tag{13.29}$$

meet immediately with $\Lambda_{k+1} = Q^{-1}(p(k+1))$. *Hence, by virtue of Schur complement, one easily finds* $\Xi(k) < 0$ *and* $\mathbb{E}\{\Delta V(k)\} < -\lambda_{\min}(-\Xi(k))\mathbb{E}\|\eta(k)\|$. *Finally, it directly follows from Lemma 1 in [186] that the exponential mean-square stability of the augmented error system (13.12) is guaranteed, and, consequently, the proof of the theorem is accomplished.*

Remark 13.3 *In Theorem 30, for the purpose of analyzing exponential stability of error system (13.12) in the mean-square sense by utilizing the time varying information of switching probability, probability-dependent Lyapunov functionals have been introduced to reduce the probable conservatism. It is worth mentioning that, in the past few years, parameter-dependent Lyapunov functionals have been usually applied to uncertain systems and time-varying parameter systems [43,154]. Furthermore, a slack variable S has been exploited in Theorem 30 to decouple the product terms between Lyapunov matrices and system matrices, and we can easily deal with the corresponding controller design problem in the latter theorem accordingly.*

13.2.2 H_∞ Synchronization Performance Analysis

In Theorem 30, a sufficient condition has been obtained to guarantee the exponential mean-square stability of the error dynamics (13.12) in terms of a sequences of LMIs (13.19). In the rest of our work in the following theorem, we focus on dealing with the H_∞ performance constraint with respect to the output error $\tilde{z}_i(k)$.

Theorem 31 *Given the disturbance attenuation level* γ, *controller parameters* K_{i0} *and* K_{ip} $(i = 1, 2, \cdots, N)$, *the exponential mean-square stability of the augmented error system (13.12) for* $\tilde{v}(k) = 0$ *and* H_∞ *constraints for all nonzero* $\tilde{v}(k)$ *under the zero initial condition are guaranteed simultaneously, subject to all* $p(k) \in [p_1, p_2]$, *if there exist matrix sequences* $Q(p(k)) > 0$, *slack matrix* S, *and two positive scalars* ε_1 *and* ε_2, *such that the following matrix* *inequalities*

$$\begin{bmatrix} \bar{\Omega}_{11} & 0 & 0 & 0 & (S^T\mathcal{A}(k))^T & 0 \\ * & -\varepsilon_1 I & 0 & 0 & (S^T\mathcal{P}(k))^T & (\theta(k)S^TR)^T \\ * & * & -\varepsilon_2 I & 0 & (S^T\bar{\mathcal{P}}(k))^T & -(\theta(k)S^TR)^T \\ * & * & * & -\gamma^2 I & (S^T\mathcal{B})^T & 0 \\ * & * & * & * & -\Gamma_{k+1} & 0 \\ * & * & * & * & * & -\theta(k)\Gamma_{k+1} \end{bmatrix} < 0 \quad (13.30)$$

hold, where $\bar{\Omega}_{11} = \Omega_{11} + \mathcal{M}^T\mathcal{M}$.

Proof 34 *Choosing the same Lyapunov functional as the one in the proof of Theorem 30 with nonzero* $\tilde{v}(k)$*, we can obtain that*

$$\begin{aligned}\mathbb{E}\Big\{\Delta V(p(k))\Big\} \leq \mathbb{E}\Big\{&\xi^T(k)\Xi(k)\xi(k) + 2\eta^T(k)\mathcal{A}^T(k)Q(p(k+1))\mathcal{B}\tilde{v}(k) \\ &+ 2\tilde{\mathcal{F}}^T(\eta(k),k)\mathcal{P}^T(k)Q(p(k+1))\mathcal{B}\tilde{v}(k) \\ &+ 2\tilde{\mathcal{G}}^T(\eta(k),k)\bar{\mathcal{P}}^T(k)Q(p(k+1))\mathcal{B}\tilde{v}(k) \\ &+ \tilde{v}^T(k)\mathcal{B}^TQ(p(k+1))\mathcal{B}\tilde{v}(k)\Big\}, \end{aligned} \quad (13.31)$$

where $\xi(k)$ *and* $\Xi(k)$ *have been defined before. Then, letting*

$$\bar{\xi}(k) = [\eta^T(k) \;\; \tilde{\mathcal{F}}^T(\eta(k),k) \;\; \tilde{\mathcal{G}}^T(\eta(k),k) \;\; \tilde{v}^T(k)]^T,$$

it can be calculated that

$$\mathbb{E}\{\Delta V(p(k))\} \leq \mathbb{E}\left\{\bar{\xi}^T(k)\begin{bmatrix} \Xi(k) & \Upsilon^T \\ \Upsilon & \mathcal{B}^TQ(p(k+1))\mathcal{B} \end{bmatrix}\bar{\xi}(k)\right\}, \quad (13.32)$$

where $\Upsilon = [\mathcal{A}^T(k)Q(p(k+1))\mathcal{B} \;\; \mathcal{P}^T(k)Q(p(k+1))\mathcal{B} \;\; \bar{\mathcal{P}}^TQ(p(k+1))\mathcal{B}]$*. For the purpose of analyzing the* H_∞ *performance, the following formula is introduced:*

$$J(s) = \mathbb{E}\left\{\sum_{k=0}^{s}\left(\tilde{z}(k)^T\tilde{z}(k) - \gamma^2\tilde{v}^T(k)\tilde{v}(k)\right)\right\}, \quad (13.33)$$

where s *is a nonnegative integer. Denote* $\hat{\Xi}(k) = \Xi(k) + \tilde{\mathcal{M}}^T\tilde{\mathcal{M}}$, $\tilde{\mathcal{M}} = [\mathcal{M} \;\; 0 \;\; 0]$*. Under the assumption of zero initial condition, adding the zero term* $\Delta V(k) - \Delta V(k)$ *to the right side of (13.33) yields*

$$\begin{aligned} J(s) &= \mathbb{E}\left\{\sum_{k=0}^{s}\Big(\tilde{z}(k)^T\tilde{z}(k) - \gamma^2\tilde{v}(k)^T\tilde{v}(k) + \Delta V(k)\Big) - \sum_{k=0}^{s}\Delta V(k)\right\} \\ &= \mathbb{E}\left\{\sum_{k=0}^{s}\bar{\xi}^T(k)\begin{bmatrix} \hat{\Xi}(k) & \Upsilon^T \\ \Upsilon & \mathcal{B}^TQ(p(k+1))\mathcal{B} - \gamma^2 I \end{bmatrix}\bar{\xi}(k)\right\} \\ &\quad - V(s+1) + V(0) \\ &\leq \mathbb{E}\left\{\sum_{k=0}^{s}\bar{\xi}^T(k)\begin{bmatrix} \hat{\Xi}(k) & \Upsilon^T \\ \Upsilon & \mathcal{B}^TQ(p(k+1))\mathcal{B} - \gamma^2 I \end{bmatrix}\bar{\xi}(k)\right\}. \end{aligned} \quad (13.34)$$

Along the similar proof line of Theorem 30, when enforcing $s \to \infty$, we can conclude $J(s) < 0$ from (13.30) immediately by confirming from Schur complement lemma. Consequently, the H_∞ performance is obtained, which completes the proof.

13.2.3 Probability-Dependent Synchronization Controller Design

Up to now, the analysis problem of H_∞ synchronization performance for the dynamical networks has been finished. Next, we are to pay our attention to the design problem of controller parameters.

Theorem 32 *Given the disturbance attenuation level $\gamma > 0$, the augmented system (13.12) is exponentially mean-square stable, and the H_∞ performance is satisfied under zero initial condition for all nonzero $\tilde{v}(k)$, if there exist a sequence of matrices $Q(p(k)) > 0$, nonsingular matrices S_1 and S_2, and two positive scalars ε_1 and ε_2, such that the following LMIs*

$$\begin{bmatrix} \bar{\Omega}_{11} & 0 & 0 & 0 & \Psi_{15}^T & 0 \\ * & -\varepsilon_1 I & 0 & 0 & \Psi_{25}^T & \Psi_{26}^T \\ * & * & -\varepsilon_2 I & 0 & \Psi_{35}^T & -\Psi_{36}^T \\ * & * & * & -\gamma^2 I & \Psi_{45}^T & 0 \\ * & * & * & * & -\bar{\Gamma}_{k+1} & 0 \\ * & * & * & * & * & -\theta(k)\bar{\Gamma}_{k+1} \end{bmatrix} < 0 \tag{13.35}$$

hold, where

$$\begin{aligned}
\Psi_{15} &= \begin{bmatrix} I \otimes S_1^T A & 0 \\ * & I \otimes S_2^T A + W \otimes S_2^T \Gamma + \bar{K}(p(k)) \end{bmatrix}, \\
\Psi_{25} &= \begin{bmatrix} 0 & 0 \\ p(k)\bar{S}_2^T & 0, \end{bmatrix}, \quad \Psi_{45} = \begin{bmatrix} 0 \\ \bar{S}_2{}^T \bar{B} \end{bmatrix}, \\
\Psi_{35} &= \begin{bmatrix} 0 & 0 \\ (1-p(k))\bar{S}_2^T & 0, \end{bmatrix}, \quad \Psi_{26} = \Psi_{36} = \begin{bmatrix} 0 & 0 \\ \theta(k)\bar{S}_2^T & 0 \end{bmatrix}, \\
\Delta &= \begin{bmatrix} I \otimes (S_1 + S_1^T) & 0 \\ 0 & I \otimes (S_2 + S_2^T) \end{bmatrix} \\
\bar{K}(p(k)) &= \mathrm{diag}\left\{\bar{K}_1(p(k))\ \bar{K}_2(p(k)), \cdots, \bar{K}_N(p(k))\right\}, \\
\bar{S}_2 &= I \otimes S_2,\ \bar{\Gamma}_{k+1} = Q(p(k)) + \Delta,
\end{aligned} \tag{13.36}$$

and other parameters have been defined previously. In this case, the controller parameters can be easily determined as

$$K_i(p(k)) = S_2^{-T} \bar{K}_i(p(k)). \tag{13.37}$$

Proof 35 *Let nonsingular matrix S be partitioned as $S = \mathrm{diag}\{I \otimes S_1, I \otimes S_2\}$, and this result can be directly obtained from Theorem 31. Therefore, the detailed proof is omitted.*

13.2.4 Admissible Solution

Because of the time-varying parameter $p(k)$ in the LMIs (13.19), (13.30), and (13.35), it will give rise to infinite numbers of LMIs to be solved, and such computation burden makes it impossible to obtain desired controller parameters. Therefore, the following theorem is proposed to convert the infinity LMIs into the admissible ones.

Theorem 33 *Given the disturbance attenuation level $\gamma > 0$, the augmented system (13.12) is exponentially mean-square stable, and the H_∞ performance is satisfied under zero initial condition for all nonzero $\tilde{v}(k)$, if there exist matrices $Q_0 > 0$, $Q_p > 0$, $\bar{K}_{i0}$, $\bar{K}_{ip}$, $(i = 1, 2, \cdots, N)$, and nonsingular matrices S_1 and S_2, and two positive scalars ε_1 and ε_2, such that the following LMIs*

$$\mathbb{M}^{hjl} = \begin{bmatrix} \bar{\Omega}_{11}^h & 0 & 0 & 0 & \Psi_{15}^{hT} & 0 \\ * & -\varepsilon_1 I & 0 & 0 & \Psi_{25}^{hT} & \Psi_{26}^{hjT} \\ * & * & -\varepsilon_2 I & 0 & \Psi_{35}^{jT} & -\Psi_{36}^{hjT} \\ * & * & * & -\gamma^2 I & \Psi_{45}^{T} & 0 \\ * & * & * & * & -\bar{\Gamma}^l & 0 \\ * & * & * & * & * & -\theta^{hj}\bar{\Gamma}^l \end{bmatrix} < 0 \tag{13.38}$$

hold, for h, j, and $l = 1, 2$, where

$$\begin{aligned} \Psi_{15}^h &= \begin{bmatrix} I \otimes S_1^T A & 0 \\ * & I \otimes S_2^T A + W \otimes S_2^T \Gamma + \bar{K}^h \end{bmatrix}, \\ \Psi_{25}^h &= \begin{bmatrix} 0 & 0 \\ p_h \bar{S}_2^T & 0, \end{bmatrix}, \ \Psi_{35}^j = \begin{bmatrix} 0 & 0 \\ (1-p_j)\bar{S}_2^T & 0, \end{bmatrix}, \\ Q^h &= Q_0 + p_h Q_p, \ \bar{K}^h = \mathrm{diag}\{\bar{K}_1^h, \ \bar{K}_2^h, \cdots, \bar{K}_N^h\}, \\ \bar{\Omega}_{11}^h &= \Omega_{11}^h + \mathcal{M}^T \mathcal{M}, \ \Omega_{11}^h = -Q^h + \varepsilon_1 \hat{G}_1^T \hat{G}_1 + \varepsilon_2 \hat{G}_2^T \hat{G}_2, \\ \bar{\Gamma}^l &= -Q^l + \Delta, \bar{K}_i^h = \bar{K}_{i0} + p_h \bar{K}_{ip}, \\ \theta^{hj} &= p_h(1-p_j), \ \Psi_{26}^{hj} = \Psi_{36}^{hj} = \begin{bmatrix} 0 & 0 \\ \theta^{hj}\bar{S}_2^T & 0, \end{bmatrix}. \end{aligned} \tag{13.39}$$

Then, the controllers are given by (13.7) and (13.8), with the parameters as follows:

$$K_{i0} = S_2^{-T}\bar{K}_{i0}, \quad K_{ip} = S_2^{-T}\bar{K}_{ip} \quad (i = 1, \ 2, \cdots, N). \tag{13.40}$$

Proof 36 *Define the Lyapunov matrix with $Q(p(k)) = Q_0 + p(k)Q_p$. In order to avoid dealing with the time-varying probability directly, we rewrite $p(k)$ with a new form as*

$$p(k) = \lambda_1(k)p_1 + \lambda_2(k)p_2, \tag{13.41}$$

with

$$\lambda_1(k) = \frac{p_2 - p(k)}{p_2 - p_1}, \quad \lambda_2(k) = \frac{p(k) - p_1}{p_2 - p_1}.$$

It easily finds $\lambda_1(k) + \lambda_2(k) = 1$. *Similarly,* $p(k+1)$ *can be rewritten as*

$$p(k+1) = \varphi_1(k)p_1 + \varphi_2(k)p_2. \tag{13.42}$$

with

$$\varphi_1(k) = \frac{p_2 - p(k+1)}{p_2 - p_1}, \quad \varphi_2(k) = \frac{p(k+1) - p_1}{p_2 - p_1},$$

and $\varphi_1(k)+\varphi_2(k) = 1$. *Hence, from the redefinition of* $p(k)$ *and* $p(k+1)$ *above, we have the following transformation:*

$$\begin{aligned} Q(p(k)) &= \sum_{h=1}^{2} \lambda_h(k)Q^h, \quad Q(p(k+1)) = \sum_{l=1}^{2} \varphi_l(k)Q^l, \\ \bar{K}(p(k)) &= \sum_{h=1}^{2} \lambda_h(k)\bar{K}^h. \end{aligned} \tag{13.43}$$

Combining with the above analysis and LMIs (13.38), one has

$$\sum_{h,\, j,\, l=1}^{2} \lambda_h(k)\lambda_j(k)\varphi_l(k)\mathbb{M}^{hjl} < 0. \tag{13.44}$$

Obviously, (13.44) is equivalent to (13.38), and, therefore, the proof of this theorem is completed.

Remark 13.4 *Theorem 33 points out the way for turning the infinite LMIs in Theorem 32 into solvable ones by rewriting* $p(k)$ *as the polyhedral form. After such transformation, the accessible LMIs only depend on the upper and lower bounds of* $p(k)$ *and then the controller parameters can be easily obtained, resorting to MATLAB Toolbox. It should be noted that such, and approach was first proposed in [192, 193] and has been employed in [86, 87, 107].*

13.3 An Illustrative Example

In this section, a simulation example is provided to demonstrate the effectiveness of our established criteria on the H_∞ synchronization problem of dynamical networks (13.1).

The system parameters are selected as follows:

$$A = \begin{bmatrix} -0.45 & 0.25 \\ 1.2 & 0.5 \end{bmatrix}, \; B = \begin{bmatrix} 0.1 & 0.02 \\ 0.8 & 0.03 \end{bmatrix}, \; M = \begin{bmatrix} 0.05 & 0 \\ 0 & 0.05 \end{bmatrix}, \tag{13.45}$$
$$\gamma = 1.5, \; p_1 = 0.1, \; p_2 = 0.65.$$

Consider the stochastic dynamical networks with three nodes, and the coupling configuration matrices and inner coupling matrix are given as

$$W = \begin{bmatrix} -0.8 & 0.8 & 0 \\ 0.8 & -1.5 & 0.7 \\ 0 & 0.7 & -0.7 \end{bmatrix}, \; \Gamma = \begin{bmatrix} 0.2 & 0 \\ 0 & 0.2 \end{bmatrix}.$$

In this example, we choose the time-varying nonlinear functions $f(x_i(k))$ and $g(x_i(k))$ as

$$f(x_i(k)) = \begin{bmatrix} \frac{0.3x_{i1}(k)}{2x_{i2}^2+1} & 0.25x_{i2}\cos(x_{i1}) \end{bmatrix},$$
$$g(x_i(k)) = \begin{bmatrix} \frac{0.03x_{i1}(k)}{|x_{i2}|+1} & 0.05x_{i2}\tanh(x_{i1}) \end{bmatrix},$$

and it is easily confirmed that the constraints (13.2) and (13.3) are satisfied simultaneously with $G_1 = \text{diag}\{0.3,\ 0.25\}$, $G_2 = \text{diag}\{0.03,\ 0.05\}$, and $\delta = \beta = 1$.

Here, we adopt the exogenous disturbances $v(k) = [\exp(-k/50) \times \text{s}(k); 2\sin(0.8k)/k + 1]$, where $\text{s}(k)$ is a random variable obeying the uniform distribution over $[0.05, 0.05]$, and set time varying probability $p(k)$ of Bernoulli distribution as $p(k) = p_1 + (p_2 - p_1)|\sin(k)|$. According to Theorem 33, the desired controller parameters are obtained as follows by solving finite LMIs:

$$K_{10} = \begin{bmatrix} 0.5496 & -0.2700 \\ -1.2367 & -0.4483 \end{bmatrix}, \; K_{20} = \begin{bmatrix} 0.6795 & -0.2643 \\ -1.2391 & -0.3096 \end{bmatrix},$$
$$K_{30} = \begin{bmatrix} 0.5367 & -0.2698 \\ -1.2363 & -0.4583 \end{bmatrix}, \; K_{1p} = \begin{bmatrix} -0.0040 & 0.0039 \\ -0.0003 & -0.0022 \end{bmatrix},$$
$$K_{2p} = \begin{bmatrix} -0.0084 & 0.0033 \\ -0.0002 & -0.0113 \end{bmatrix}, \; K_{3p} = \begin{bmatrix} -0.0112 & 0.0004 \\ -0.0070 & -0.0218 \end{bmatrix}.$$

Figure 13.1 and Figure 13.2 depict the state trajectories of system (13.1) and (13.4) as well as Figure 13.3, and Figure 13.4 give the synchronization errors of $x_i(k), (i = 1, 2, 3)$ and $s(k)$, and Figure 13.5 is the curve of time-varying probability $p(k)$, from which we can obviously find out that the gain-scheduled controllers play very well, and the H_∞ synchronization performance is guaranteed with a prescribed attenuation level γ.

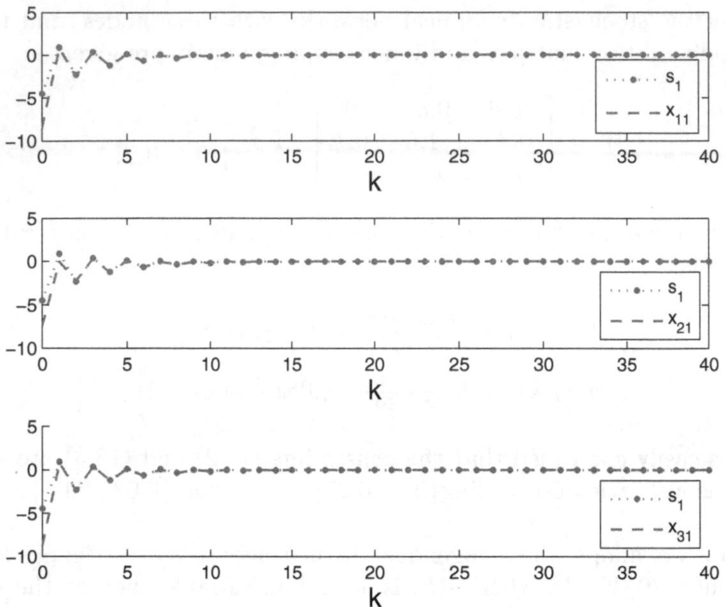

FIGURE 13.1
The state trajectories of $x_{i1}(k), (i = 1, 2, 3)$, and $s_1(k)$.

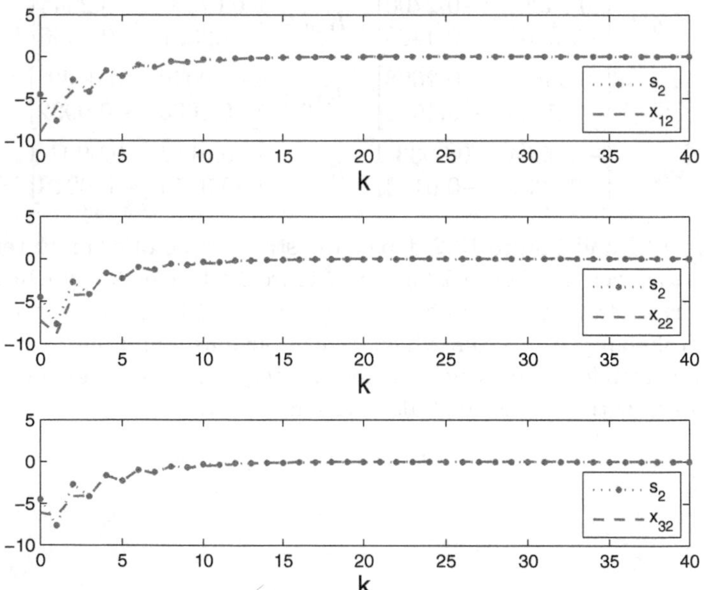

FIGURE 13.2
The state trajectories of $x_{i2}(k), (i = 1, 2, 3)$, and $s_2(k)$.

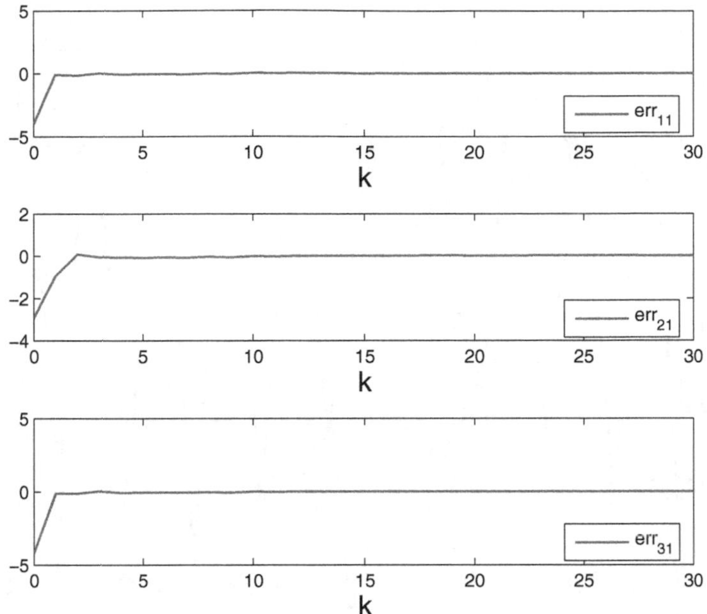

FIGURE 13.3
The synchronization error of $x_{i1}(k), (i = 1, 2, 3)$, and $s_1(k)$.

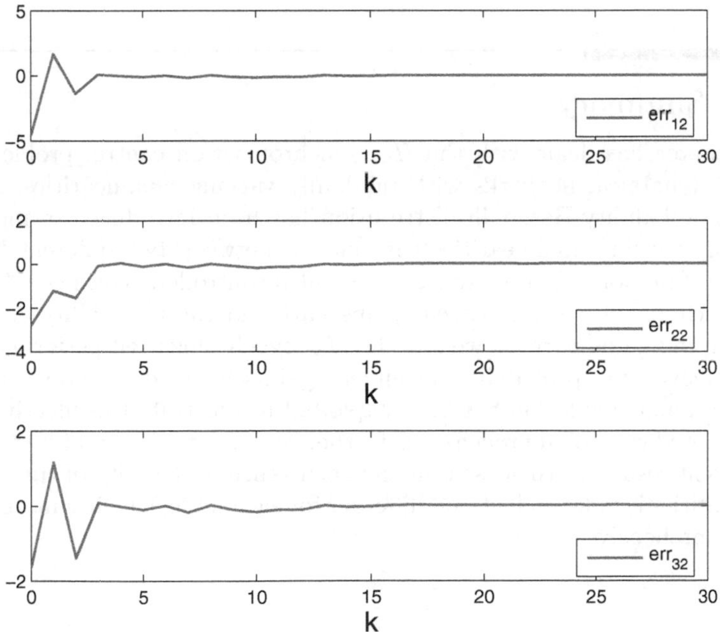

FIGURE 13.4
The synchronization error of $x_{i2}(k), (i = 1, 2, 3)$, and $s_2(k)$.

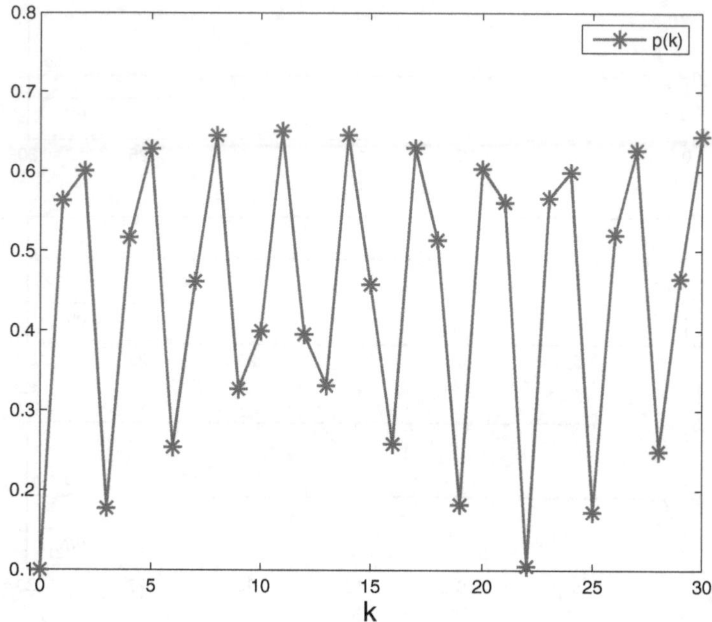

FIGURE 13.5
Time varying probability $p(k)$.

13.4 Summary

This chapter has dealt with the H_∞ synchronization control problem for a class of dynamical networks with randomly varying nonlinearities. A time-varying probability Bernoulli distribution has been introduced to model the randomly varying nonlinearities that switch between two different kinds of nonlinear functions. A series of gain-scheduled controllers have been designed, which include the constant(fixed) gains part and the probability-dependent time-varying gains part, as well as the H_∞ synchronization performance has been achieved via probability-dependent gain-scheduled method. Finally, a computational simulation has been exploited to illustrate the effectiveness of our proposed control approach. Our further research topic would be to extend the current results to the case that the occurrence probability of time varying nonlinearities is independent and different for each node, which is more general and comprehensive.

Bibliography

[1] A. Abdullah and M. Zribi. Sensor fault tolerant control for a class of linear parameter varying systems with practical examples. *IEEE Transactions on Industrial Electronics*, 60(11):5239–5251, 2013.

[2] H. Ahmad and T. Namerikawa. Extended Kalman filter-based mobile robot localization with intermittent measurements. *Systems Science & Control Engineering: An Open Access Journal*, 1(1):113–126, 2013.

[3] P. Apkarian and P. Gahinet. A convex characterization of gain-scheduled H_∞ controllers. *IEEE Transactions on Automatic Control*, 40(5):853–864, 1995.

[4] P. Apkarian, P. Gahinet, and G. Becker. Self-scheduled H_∞ control of linear parameter-varying systems: A design example. *Automatica*, 31(9):1251–1261, 1995.

[5] P. Apkarian, P. Pellanda, and H. Tuan. Mixed H_2/H_∞ multi-channel linear parameter-varying control in discrete time. *Systems & Control Letters*, 41(5):333–346, 2000.

[6] L. Arnold. *Stochastic Differential Equations: Theory and Applications*. John Wiley Sons, 1972.

[7] K. J. Åström. *Introduction to Stochastic Control Theory*. Academic Press, 1970.

[8] K. J. Åström and P. Kumar. Control: A perspective. *Automatica*, 50(1):3–43, 2014.

[9] A. L. Barabasi and R. Albert. Emergence of scaling in random networks. *Science*, 286(5439):509–512, 1999.

[10] N. Berman and U. Shaked. h_∞ for nonlinear stochastic systems. In *Proceedings of 42nd IEEE conference on decision and control, Maui, HI*, volume 5, pages 5025–5030. Springer, 2003.

[11] H. Bolouri and E. Davidson. Modeling transcriptional regulatory networks. *BioEssay*, 24(12):1118–1129, 2002.

[12] E. K. Boukas and Z. K. Liu. *Deterministic and Stochastic Time-Delay Systems*. Birkhauser, Boston, 2002.

[13] S. Boyd, L. EI Ghaoui, E. Feron, and V. Balakrishnan. *Linear Matrix Inequalities in Systems and Control Theory.* SIAM, Philadelphia, PA, 1994.

[14] Y. Cao, J. Lam, and L. Hu. Delay-dependent stochastic stability and H_∞ analysis for time-delay systems with Markovian jumping parameters. *Journal of the Franklin Institute*, 340(6–7):423–434, 2003.

[15] Y. Cao, Z. Lin, and B. M. Chen. An output feedback H_∞ controller design for linear systems subject to sensor nonlinearities. *IEEE Transactions on Circuits and Systems I: Fundamental Theory and Applications*, 50(7):914–921, 2003.

[16] Y. Cao, Z. Lin, and Y. Shamash. Set invariance analysis and gain-scheduling control for LPV systems subject to actuator saturation. *Systems & Control Letters*, 46(2):137–151, 2002.

[17] B. S. Chen and Y. Wang. On the attenuation and amplification of molecular noise in genetic regulatory networks. *BMC Bioinformatics*, 7(1):52, 2006.

[18] G. Chen. A simple treatment for suboptimal Kalman filtering in case of measurement data missing. *IEEE Transactions on Aerospace and Electronic Systems*, 26(2):413–415, 1990.

[19] L. Chen and K. Aihara. Stability of genetic regulatory networks with time delay. *IEEE Transactions on Circuits and Systems I-Fundamental Theory and Applycations*, 49(5):602–608, 2002.

[20] T. Chen, H. He, and G. Church. Modeling gene expression with differential equations. *Proceedings of the Pacific Symposium on Biocomputing*, 4:29–40, 1999.

[21] W. Chen, Z. Guan, and X. Liu. Delay-dependent exponential stability of uncertain stochastic systems with mutiple delays: An LMI approach. *Systems & Control Letters*, 54(6):547–555, 2005.

[22] T. M. Cheng and A. V. Savkin. Output feedback stabilization of nonlinear networked control systems with non-decreasing nonlinearities: A matrix inequalities approach. *International Journal of Robust and Nonlinear Control*, 17(5-6):387–404, 2007.

[23] D. Cook, A. Gerber, and S. Tapscott. Modeling stochastic gene expression: Implications for haploinsufficiency. *Proceedings of the National Academy of Science*, 95(26):15641–15646, 1998.

[24] D. F. Coutinho, C. E. de Souza, and K. A. Barbosa. Robust H_∞ filter design for a class of discrete-time parameter varying systems. *Automatica*, 45(12):2946–2954, 2009.

[25] M. de Hoon, S. Imoto, K. Kobayashi, N. Ogasawara, and S. Miyano. Infering gene regulatory networks from time-ordered gene expression data of bacillus using differential equations. *Proceedings of the Pacific Symposium on Biocomputing*, 8:17–28, 2003.

[26] P. D'haeseleer, X. Wen, S. Fuhrman, and R. Somogyi. Linear modeling of mRNA expression levels during CNS development and injury. *Proceedings of the Pacific Symposium on Biocomputing*, 4:41–52, 1999.

[27] E. Diller, S. Floyd, C. Pawashe, and M. Sitti. Control of multiple heterogeneous magnetic microrobots in two dimensions on nonspecialized surfaces. *IEEE Transactions on Robotics*, 28(1):172–182, 2012.

[28] D. Ding, Z. Wang, J. Hu, and H. Shu. Dissipative control for state-saturated discrete time-varying systems with randomly occurring nonlinearities and missing measurements. *International Journal of Control*, 86(4):674–688, 2013.

[29] D. Ding, Z. Wang, B. Shen, and H. Shu. H_∞ state estimation for discrete-time complex networks with randomly occurring sensor saturations and randomly varying sensor delays. *IEEE Transactions on Neural Networks and Learning Systems*, 23(5):725–736, 2012.

[30] D. Ding, Z. Wang, B. Shen, and H. Shu. H_∞ state estimation for discrete-time complex networks with randomly occurring sensor saturations and randomly varying sensor delayss. *IEEE Transactions Neural Networks and Learning Systems*, 23(5):725–736, 2012.

[31] H. Dong, Z. Wang, X. Chen, and H. Gao. A review on analysis and synthesis of nonlinear stochastic systems with randomly occurring incomplete information. *Mathematical Problems in Engineering*, 2012(416358), 2012.

[32] H. Dong, Z. Wang, and H. Gao. Fault detection for Markovian jump systems with sensor saturations and randomly varying nonlinearities. *IEEE Transactions on Circuits and Systems I: Regular Papers*, 59(10):2354–2362, 2012.

[33] H. Dong, Z. Wang, and H. Gao. Distributed H_∞ filtering for a class of Markovian jump nonlinear time-delay systems over lossy sensor networks. *IEEE Transactions on Industrial Electronics*, 60(10):4665–4672, 2013.

[34] J. Doyle, K. Glover, P. Khargonekar, and B. Francis. State-space solutions to the standard H_2 and H_∞ control problems. *IEEE Transactions Automatic Control*, 34(8):31–47, 1989.

[35] A. El-Gohary and M. Yassen. Optimal control and synchronization of Lotka–Volterra model. *Chaos, Solitons and Fractals*, 12(11):2087–2093, 2001.

[36] M. B. Elowitz and S. Leibler. A synthetic oscillatory network of transcriptional regulators. *Nature*, 403(6767):335–338, 2000.

[37] E. Fridman and U. Shaked. A descriptor system approach to H_∞ control of linear time-delay systems. *IEEE Transactions on Automatic Control*, 47(2):253–270, 2002.

[38] A. Friedman. *Stochastic Differential Equations and their Applications*, volume 2. Academic Press, New York, 1976.

[39] H. Gao, J. Lam, and G. Chen. New criteria for synchronization stability of general complex dynamical networks with coupling delays. *Physics Letters A*, 360(2):263–273, 2006.

[40] H. Gao, J. Lam, and C. Wang. Robust energy-to-peak filter design for stochastic time-delay systems. *Systems & Control Letters*, 55(2):101–111, 2006.

[41] H. Gao, J. Lam, C. Wang, and S. Xu. Robust H_∞ filtering for 2D stochastic systems. *Circuits, Systems and Signal Processing*, 23(6):479–505, 2004.

[42] H. Gao, X. Meng, and T. Chen. A parameter-dependent approach to robust H_∞ filtering for time-delay systems. *IEEE Transactions on Automatic Control*, 53(10):2420–2425, 2008.

[43] H. Gao, P. Shi, and J. Wang. Parameter-dependent robust stability of uncertain time-delay systems. *Journal of Computational and Applied Mathematics*, 206(1):366–373, 2007.

[44] H. Gao and C. Wang. Delay-Dependent Robust H_∞ and L_2-L_∞ filtering for a class of uncertain nonlinear time-delay systems. *IEEE Transactions on Automatic Control*, 48(9):1661–1666, 2003.

[45] H. Gao and C. Wang. Robust L_2-L_∞ filtering for uncertain systems with mutiple time-varing state delays. *IEEE Transactions on Circuits and Systems I: Fundamental Theory and Applications*, 50(4):594–599, 2003.

[46] H. Gao and C. Wang. A delay-dependent approach to robust H_∞ filtering for uncertain discrete-time state-delayed systems. *IEEE Transactions on Signal Processing*, 52(6):1631–1640, 2004.

[47] H. Gao, C. Wang, and J. Wang. On H_∞ performance analysis for continuous-time stochastic systems with polytopic uncertainties. *Circuits, Systems and Signal Processing*, 24(4):415–429, 2005.

[48] J. Gao, B. Huang, and Z. Wang. LMI-based H_∞ control of uncertain linear jump systems with time-delays. *Automatica*, 37(7):1141–1146, 2001.

[49] T. Gardner, C. Cantor, and J. Collins. Construction of a genetic toggle switch in Escherichia coli. *Nature*, 403(6767):339–342, 2000.

[50] C. A. C. Gonzaga and O. L. V. Costa. Stochastic stabilization and induced L_2-gain for discrete-time Markov jump Lur'e systems with control saturation. *Automatica*, 50(9):2397–2404, 2014.

[51] G. C. Goodwin, H. Haimovich, D. E. Quevedo, and J. S. Welsh. A moving horizon approach to Networked Control system design. *IEEE Transactions on Automatic Control*, 49(9):1427–1445, 2004.

[52] L. Guo and H. Wang. Fault detection and diagnosis for general stochastic systems using B-spline expansions and nonlinear filters. *IEEE Transactions on Circuits and Systems I: Regular Papers*, 52(8):1644–1652, 2005.

[53] L. Guo, F. Yang, and J. Fang. Multiobjective filtering for nonlinear time-delay systems with nonzero initial conditions based on convex optimizations. *Circuits, Systems and Signal Processing*, 25(5):591–607, 2006.

[54] Q. L. Han. Absolute stability of time delay systems with sector-bounded nonlinearity. *Automatica*, 41(12):2171–2176, 2005.

[55] J. Hasty, J. Pradines, M. Dolink, and J. Collins. Biophysics noise-based switches and amplifiers for gene expression. In *Proceedings of the National Academy of Sciences*, volume 97, 2075–2080, 2000.

[56] J. He. Mysterious Pi and a possible link to DNA sequencing. *International Journal of Nonlinear Sciences and Numerical Simulation*, 5(3):263–274, 2004.

[57] X. He, Z. Wang, and D. Zhou. Robust H_∞ filtering for networked systems with multiple state delays. *International Journal of Control*, 80(8):1217–1232, 2007.

[58] Y. He, Q. G. Wang, L. Xie, and C. Lin. Further improvement of free-weighting matrices technique for systems with time-varying delay. *IEEE Transactions on Automatic Control*, 52(2):293–299, 2007.

[59] Y. He, M. Wu, J. She, and G. Liu. Delay-dependent robust stability criteria for uncertain neutral systems with mixed delays. *Systems & Control Letters*, 51(1):57–65, 2004.

[60] Y. He, M. Wu, J. She, and G. Liu. Parameter-dependent Lyapunov functional for stability of time-delay systems with polytopic-type uncertainties. *IEEE Transactions on Automatic Control*, 49(5):828–832, 2004.

[61] N. Hoang, H. Tuan, P. Apkarian, and S. Hosoe. Gain-Scheduled Filtering for Time-Varying Discrete Systems. *IEEE Transactions on Signal Processing*, 52(9):2464–2476, 2004.

[62] F C. Hoppensteadt and E. M. IzhikevichPattern. Recognition via synchronization in phase-locked loop neural networks. *IEEE Transactions on Neural Networks*, 11(3):734–738, 2000.

[63] F. O. Hounkpevi and E. E. Yaz. Minimum variance generalized state estimators for multiple sensors with different delay rates. *Signal Processing*, 87(4):602–613, 2007.

[64] F. O. Hounkpevi and E. E. Yaz. Robust minimum variance linear state estimators for multiple sensors with different failure rates. *Auotmatica*, 43(7):1274–1280, 2007.

[65] J. Hu, Z. Wang, H. Gao, and L. Stergioulas. Probability-guaranteed H_∞ finite-horizon filtering for a class of nonlinear time-varying systems with sensor saturations. *System & Control Letters*, 61:477–484, 2012.

[66] J. Hu, Z. Wang, B. Shen, and H. Gao. Quantized recursive filtering for a class of nonlinear systems with multiplicative noises and missing measurements. *International Journal of Control*, 86(4):650–663, 2013.

[67] S. Hu and J. Wang. On stabilization of a new class of linear time-invariant interval systems via constant state feedback control. *IEEE Transactions on Automatic Control*, 45(11):21.6–2111, 2000.

[68] S. Hu and Q. Zhu. Stochastic optimal control and analysis of stability of networked control systems with long delay. *Automatica*, 39(11):1877–1884, 2003.

[69] S. Huang. Gene expression profiling, genetic networks, and cellular states: An integrating concept for tumorigenesis and drug discovery. *Journal of Molecular Medicine*, 77:469–480, 1999.

[70] Y. Ji and H. J. Chizeck. Controllability, stabilizability, and continuous-time Markovian jump linear quadratic control. *IEEE Transactions on Automatic Control*, 35(3):777–788, 1990.

[71] J. Jost and M. P. Joy. Spectral properties and synchronization in coupled map lattices. *Physical Review E*, 65(1):061201, 2002.

[72] T. Kaczorek. *Two-dimensional Linear Systems*. Springer, 1985.

[73] R. E. Kalman. A new approach to linear filtering and prediction problems. *Journal of Fluids Engineering*, 82(1):35–45, 1960.

[74] Y. Kang, Y. Xie, and J. Xu. Observing nonlinear stochastic piecewise constant driving forces by the method of moments. *Chaos, Solitons and Fractals*, 27:715–721, 2006.

[75] H. K. Khalil. *Nonlinear Systems (Third Edition).* New Jersey, Prentice Hall, 2001.

[76] P. P. Khargonekar, I. R. Petersen, and K. Zhou. Robust stabilization of uncertain linear systems: Quadratic stabilizability and H_∞ control theory. *IEEE Transactions on Automatic Control*, 35:356–361, 1990.

[77] S. J. Kim and I. J. Ha. A state-space approach to analysis of almost periodic nonlinear systems with sector nonlinearities. *IEEE Transactions on Automatic Control*, 44:66–70, 1999.

[78] G. Kreisselmeier. Stabilization of linear systems in the presence of output measurement saturation. *Systems Control Letters*, 29:27–30, 1996.

[79] V. I. Krinsky, V. N. Biktashev, and I. R. Efimov. Autowave principles for parallel image processing. *Physica D: Nonlinear Phenomena*, 49:247–253, 1991.

[80] R. Krtolica, U. Ozguncr, H. Chan, H. Goktas, J. Winkelman, and M. Liubakka. Stability of linear feedback systems withn random communication delays. *International Journal of Control*, 59:925–953, 1994.

[81] H. Kushur. *Stochastic Stability and Control.* Academic Press, New York, 1976.

[82] L. Kuzmenkov, S. Maximov, and J. Guardado. On the asymptotic solutions of the coupled quasiparticle-oscillator system. *Chaos, Solitons and Fractals*, 15:597–610, 2003.

[83] J. Lam, H. Gao, S. Xu, and C. Wang. H_∞ and L_2 model reduction for system input with sector nonlinearities. *Journal of Optimization Theory and Applications*, 125:137–155, 2005.

[84] G. Lausterer and W. H. Ray. Distributed parameter state estimation and optimal feedback control—An experimental study in two space dimensions. *IEEE Transactions on Automatic Control*, 24(2):179–190, 1979.

[85] C. Li, L. Chen, and K. Aihara. Synchronization of coupled nonidentical genetic oscillators. *Physical Biology*, 3:37–44, 2006.

[86] W. Li, G. Wei, K. H. Reza, and X. Liu. Nonfragile gain-scheduled control for discrete time stochastic systems with randomly occurring sensor saturations. *Abstract and Applied Analysis*, 2013, 2013.

[87] W. Li, G. Wei, and L. Wang. Probability-dependent static output feedback control for discrete-time nonlinear stochastic systems with missing measurements. *Mathematical Problems in Engineering*, 2012, 2012.

[88] W. Li, W. Xu, J. Zhao, and Y. Jin. Probability-dependent static output feedback control for discrete-time nonlinear stochastic systems with missing measurements. First passage problem for stochastic dynamical systems. *Chaos, Solitons and Fractals*, 28:414–421, 2006.

[89] X. Li, X. Wang, and G. Chen. Pinning a complex dynamical network to its equilibrium. *IEEE Transactions on Circuits and Systems I: Fundamental Theory and Applications, Regular Papers*, 51:2074–2087, 2004.

[90] J. Liang, F. Sun, and X. Liu. Finite-horizon H_∞ filtering for time-varying delay systems with randomly varying nonlinearities and sensor saturations. *Systems Science & Control Engineering: An Open Access Journal*, 2:108–118, 2004.

[91] J. Liang, Z. Wang, and X. Liu. State estimation for coupled uncertain stochastic networks with missing measurements and time-varying delays: The discrete-time case. *IEEE Transactions on Neural Networks*, 20:781–793, 2009.

[92] J. Liang, Z. Wang, and X. Liu. Robust state estimation for two-dimensional stochastic time-delay systems with missing measurements and sensor saturation. *Multidimention Systems and Signal Processing*, 25(1):157–177, 2014.

[93] J. Liang, Z. Wang, X. Liu, and P. Louvieris. Robust synchronization for 2-D discrete-time coupled dynamical networks. *IEEE Transactions on Neural Networks and Learning Systems*, 23(6):942–953, 2012.

[94] J. Liang, Z. Wang, and Y. Liu. Distributed state estimation for discrete-time sensor networks with randomly varying nonlinearities and missing measurements. *IEEE Transactions on Neural Networks*, 22:486–496, 2011.

[95] J. Liang, Z. Wang, Y. Liu, and X. Liu. Robust synchronization of an array of coupled stochastic discrete-time delayed neural networks. *IEEE Transactions on Neural Networks*, 19:1910–1921, 2008.

[96] J. Liang, Z. Wang, B. Shen, and X. Liu. Distributed state estimation in sensor networks with randomly occurring nonlinearities subject to time-delays. *ACM Transactions on Sensor Networks*, 19(1), 2012.

[97] Z. Lin and T. Hu. Semi-global stabilization of linear systems subject to output saturation. *Systems Control letters*, 43:211–217, 2001.

[98] H. Liu, F. Sun, K. He, and Z. Sun. Design of reduced-order H_∞ filter for Markovian jumping systems with time delay. *IEEE Transactions on Circuits and Systems II: Analog and Digital Signal Processing*, 51:607–612, 2004.

[99] Y. Liu, Z. Wang, J. Liang, and X. Liu. Synchronization and state estimation for discrete-time complex networks with distributed delays. *IEEE Transactions on Systems, Man, and Cybernetics Part B-Cybernetics*, 38:1314–1325, 2008.

[100] Y. Liu, Z. Wang, and X. Liu. Robust H_∞ control for a class of nonlinear stochastic systems with mixed time-delay. *International Journal of Robust and Nonlinear Control*, 17:1525–1551, 2007.

[101] J. Lu and D. W. C. Ho. Globally exponential synchronization and synchronizability for general dynamical networks. *IEEE Transaction on Systems, Man, and Cybernetics Part B-Cybernetics*, 40:350–361, 2010.

[102] J. Lu, D. W. C. Ho, and Z. Wang. Pinning stabilization of linearly coupled stochastic neural networks via minimum number of controllers. *IEEE Transactions on Neural Networks*, 20:1617–1629, 2009.

[103] W. Lu and T. Chen. Synchronization of coupled connected neural networks with delays. *IEEE Transaction Circuits and Systems I: Fundamental Theory and Applications*, 51:2491–2503, 2004.

[104] W. Lu and T. Chen. Global synchronization of discrete-time dynamical network with a directed graph. *IEEE Transaction on Circuits Systems II: Analog and Digital Signal Processing*, 54:136–140, 2007.

[105] W. Lu and E. B. Lee. Stability analysis for two-dimensional systems via a Lyapunov approach. *IEEE Transactions on Circuits and Systems*, 32(1):61–68, 1985.

[106] X. Lu, L. Xie, H. Zhang, and W. Wang. Robust Kalman filtering for discrete-time systems with measurement delay. *IEEE Transactions on Circuits and Systems II: Analog and Digital Signal Processing*, 54:522–526, 2007.

[107] Y. Luo, G. Wei, H. R. Karimi, and L. Wang. Deconvolution filtering for nonlinear stochastic systems with randomly occurring sensor delays via probability-dependent method. *Abstract and Applied Analysis*, 2013, 2013.

[108] L. Ma, Z. Wang, Y. Bo, and Z. Guo. A game theory approach to mixed H_2/H_∞ control for a class of stochastic time-varying systems with randomly occurring nonlinearities. *Systems & Control Letters*, 60:1009–1015, 2011.

[109] X. Mao. *Stability of Stochastic Differential Equations with Respect to Semi-Martingales.* Longman Scientific and Technical. Horwood, New York, 1991.

[110] X. Mao. *Exponential Stability of Stochastic Differential Equations.* Marcel Dekker, New York, 1994.

[111] X. Mao. Razumikhin-type theorems on exponential stability of stochastic functional differential equations. *Stochastic Process Application*, 5:233–250, 1996.

[112] X. Mao. Razumikhin-type theorems on exponential stability of neutral stochastic functional differential equations. *SIAM Journal on Mathematical Analysis*, 28:389–401, 1997.

[113] X. Mao. *Stochastic Differential Equations and Applications (second edition)*. Woodhead Publishing, Cambridge, 2007.

[114] X. Mao and J. Chu. Quadratic stability and stabilization of dynamic interval systems. *IEEE Transactions on Automatic Control*, 48:1007–1012, 2003.

[115] X. Mao, J. Lam, S. Xu, and H. Gao. Razumikhin method and exponential stability of hybrid stochastic delay interval systems. *Journal of Mathematical Analysis and Applications*, 314:45–66, 2006.

[116] X. Mao and S. Sabanis. Numerical solutions of stochastic differential delay equations under local Lipschitz condition. *Journal of Computational and Applied Mathematics*, 51:215–227, 2003.

[117] A. S. Matveev and A. V. Savkin. Optimal control of networked systems via asynchronous communication channels with irregular delays. *Proceedings of the 40th IEEE Conference on Decision and Control*, pages 2323–2332, 2001.

[118] S. K. Mitra, A. Sagar, and N. Pendergrass. Realizations of two-dimensional recursive digital filters. *IEEE Transactions on Circuits and Systems*, 22(3):177–184, 1975.

[119] N. A. M. Monk. Oscillatory expression of Hes1, p53, and NF-κB driven by transcriptional time delays. *Current Biology*, 13:1409–1413, 2003.

[120] Y. S. Moon, P. Park, W. H. Kwon, and Y. S. Lee. Delay-dependent robust stabilization of uncertain state-delayed systems. *International Journal of Control*, 74:1447–1455, 2001.

[121] Y. S. Moon, P. G. Park, and W. H. Kwon. Robust stabilization of uncertain input-delayed systems using reduction method. *Automatica*, 37:307–312, 2001.

[122] N. Nahi. Optimal recursive estimation with uncertain observation. *IEEE Transactions on Information Theory*, 15:457–462, 1969.

[123] S. Niculescu, E. Verriest, L. Dugard, and J. Dion. Stability and robust stability of time-delay systems: A guided tour. In *Stability and Control of Time-Delay Systems*, pages 1–71, 1998.

[124] J. Nilsson, B. Bernhardsson, and B. Wittenmark. Stochastic analysis and control of real-time systems with random time delays. *Automatica*, 34:57–64, 1998.

[125] Y. Niu, D. W. C. Ho, and J. Lam. Robust integral sliding mode control for uncertain stochastic systems with time-varying delay. *Automatica*, 41:873–880, 2005.

[126] J. Paulsson. Summing up the noise in gene networks. *Nature*, 427:415–418, 2004.

[127] R. Pearson. Gray-box modeling of nonideal sensors. *Proceedings of American Control Conference*, pages 4404–4409, 2001.

[128] A. Pease, D. Solas, E. Sullivan, M. Cronin, C. Holmes, and S. Fodor. Light-generated oligonucleotide arrays for rapid DNA sequense analysis. *Proceedings of the National Academy of Sciences of the United States of America*, 91:5022–5026, 1994.

[129] V. Perez-Munuzuri, V. Perez-Villar, and L. O. Chua. Autowaves for image processing on a 2-D CNN array of excitable nonlinear circuits: Flat and wrinkled labyrinths. *IEEE Transactions on Circuits and Systems I: Fundamental Theory and Applications*, 40:174–181, 1993.

[130] S. R. Powell and L. M. Silvermann. Modeling of two-dimensional covariance functions with applications to image restoration. *IEEE Transactions on Automatic Control*, 19(1):8–13, 1974.

[131] J. Qiu, G. Feng, and H. Gao. Fuzzy-model-based piecewise H_∞ static output feedback controller design for networked nonlinear systems. *IEEE Transactions on Fuzzy Systems*, 18(5):919–934, 2010.

[132] A. Raouf and G. Moatimid. Nonlinear dynamics and stability of two streaming magnetic fluids. *Chaos and Solitons and Fractals*, 12:1207–1216, 2001.

[133] A. Ray. Output feedback control under randomly varying distributed delays. *Journal of Guidance Control and Dynamics*, 17:701–711, 1994.

[134] F. Ren and J. Cao. Asymptotic and robust stability of genetic regulatory networks with time-varying delays. *Neurocomputing*, 71:834–842, 2008.

[135] R. P. Roesser. A discrete state-space model for linear image processing. *IEEE Transactions on Automatic Control*, 20(1):1–10, 1975.

[136] W. J. Rugh and J. S. Shamma. Research on gain scheduling. *Automatica*, 36:1401–1425, 2000.

[137] A. V. Savkin and T. M. Cheng. Detectability and output feedback stability of nonlinear networked control systems. *IEEE Transactions on Automatic Control*, 52:730–735, 2007.

[138] A. V. Savkin, I. R. Peterson, and S. O. R. Moheimani. Model validation and state estimation for uncertain continuous-time systems with missing discrete-continuous data. *Computers & Electrical Engineering*, 25:29–43, 1999.

[139] B. Schenato, B. Sinopoll, M. Franceschettti, K. Poolla, and S. S. Sastaty. Foundations of control and estimation over lossy networks. *Proceedings of the IEEE*, 95:163–187, 2007.

[140] P. Seiler and R. Sengupta. An H approach to networked control. *IEEE Transactions on Automatic Control*, 50:356–361, 2005.

[141] B. Shen, Z. Wang, D. Ding, and H. Shu. H_∞ state estimation for complex networks with uncertain inner coupling and incomplete measurements. *IEEE Transactions on Neural Networks and Learning Systems*, 24(12):2027–2037, 2013.

[142] B. Shen, Z. Wang, J. Liang, and Y. Liu. Recent advances on filtering and control for nonlinear stochastic complex systems with incomplete information: A survey. *Mathematical Problems in Engineering*, 2012, 2012.

[143] B. Shen, Z. Wang, H. Shu, and G. Wei. Robust H_∞ finite-horizon filtering with randomly occurred nonlinearities and quantization effects. *Automatica*, 46:1743–1751, 2010.

[144] P. Shi, E. K. Boukas, and R. K. Agawal. Kalman filtering for continuous-time uncertain systems with Markovian jumping parameters. *IEEE Transactions on Automatic Control*, 44:1592–1597, 1999.

[145] P. Shi, M. Mahmoud, S. K. Nguang, and A. Ismail. Robust filtering for jumping systems with mode-dependent delays. *Signal Processing*, 86:140–152, 2006.

[146] P. Shi, M. Mahmoud, J. Yi, and A. Ismail. Worst case control of uncertain jumping systems with multi-state and input delay information. *Information Sciences*, 176:186–200, 2005.

[147] H. Shu and G. Wei. H_∞ analysis of nonlinear stochastic time-delay systems. *Chaos, Solitons and Fractals*, 26:637–647, 2005.

[148] Z. Shu and J. Lam. Delay-dependent exponential estimates of stochastic neural networks with time delay. *Proceedings of the International Conference on Neural Information Processing Part I (Lecture Notes in Computer Science)*, 4232:332–341, 2006.

[149] B. Sinopoli, L. Schenato, M. Franceschetti, K. Poolla, M. I. Jordan, and S. S. Sastry. Kalman filtering with intermittent observations. *IEEE Transactions on Automatic Control*, 49:1453–1464, 2004.

[150] P. Smolen, D. Baxter, and J. Byrne. Mathematical modeling of gene networks. *Neuron*, 26:567–580, 2000.

[151] P. Smolen, D. Baxter, and J. Byrne. Modeling circadian oscillations with interlocking positive and negative feedback loops. *Journal of Neuroscience*, 21:6644–6656, 2001.

[152] Y. Song, S. Liu, and G. Wei. Constrained robust distributed model predictive control for uncertain discrete-time Markovian jump linear system. *Journal of the Franklin Institute-Engineering and Applied Mathematics*, 352:73–92, 2015.

[153] C. E. Souza and A. Trofino. Gain-scheduled H_2 controller synthesis for linear parameter varying systems via parameter-dependent Lyapunov functions. *International Journal of Robust and Nonlinear Control*, 16(5):243–257, 2006.

[154] C. E. De Souza and A. Trofino. Gain-scheduled H_2 controller synthesis for linear parameter varying systems via parameter-dependent Lyapunov functions. *International Journal of Robust Nonlinear Control*, 16(5):243–257, 2006.

[155] D. J. Stilwell and J. R. Wilson. Stability and L_2 gain properties of LPV systems. *Automatica*, 38(9):1601–1606, 2002.

[156] M. Taksar, A. Poinyak, and A. Iparraguirre. Robust output feedback control for linear stochastic systems in continuous-time with time-varying parameters. *IEEE Transactions on Automatic Control*, 43(8):1133–1137, 1998.

[157] Y. Tang, S. Y. S. Leung, W. K. Wong, and J. Fang. Impulsive pinning synchronization of stochastic discrete-time networks. *Neurocomputing*, 73(10–12):2132–2139, 2010.

[158] A. Teel. Connections between Razumikhin-type theorems and the ISS nonlinear small gain theorem. *IEEE Transactions on Automatic Control*, 43(7):960–964, 1998.

[159] E. Tian, P. Chen, and G. Zhou. Fault tolerant control for discrete networked control systems with random faults. *International Journal of Control, Automation and Systems*, 10(2):444–448, 2012.

[160] E. Tian and C. Peng. Delay-dependent stability and synthesis of uncertain T-S fuzzy systems with time-delay. *Fuzzy Sets & Systems*, 157(4):544–559, 2005.

[161] T. Tian and K. Burrage. Stochastic neural network models for gene regulatory networks. In *Proceedings of IEEE Congress on Evolutionary Computation*, pages 162–169. ACT, 2003.

[162] Y. Tipsuwan and M. Y. Chow. Control methodologies in networked control systems. *Control Engineering Practice*, 11(10):1099–1111, 2003.

[163] O. Toker. On the complexity of the robust stability problem for linear parameter varying systems. *Automatica*, 33(11):2015–2017, 1997.

[164] N. Tsai and A. Ray. Stochastic optimal control under randomly varying delays. *International Journal of Control*, 69(5):1179–1202, 1997.

[165] G. C. Walsh, O. Beldiman, and L. G. Bushnell. Asymptotic behavior of nonlinear networked control. *IEEE Transactions on Automatic Control*, 46(7):1093–1097, 2001.

[166] F. Wang and V. Balakrishnan. Improved stability analysis and gain-scheduled controller synthesis for parameter-dependent systems. *IEEE Transactions on Automatic Control*, 47(5):720–734, 2002.

[167] L. Wang, G. Wei, and W. Li. Probability-dependent H_∞ synchronization control for dynamical networks with randomly varying nonlinearities. *Neurocomputing*, 33:369–376, 2014.

[168] L. Wang, G. Wei, and H. Shu. State estimation for complex networks with randomly occurring coupling delays. *Neurocomputing*, 122:513–520, 2013.

[169] Y. Wang, Z. Wang, and J. Liang. Global synchronization for delayed complex networks with randomly occurring nonlinearities and multiple stochastic disturbances. *Journal of Physics A: Mathematical and Theoretical*, (13):135–101.

[170] Y. Wang, L. Xie, and C. E. De Souza. Robust control of a class of uncertain nonlinear systems. *Systems & Control Letters*, (2):139–149.

[171] Z. Wang and K. Burnham. LMI approach to output feedback control for linear uncertain systems with D-stability constraints. *Journal of Optimization Theory and Application*, 113(2):357–372, 2002.

[172] Z. Wang, H. Dong, B. Shen, and H. Gao. Finite-horizon H_∞ filtering with missing measurements and quantization effects. *IEEE Transactions on Automatic Control*, 58(7):1707–1718, 2013.

[173] Z. Wang, D. Goodall, and K. Burnham. On designing observers for time-delay systems with nonlinear disturbances. *International Journal of Control*, 75(11):803–811, 2002.

[174] Z. Wang and D. W. C. Ho. Robust filtering under randomly varying sensor delays with variance constraints. *IEEE Transactions Circuits and Systems: Express Briefs*, 51(6):320–326, 2004.

[175] Z. Wang, D. W. C. Ho, and X. Liu. Robust filtering under randomly varying sensor delay with variance constraints. *IEEE Transactions on Circuits and Systems II: Analog and Digital Signal Processing*, 51(6):320–326, 2004.

[176] Z. Wang, D. W. C. Ho, and X. Liu. State estimation for delayed neural networks. *IEEE Transactions on Neural Networks*, 16(1):279–284, 2005.

[177] Z. Wang, J. Lam, and X. Liu. Exponential filtering for uncertain Markovian jump time-delay systems with nonlinear disturbances. *IEEE Transactions on Circuits and Systems II: Analog and Digital Signal Processing*, 51(5):262–268, 2004.

[178] Z. Wang, J. Lam, G. Wei, K. Fraser, and X. Liu. Filtering for nonlinear genetic regulatory networks with stochastic disturbances. *IEEE Transactions on Automatic Control*, 53(10):2448–2457, 2008.

[179] Z. Wang, Y. Liu, M. Li, and X. Liu. Stability analysis for stochastic Cohen-Grossberg neural networks with mixed time delays. *IEEE Transactions on Neural Networks*, 17(3):814–820, 2006.

[180] Z. Wang, Y. Liu, and X. Liu. H_∞ filtering for uncertain stochastic time-delay systems with sector-bounded nonlinearities. *Automatica*, 44(5):1268–1277, 2008.

[181] Z. Wang, H. Qiao, and K. Burnham. On stabilization of bilinear uncertain time-delay stochastic systems with Markovian jumping parameters. *IEEE Transactions on Automatic Control*, 47(4):640–662, 2002.

[182] Z. Wang, B. Shen, and X. Liu. H_∞ filtering with randomly occurring sensor saturations and missing measurements. *Automatica*, 48(3):556–562, 2012.

[183] Z. Wang, B. Shen, H. Shu, and G. Wei. Quantized H_∞ control for nonlinear stochastic time-delay systems with missing measurements. *IEEE Transactions on Automatic Control*, 57(6):1431–1444, 2012.

[184] Z. Wang, H. Shu, and X. Liu. Reliable stabilization of stochastic time-delay systems with nonlinear disturbances. *International Journal of General Systems*, 34(5):523–535, 2005.

[185] Z. Wang, Y. Wang, and Y. Liu. Global synchronization for discrete-time stochastic complex networks with randomly occurred nonlinearities and mixed time-delays. *IEEE Transactions on Neural Networks*, 21(1):11–25, 2010.

[186] Z. Wang, F. Yang, D. W. C. Ho, and X. Liu. Robust H_∞ filtering for stochastic time-delay systems with missing measurements. *IEEE Transactions on Signal Processing*, 54(7):2579–2587, 2006.

[187] Z. Wang, F. Yang, D. W. C. Ho, and X. Liu. Robust H_∞ control for networked systems with random packet losses. *IEEE Transactions on Systems, Man, and Cybernetics Part B-Cybernetics*, 37(4):916–924, 2007.

[188] Z. Wang, F. Yang, and X. Liu. Robust filtering for systems with stochastic nonlinearities and deterministic uncertainties. *Proceedings of the Institution of Mechanical Engineers, Part I — Journal of Systems and Control Engineering*, 220(3):171–182, 2006.

[189] G. Wei, S. Liu, Y. Song, and Y. Liu. Probability-guaranteed set-membership filtering for systems with incomplete measurements. *Automatica*, 60:12–16, 2015.

[190] G. Wei, L. Wang, and W. Li. Fault-tolerant control for discrete-time stochastic systems with randomly occurring faults. In *Proceedings of the 31st Chinese Control Conference*, pages 1535–1540, 2012.

[191] G. Wei, Z. Wang, X. He, and H. Shu. Filtering for networked stochastic time-delay systems with sector nonlinearity. *IEEE Transactions on Circuits and Systems II: Express Briefs*, 56(1):71–75, 2009.

[192] G. Wei, Z. Wang, and B. Shen. Probability-dependent gain-scheduled control for discrete stochastic delayed systems with randomly occurring nonlinearities. *International Journal of Robust and Nonlinear Control*, 23(7):815–826, 2013.

[193] G. Wei, Z. Wang, B. Shen, and M. Li. Probability-dependent gain-scheduled filtering for stochastic systems with missing measurements. *IEEE Transactions on Circuits and Systems-II: Express Briefs*, 58(11):753–757, 2011.

[194] G. Wei, Z. Wang, and H. Shu. Nonlinear H_∞ control of stochastic delayed systems with Markovian switching. *Chaos, Solitons and Fractals*, 35(3):442–451, 2008.

[195] G. Wei, Z. Wang, and H. Shu. Robust filtering with stochastic nonlinearities and multiple missing measurements. *Automatica*, 45(3):836–841, 2009.

[196] G. Wei, Z. Wang, H. Shu, K. Fraser, and X. Liu. Robust filtering for gene expression time series data with variance constraints. *International Journal of Computer Mathmatics*, 84(5):619–633, 2007.

[197] F. Wu and K. M. Grigoriadis. LPV systems with parameter-varying time delays: Analysis and control. *Automatica*, 37(2):221–229, 2001.

[198] L. Wu, X. Su, and P. Shi. Mixed H_2/H_∞ approach to fault detection of discrete linear repetitive processes. *Journal of the Franklin Institute*, 348(2):393–414, 2011.

[199] M. Wu, Y. He, and J. She. Stability analysis and robust control of time-delay systems. *Springer*, 2010.

[200] L. Xie and C. E. de Souza. Robust H_∞ control for linear systems with norm-bounded time-varying uncertainty. *IEEE Transactions on Automatic Control*, 37(8):1188–1191, 1992.

[201] L. Xie, E. Fridman, and U. Shaked. Robust H_∞ control of distributed delay systems with application to combustion control. *IEEE Transactions on Automatic Control*, 46(12):1930–1935, 2001.

[202] L. Xie, Y. C. Soh, and C. E. de Souza. Robust Kalman filtering for uncertain discrete-time systems. *IEEE Transactions on Automatic Control*, 39(6):1310–1314, 1994.

[203] B. Xu and Y. Tao. External noise and feedback regulation: Steady-state statistics of auto-regulatory genetic network. *Journal of Theoretical Biology*, 243(2):214–221, 2006.

[204] S. Xu and T. Chen. Robust H_∞ control for uncertain stochastic systems with state delay. *IEEE Transactions on Automatic Control*, 47(12):2089–2094, 2002.

[205] S. Xu and T. Chen. H_∞ output feedback control for uncertain stochastic systems with time-varying delays. *Automatica*, 40(3):343–354, 2004.

[206] S. Xu and J. Lam. Improved delay-dependent stability results for time-delay systems. *IEEE Transactions on Automatic Control*, 50(3):384–387, 2005.

[207] S. Xu, J. Lam, X. Mao, and Y. Zou. A new LMI condition for delay-dependent robust stability of stochastic time-delay systems. *Asian Journal of Control*, 7(4):419–423, 2005.

[208] I. Yaesh, S. Boyarski, and U. Shaked. Probability-guaranteed robust H_∞ performance analysis and state-feedback design. *Systems & Control Letters*, 48(5):351–364, 2003.

[209] F. Yang, Y. Li, and X. Liu. Robust error square constrained filter design for systems with non-Gaussian noises. *IEEE Signal Processing Letters*, 15:930–933, 2008.

[210] F. Yang, Z. Wang, D. W. C. Ho, and M. Gani. Robust H_∞ control with missing measurements and time-delays. *IEEE Transactions on Automatic Control*, 52(9):1666–1672, 2007.

[211] F. Yang, Z. Wang, D. W. C. Ho, and X. Liu. Robust H_2 filtering for a class of systems with stochastic nonlinearities. *IEEE Transactions on Circuits and Systems II-Express Briefs*, 53(3):235–239, 2006.

[212] F. Yang, Z. Wang, and Y. S. Hung. Robust kalman filtering for discrete time-varying uncertain systems with multiplicative noises. *IEEE Transactions on Automatic Control*, 47(7):1179–1183, 2002.

[213] F. Yang, Z. Wang, Y. S. Hung, and M. Gani. H_∞ control for networked systems with random communication delays. *IEEE Transactions on Automatic Control*, 51(3):511–518, 2006.

[214] F. Yang, Z. Wang, Y. S. Hung, and M. Gani. H_∞ control for networked systems with random communication delays. *IEEE Transactions on Automatic Control*, 51(3):511–518, 2006.

[215] R. Yang, P. Shi, and G. Liu. Filtering for discrete-time networked nonlinear systems with mixed random delays and packet dropouts. *IEEE Transactions on Automatic Control*, 56(11):2655–2660, 2011.

[216] R. Yang, Z. Zhang, and P. Shi. Exponential stability on stochastic neural networks with discrete interval and distributed delays. *IEEE Transactions on Neural Networks*, 21(1):169–175, 2010.

[217] J. Yao, W. Wang, and S. Ye. Robust stabilization of 2-D state-delayed systems with stochastic perturbation. In *Proceedings of the 11th International Conference on Control, Automation, Robotics and Vision*, pages 1951–1956, 2010.

[218] E. Yaz and A. Ray. Linear unbiased state estimation for random models with sensor delay. In *Proceedings of 35th IEEE Conference on Decision and Control*, pages 47–52, 1996.

[219] E. E. Yaz and Y. I. Yaz. State estimation of uncertain nonlinear stochastic systems with general criteria. *Applied Mathematics Letters*, 14(5):605–610, 2001.

[220] C. Yuan and X. Mao. Robust stability and controllability of stochastic differential delay equations with Markovian switching. *Automatica*, 40(3):343–354, 2004.

[221] D. Yue and Q. Han. Delay-dependent exponential stability of stochastic systems with time-varing delay, nonlinearity and Markovian switching. *IEEE Transactions on Automatic Control*, 50(2):217–222, 2005.

[222] D. Yue and Q. Han. Delay-dependent robust H_∞ controller design for uncertain descriptor systems with time-varying discrete and distributed delays. In *IEE Proceedings—Control Theory and Applications*, pages 628–638, 2005.

[223] D. Yue, Q. Han, and C. Peng. State feedback controller design of networked control systems. *IEEE Transactions on Circuits and Systems II: Analog and Digital Signal Processing*, 51(11):640–644, 2004.

[224] D. Yue and J. Lam. Non-fragile guaranteed cost control for uncertain descriptor systems with time-varying state and input delays. *Optimal Control Applications & Methods*, 26(2):85–105, 2005.

[225] G. Zames. Feedback and optimal sensitivity: Model reference transformations, multiplicative seminorms, and approximate inverses. *IEEE Transactions on Automatic Control*, 26(2):301–320, 1981.

[226] W. Zhang, M. S. Branicky, and S. M. Phillips. Stability of networked control systems. *IEEE Control System Magzine*, 21(1):81–99, 2004.

[227] A. L. Zheleznyak and L. O. Chua. Coexistence of low- and high-dimensional spatio-temporal chaos in a chain of dissipatively coupled Chuas circuits. *International Journal of Bifurcation and Chaos*, 4(3):639–674, 1994.

[228] S. Zhou and G. Feng. H_∞ filtering for discrete-time systems with randomly varying sensor delays. *Automatica*, 44(7):1918–1922, 2008.

[229] S. Zhou, B. Zhang, and W. Zheng. Gain-scheduled H_∞ filtering of parameter-varying discrete-time systems via parameter-dependent Lyapunov functions. *International Journal of Control, Automation and Systems*, 7(3):475–479.

Index

Milton Keynes UK
Ingram Content Group UK Ltd.
UKHW040109071024
449327UK00019B/918

9 780367 574581